Advanced Concepts and Technologies for Electric Aircraft

This edited volume with expert contributions addresses the critical engineering challenges in aircraft electrification. It focuses on the power electronics, electric motors, insulation systems, and energy storage essential for achieving sustainable aviation and meeting global carbon reduction targets.

Throughout this book, expert chapter contributors provide comprehensive coverage of electric aircraft technologies, from propulsion system architectures to advanced power conversion solutions. Beginning with an overview of the aviation industry's decarbonisation journey, it examines hybrid-electric, fully electric, and hydrogen-powered alternatives across different market segments. Detailed technical discussions encompass power electronics fundamentals, including novel converter topologies such as zigzag transformer-based systems and switched-midpoint boost inverters. The book explores DC–DC converter technologies, high-speed permanent magnet synchronous motor drives, and magnetic geared machines optimised for aerospace applications. Advanced topics include prognostics and health management systems for predictive maintenance, wide bandgap semiconductor technologies, and critical insulation challenges in high-voltage, high-altitude environments. Quality assurance protocols and testing methodologies for motor insulation systems under aerospace conditions are thoroughly examined. This in-depth approach enables readers to apply their knowledge to design fully electric aircraft, including their motor drives, control systems, energy management, and insulation.

This book is a vital resource for anyone researching the development of electric aircraft systems and sustainable aviation technologies. It is particularly useful for postgraduate students of electrical engineering, aerospace engineering, and power electronics and researchers and engineers in R&D departments in aircraft and propulsion systems manufacturers.

Akshay Kumar Rathore, an IEEE Fellow, is Professor and Director of Power and Energy Systems Research Area in the Department of Electrical and Computer Engineering, National University of Singapore. He obtained his PhD degree from the University of Victoria, BC, Canada, and MTech degree from the Indian Institute of Technology (BHU), Varanasi, India.

Advanced Concepts and Technologies for Electric Aircraft

Power Systems and Energy Management

Edited by
Akshay Kumar Rathore

CRC Press
Taylor & Francis Group
Boca Raton London New York

CRC Press is an imprint of the
Taylor & Francis Group, an **informa** business

Designed cover image: iStock

First edition published 2027
by CRC Press
2385 NW Executive Center Drive, Suite 320, Boca Raton FL 33431

and by CRC Press
4 Park Square, Milton Park, Abingdon, Oxon, OX14 4RN

CRC Press is an imprint of Taylor & Francis Group, LLC

ISBN: 978-1-041-25760-8 (hbk)
ISBN: 978-1-041-25762-2 (pbk)
ISBN: 978-1-003-74799-4 (ebk)

DOI: 10.1201/9781003747994

Typeset in Palatino
by Apex CoVantage, LLC

Contents

About the Editor

Akshay Kumar Rathore, an IEEE Fellow, is Professor and Director of Power and Energy Systems Research Area in the Department of Electrical and Computer Engineering, National University of Singapore. He obtained his PhD degree from the University of Victoria, BC, Canada, and MTech degree from the Indian Institute of Technology (BHU), Varanasi, India. He is a recipient of several prestigious awards, such as the IEEE IAS Andrew W Smith Outstanding Young Member Achievement Award, the Isao Takahashi Power Electronics Award, the IEEE David Irwin Early Career Award, the IEEE Bimal Bose Award for Industrial Electronics Applications in Energy Systems, the Nagamori Award, and the distinguished alumnus award (young achiever) from IIT (BHU) Varanasi, India. During his tenure at Concordia University in Canada, he was listed in the Provost Circle of Distinction in 2021.

Rathore has contributed significantly to the design of novel power electronic converter topologies and modulation strategies, publishing about 300 research articles, including about 105 IEEE Transactions. He edited a book on advanced concepts and technologies for electric transportation, published by CRC Press in 2022. He has delivered over 200 invited talks, keynotes, and tutorials in several events across the globe. He is currently Distinguished Lecturer of the IEEE Power Electronics Society, Chair of the IEEE IAS Andrew W Smith Outstanding Young Member Achievement Award Committee, and serves on the committees for the IEEE Medal in Power Engineering, the IEEE Nikola Tesla Award, and the IEEE Fellow Committee. He has chaired three technical committees in the IEEE, served on the IAS Executive Board during 2017–2025 and IES AdCom during 2016–2017 and 2021–2026.

Contributors

Altaf Q. H. Badar is currently Assistant Professor in the Department of Electrical Engineering at the NIT, Warangal, India. He received his BE and MTech from RTM Nagpur University, Nagpur, and later earned his PhD from VNIT, Nagpur. His research focuses on integrating intelligent computational techniques into modern power and energy systems. His primary areas of interest include energy trading, artificial intelligence, and machine learning applications in power systems, energy management, techno-economic analysis, hybrid renewable energy systems, and evolutionary optimisation algorithms. He has published more than 50 research papers in reputable international journals and conferences. In addition to his engineering qualifications, he has also completed a Diploma in Business Management, reflecting his interest in bridging technical innovation with managerial and economic perspectives in the energy sector. He is a senior member of the IEEE.

Sheron Figarado has completed his BTech from Cochin University of Science and Technology in 2004. He received his PhD degree in power electronics from the Centre for Electronic Design and Technology, Indian Institute of Science, Bengaluru, India, in 2009. He had worked as Assistant Professor at BITS Pilani KK Birla Goa Campus during 2009–2015 and later at National Institute of Technology Karnataka from 2015 to 2018. He joined IIT Goa in 2018 and is currently working as Associate Professor at the School of Electrical Sciences, Indian Institute of Technology Goa, India. His research interests include design and control of multilevel inverters for induction motor drives, inductive power transfer for EV applications, and PWM techniques for multilevel inverters.

Felipe B. Grigoletto received his BSc, MSc, and DSc degrees in electrical engineering from the Federal University of Santa Maria, Brazil, in 2007, 2009, and 2013 respectively. He is currently Professor at the Federal University of Pampa and Researcher at the Laboratory of Power Electronics and Control, Alegrete, Brazil. His research interests include modulation and control strategies for power converters, modelling and design of power electronics circuits, and renewable energy conversion systems. Dr Grigoletto was the recipient of the First Prize Paper Award at the IEEE Innovative Smart Grid Technologies Latin America (ISGTLA 2019) and at the 14th Seminar on Power Electronics and Control (SEPOC 2022). He is also a frequent reviewer of several IEEE journals in the area of power electronics.

Sara Hasanpour received the B.S. degree in Electronic Engineering from Azad Islamic University, Lahijan Branch, Iran, in 2002, the M.S. degree in Electrical Engineering from the Isfahan University of Technology, Iran, in 2005, and the PhD degree in Power Electronics Engineering from the University of Guilan, Iran, in 2019. She is currently Assistant Professor with Azad Islamic University, Iran. Dr Hasanpour is a frequent reviewer of the *IEEE Transactions on Industrial Electronics, IEEE Transactions on Power Electronics*, and *IET Power Electronics*. She is also Associate Editor of *IEEE Journal of Emerging and Selected Topics in Industrial Electronics.*

Ananda Shankar Hati received his BTech in 2007 and MTech in 2009 from WBUT, India, and earned his PhD in Electrical Engineering from IIT (ISM) Dhanbad, India, in 2016. He began his academic career as Lecturer at BITM, Shantiniketan, followed by doctoral research at IIT (ISM) Dhanbad, and a brief tenure as Assistant Professor at the NIT, Arunachal Pradesh, India. Since 2013, he has been serving as Assistant Professor in the Department of Electrical Engineering at IIT (ISM) Dhanbad. He also completed his post-doctoral research at Deakin University, Australia, in 2021. He has authored or co-authored more than 50 publications and holds 38 patents. His funded research includes intelligent mine ventilation and smart helmet technologies for miner safety. He has delivered professional training programmes for organisations such as CIL, SCCL, NALCO, and UCIL. He is Fellow of the Royal Society of Arts and IETE, a Chartered Engineer (India). His research interests include machine intelligence, Internet of Things, sensor networks, prognostics and health management, and energy-efficient electric machines.

Md. Asif Hussain received his BTech in 2016 from MAKAUT, India, and MTech from IIT (ISM), Dhanbad, India, in 2020. He is currently pursuing his PhD from IIT (ISM), Dhanbad. His research focuses on advancing efficient, reliable electrical systems for industrial and mining applications. His areas of interest include electrical drives and control, mining ventilation systems, and power electronic converters, with particular emphasis on improving performance, safety, and energy efficiency in modern electrical engineering systems.

Shesha H. Jayaram is Professor and Director of the High Voltage Engineering Laboratory in the Department of Electrical and Computer Engineering at the University of Waterloo. She previously held the University Research Chair and is also Adjunct Professor at both the University of Guelph and McMaster University. Dr Jayaram earned her BASc degree in Electrical Engineering from Bangalore University, India, in 1980, MASc degree in High Voltage Engineering from the Indian Institute of Science, Bangalore, in 1983, and PhD degree in Electrical Engineering from the University of Waterloo in 1990. Prior to joining the University of Waterloo, she was Assistant Professor at the University of Western Ontario from 1990 to 1992 and Adjunct Professor from 1992 to 2003. Dr Jayaram's research focuses on high-voltage engineering, insulation diagnostics, solution-based outputs, nanocomposite materials, and pulse power applications in biotechnology and environmental engineering. She has made significant contributions to electrical insulation integrity in power grids, and her research has facilitated technology transfer to industries globally. An active member of the IEEE Dielectrics and Electrical Insulation and Industry Applications Societies, as well as the Electrostatic Society of America, Dr Jayaram is a Registered Professional Engineer in the Province of Ontario, Canada.

Kalpana R. (Senior Member, IEEE) received her PhD degree in Electrical Engineering from the Indian Institute of Technology Delhi, India. She joined as Assistant Professor in the Department of Electrical and Electronics Engineering, National Institute of Technology Karnataka (NITK), India, in 2013. She is presently working as Associate Professor in the Department of Energy and Environment, National Institute of Technology Tiruchirapalli, India. She has more than 12 years of teaching and research experience and she has demonstrated excellence in securing research funding, having completed four projects with a total grant amount of Rs. 90 lakhs. She has supervised eight PhD students, published 50 journal/transactions papers, 100 conference papers, and holds four Indian patents and filed one Indian patent. She has been elected as IEEE IAS Member at Large Executive Board Member 2025–2026.

Prashant Kumar received his BTech degree in 2014 and subsequently completed his MTech and PhD from IIT (ISM) Dhanbad, India, in 2017 and 2022, respectively. He is currently serving as Assistant Professor in the Department of AI and Big Data at Woosong University, Republic of Korea. His research focuses on the application of artificial intelligence to modern industrial systems, particularly in the prognostics and health management (PHM) of rotating machines and industrial robots. His research interests include electrical drives, smart manufacturing, and emerging e-mobility systems, with an emphasis on improving reliability, efficiency, and intelligent monitoring of complex engineering systems.

Sze Sing Lee received the B.Eng. (Hons.) and PhD degrees in Electrical Engineering from Universiti Sains Malaysia, Malaysia, in 2010 and 2013, respectively. From 2014 to 2019, he was Lecturer/Assistant Professor at the University of Southampton Malaysia Campus. From 2018 to 2019, he was Visiting Research Professor at Ajou University, South Korea. He is currently Assistant Professor and Degree Program Director of Electrical Power Engineering at Newcastle University in Singapore. His research interests include power converter/inverter topologies and their control strategies. Dr Lee is Associate Editor of the *IEEE Transactions on Power Electronics* and the *IEEE Transactions on Industrial Electronics*. He is a Chartered Engineer registered with the Engineering Council, UK, and currently serves as Professional Review Interviewer and International Professional Registration Advisor.

Xiaodong Li is Professor in the Faculty of Innovation Engineering, Macau University of Science and Technology. He completed his PhD from the University of Victoria, British Columbia, Canada. His research interests include power converters in transportation electrification and AI applications in smart grids. He has published more than 100 journals and conference papers with over 5,000 citations (data from Google Scholar). He also holds four US patents and five Australia Innovation Patents. He is on the list of "the World's Top 2% Scientists" by Elsevier and Stanford University since 2022. He was a recipient of Industry Postgraduate Scholarship (IPS) from the Natural Sciences and Engineering Research Council of Canada (NSERC), the IEEE Power and Energy Society Best Paper Prize in 2007, and the BOC Excellent Research Award from the Macau University of Science and Technology in 2013. Dr Li is a senior member of IEEE, Chair of IEEE Macau Section in 2022–2026.

Nishant Mishra received his BTech degree in 2017 from GGSI University, India, and MTech degree in 2022 from the National Institute of Technology, India. He is currently pursuing his PhD at IIT(ISM) Dhanbad, India. His research interests lie in developing intelligent and energy-efficient solutions for modern building and industrial systems. His work focuses on HVAC systems, building energy management, machine intelligence, and advanced energy-efficient control strategies to improve sustainability and operational performance in smart infrastructure.

Sivakumar Nadarajan is currently Associate Professor in the Department of Electrical Engineering at the Singapore Institute of Technology (SIT), Singapore. He received the BTech degree in Electrical and Electronics Engineering and the MTech degree in Electrical Drives and Controls from Pondicherry Engineering College, India, and the PhD degree in Electrical Engineering from the National University of Singapore. Prior to joining SIT, he accumulated around 19 years of industrial experience across the aerospace, marine, and automotive sectors, including service at Rolls-Royce Singapore Pte. Ltd. from 2009 to 2024 as Technical Manager and Head of the Centre of Excellence for Electrical Equipment Health Monitoring. His research and teaching interests include condition monitoring, failure prognosis, electrical machines and drives, power electronics, machine modelling, and marine EV charging. His notable achievements include filing over 40 patents in the United States and Europe, publishing more than 50 articles, attaining IET Chartered Engineer status, and received awards such as the 2023 ASEAN Outstanding Engineering Achievement Award, the 2023 IES Prestigious Engineering Award (Singapore), and the Sir Henry Royce Award for innovation and engineering excellence from Rolls-Royce.

Kranthi Panuganti received his BTech degree in electrical and electronics engineering from the National Institute of Technology Goa, India, in 2019. He is currently pursuing the PhD degree in Electrical Engineering with the School of Electrical Sciences, Indian Institute of Technology Goa, Goa.

His research interests include modular multilevel converters, electric motor drives, pulse width modulation schemes, multilevel DC–DC converters, and model predictive control schemes.

Kaushik Paul is currently serving as Assistant Professor in the Department of Electrical Engineering at BIT Sindri, India. He obtained his highest degree, PhD, from NIT Jamshedpur. His research and teaching interests include congestion management and related areas in electrical engineering. Among his noteworthy achievements are his academic and professional contributions through teaching, research, and scholarly activities in his area of specialisation.

Saravana Prakash P. (Senior Member, IEEE) received his PhD degree in Electrical and Electronics Engineering from the National Institute of Technology Karnataka (NITK), India, in 2019. He has worked as Assistant Professor in the Department of Electrical Engineering, Malaviya National Institute of Technology (MNIT), India, from March 2020 to January 2025. He is currently working as Assistant Professor in the Department of Electrical and Electronics Engineering, National Institute of Technology Puducherry (NITPY), India. He has authored more than 50 technical papers in *IEEE Transactions/IET Transactions*, conferences, and holds three Indian patents. His research interests include power electronics, power quality improvement in AC–DC converters, SMPS, and electric vehicles.

Sashidhar Sampathirao (Senior Member, IEEE) received his PhD degree in Electrical Engineering from the Indian Institute of Technology (IIT) Bombay, India, in 2017. He is currently working as Associate Professor with the School of Electrical Sciences, IIT Goa, India. He was a research assistant with the Naval Science and Technology Laboratory, DRDO, for his M.E. project until 2011. In 2012, he joined the National Centre for Photovoltaic Research and Education, IIT Bombay, as Senior Research Assistant. He was Project Research Scientist with the Electrical Engineering Department, IIT Bombay, until 2018. His research interests include design and finite-element modelling of electrical machines and motor drives.

Shakthi Prasad D. completed his B.E. in Electrical Engineering from Malnad College of Engineering, and his MTech (Research) in Power and Energy Systems from the National Institute of Technology Karnataka, India. He pursued his PhD in High Voltage Engineering at the Indian Institute of Science, Bangalore. He is currently Associate Professor in the School of Electrical Sciences at the Indian Institute of Technology Goa. His areas of interest include high-voltage engineering, insulation engineering, computational electromagnetics, and plasma applications. He serves as Associate Editor for *IET Nanodielectrics*. He was elected as an INAE Young Associate in 2024 and received the Excellence in Teaching Award at IIT Goa in 2025.

Neel Shrivastava received his MTech degree in Electrical Engineering from the Indian Institute of Technology (IIT) Goa, India, in 2022. He is currently pursuing a PhD degree in Electrical Engineering at IIT Goa, India. His research interests include the design, finite-element modelling, and analysis of electrical machines, including field-modulated linear and rotary machines, magnetic-geared electrical machines, and generators for renewable energy applications.

Pampa Sinha is currently associated with KIIT University, India, where she is engaged in teaching, research, and academic development in the field of electrical engineering. She holds a PhD in Electrical Engineering and has more than ten years of teaching and research experience. Her research and teaching interests include power systems, power quality monitoring, fault detection, smart grids, signal processing, and AI-based analysis for modern energy systems. Her noteworthy achievements include significant contributions to advanced fault detection, power quality assessment, and intelligent monitoring frameworks for next-generation power and energy systems.

Snehalika is currently serving as Assistant Professor at KIIT University, India, and is actively engaged in academic and research activities in the field of power electronics. She holds a PhD, and has been involved in teaching and research in her area of specialisation. Her research and teaching interests include power electronics, power systems, control techniques, and related emerging technologies. She is also associated with IEEE-related academic and professional activities. Her noteworthy achievements include her contributions to academics, research, and professional practice through teaching, scholarly work, and active participation in her domain.

Kazem Varesi received his BSc in Electrical Engineering (Power Engineering) from the University of Tabriz in 2008, followed by MSc with Honours in Power Engineering (Power Electronics and Electric Machines) from K. N. Toosi University of Technology, Iran, in 2011. He completed his PhD in Power Engineering (Power Electronics) at the University of Tabriz in 2017. That same year, he joined Sahand University of Technology, Iran, as Assistant Professor in the Faculty of Electrical and Computer Engineering and was promoted to Associate Professor in March 2023. His research interests include DC–DC converters, multilevel inverters, and renewable energy systems. In 2023, he received the Best Researcher Award from Sahand University of Technology's Faculty of Electrical and Computer Engineering, followed by the Best Young Researcher Award of East Azerbaijan Province in 2024. He has been recognised in Stanford University's (Scopus) ranking of the World's Top 2% Most Cited Researchers in 2022, 2023, and 2024.

Acknowledgement

This book is inspired by a joint research project, "More Electric and All Electric Aircrafts towards Resilient Decarbonized Communities," funded under the Scheme for Promotion of Academic and Research Collaboration (SPARC) by the Government of India. The project was hosted by the Indian Institute of Technology (IIT) Goa, India, in collaboration with the University of Waterloo, Canada; Singapore Institute of Technology, Singapore; and the National University of Singapore.

1

Net-Zero and Aircraft Electrification

Sivakumar Nadarajan

1.1 Introduction

The aviation industry is undergoing a transformation driven by environmental sustainability and technological innovation. As industries worldwide strive to achieve global net-zero targets, the aviation industry is intensifying its focus on reducing emissions from their products. The aviation sector currently contributes around 3% of global carbon dioxide emissions. Emission is projected to increase significantly in coming decades due to the coming passenger demand. This would make the aviation industry as one of the major contributors of global emission, hence it is important for the industry to focus on measures to reduce the emission from their products and augment with carbon offsetting mechanisms to strike balance to achieve net zero.

Aviation industry is currently looking at various avenues to reduce the emissions which include electrification, alternative fuel, innovative aircraft design, and novel materials. As aviation industry has various market segments, a tailored approach is required to reduce emission from each market segment and aircraft type. Aircraft electrification includes full-electric aircraft, hybrid-electric aircraft, and more-electric aircraft. The full-electric aircraft is a zero-emission option. The hybrid-electric aircraft relies on both the gas turbine engine and electrical systems for power and propulsion; it is not a zero-emission option but assists to reduce the emission significantly. The more-electric aircraft focuses on incremental improvement in energy efficiency and performance by replacing conventional mechanical, hydraulic, and pneumatic systems. In general, electrification offers profound benefits of improved energy efficiency, reduced system weight, reduced noise, better system control, and reduced operational costs. In addition, the full-electric aircraft enables the new mobility paradigms such as urban air mobility (UAM) and reduces the environmental carbon footprint.

DOI: 10.1201/9781003747994-1

1.2 Net-Zero Target

The global greenhouse gas emission is increasing exponentially since the beginning of the industrial revolution, due to the global average temperature increase of about 1.5°C compared to the preindustrial era, and the temperature is continuously increasing. Figure 1.1 shows the global average temperature since 1850 [1].

The concept of net-zero emissions emerges after the Paris Agreement adopted in December 2015 at COP21 [2]. The agreement set the ambitious target of limiting the global average temperature increase below 2 °C and the stretched target of limit below 1.5°C by going net zero no later than 2050.

The definition of net zero revolves around balancing the GHG emission with the CO_2 removal, which results in net-zero emission; but this is unrealistic to achieve with the current level of GHG emission, that is, >55GT CO_2e, because there are insufficient natural and artificial carbon-removal mechanisms of the current global GHG emission. Hence, the first step towards net zero is to significantly reduce the global GHG emission in coming decades to the level where the natural and artificial CO_2 removal is possible [3].

Globally, over 100 countries have already pledged to go net zero by mid- or second-half of the century, through implementing national climate action plans, long-term strategies, and policies. These countries collectively account for more than 80% of the GHG emission.

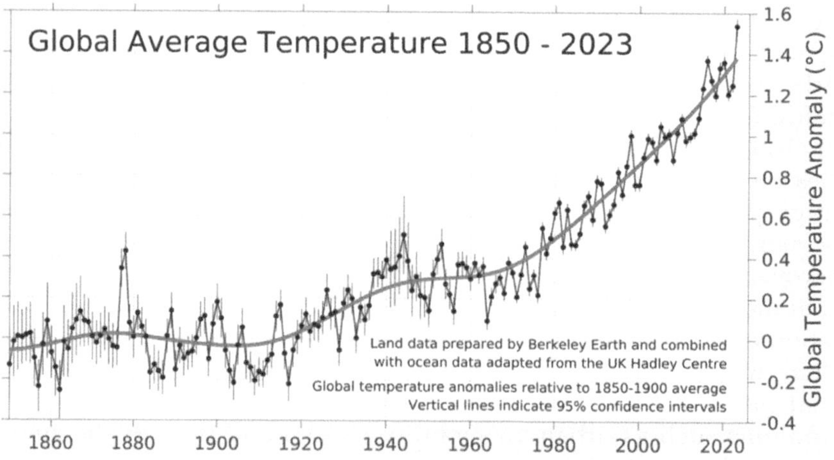

FIGURE 1.1
Global average temperature anomaly since preindustrial era.

Achieving net-zero goal necessitates initiatives to reduce emissions across all sectors, including aviation by second half of the twenty-first century. In the aviation sector, the International Civil Aviation Organization (ICAO) has endorsed the long-term aspiration of achieving net zero by 2050. The aviation industry is pursuing various strategies and initiatives to march towards the net zero, which includes sustainable aviation fuels (SAF), improvement in operation efficiency, and innovative propulsion technologies.

1.3 Net-Zero Progress Status

Although countries and industries have pledged to achieve the net-zero goals through their products, policies, and actions, the progress of achieving the net-zero target is still behind the schedule. In 2023, a comprehensive analysis was conducted to evaluate the progress through 42 climate action indicators across sectors. The analysis revealed that only the adoption of electric vehicles is currently on track to meet the 2030 targets [4, 5]. The other indicators are either off-track or progressing in a wrong direction, as shown in Figure 1.2. In the aviation sector, the production of SAF shows incremental growth, but the challenges still exist on scalability, feedstock availability, and fuel cost. The development of electric aircraft has gained momentum, several companies have developed and tested their prototypes, and regulatory bodies have started developing certification framework and guidelines to accommodate electric aircraft.

1.4 Aviation Industry GHG Emission

The transportation industry is responsible for 15% of the global GHG emissions [6], and the aviation industry has contributed about 12% of the total emission from the transportation industry [7]; hence the aviation industry accounts for approximately 2–3% of global GHG emission [8], as shown in Figure 1.3. Figure 1.4 shows the emission from the commuter aviation.

The projected growth of aviation industry is about 4.3% per annum [9] and it is expected that by 2050 the number of passengers travelling by air would reach 10 billion annually [10]. This is due to increase in individual income, urbanisation, and globalisation. Figure 1.5 shows the projected aviation emission till 2050. This necessitates the business case for rapid decarbonisation of the aviation industry as no intervention in this regard would lead the aviation emissions to triple by 2050. The emissions from aviation industry have arisen from multiple sources, including flight operations, ground-handling, and maintenance. The

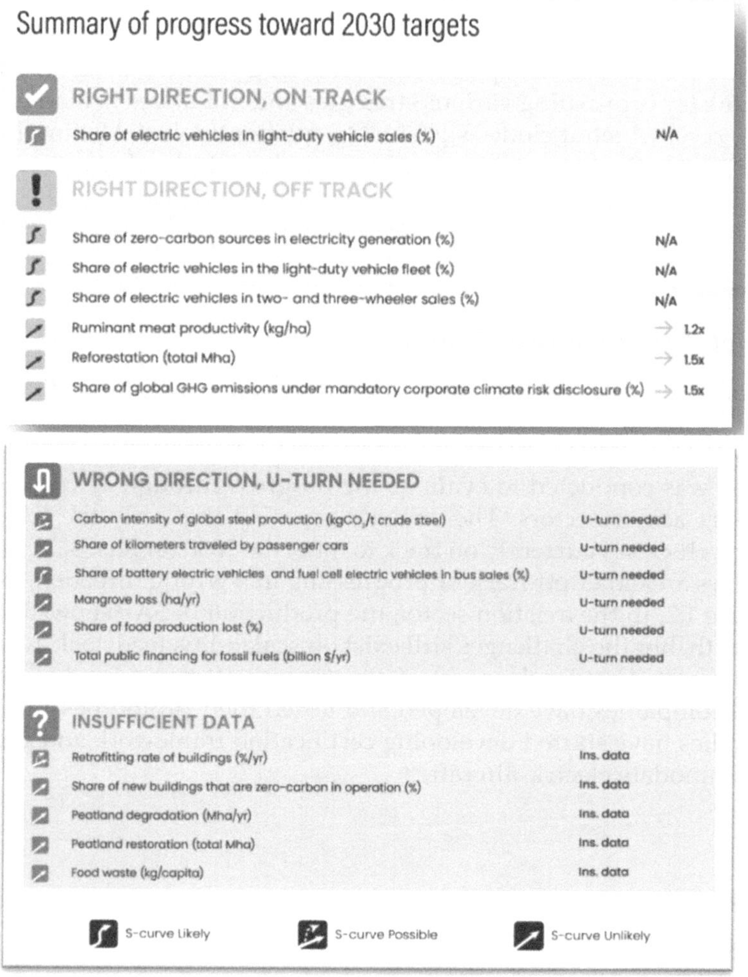

FIGURE 1.2
Progress of aviation sector indicators towards 2030 climate goals.

wide and narrowbody aircraft contribute a significant portion of the emission [11]; the United Nations have predicted that by 2050, two-thirds of the world population will live in cities [12] and the short-distance air taxi market would pick up as the urbanisation make land transport challenging. Hence, the emission from the intercity air taxi would increase significantly.

Aviation industry is looking at four levers to reduce the GHG emissions: technology innovations, improvements in operations and infrastructure, sustainable aviation fuel (SAF), and out-of-sector carbon reduction market [10].

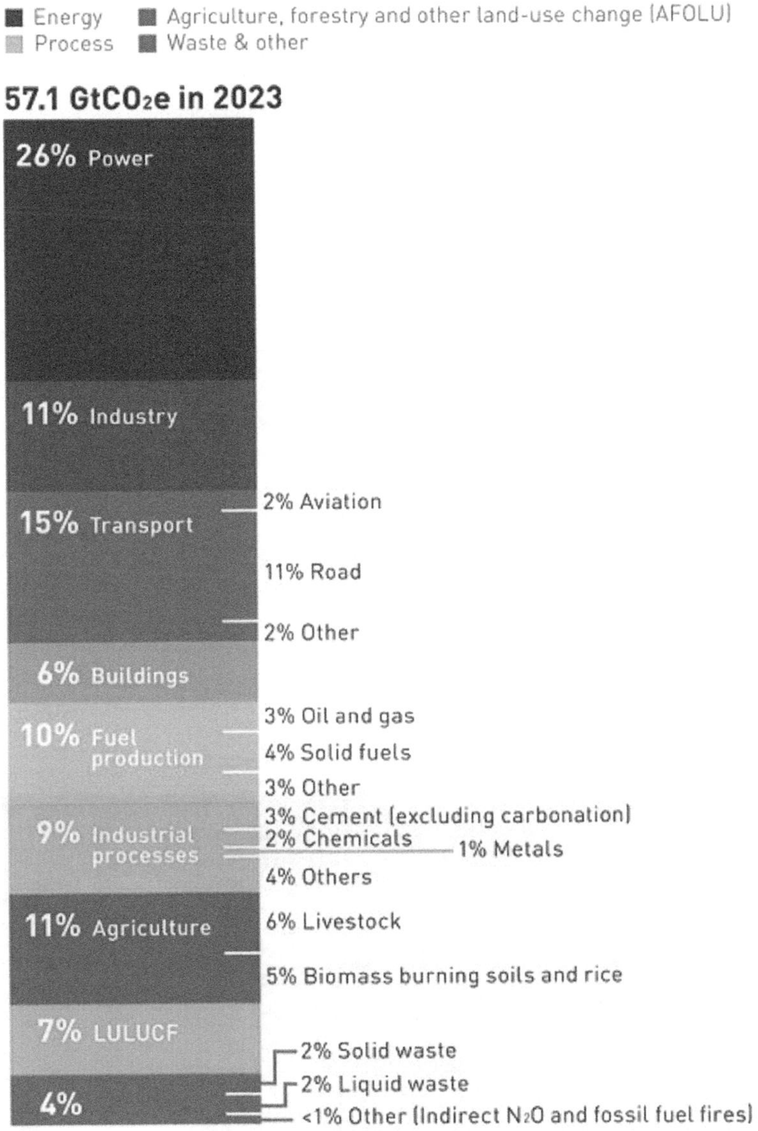

FIGURE 1.3
Share of aviation CO_2 emissions compared with other parts of the economy [8].

The technology innovations focus on leveraging innovative or next-generation technologies to improve the overall fuel efficiency of the existing aircraft and develop innovative aircraft design and propulsion concepts for next-generation aircraft. The aircraft engine efficiency improves over the

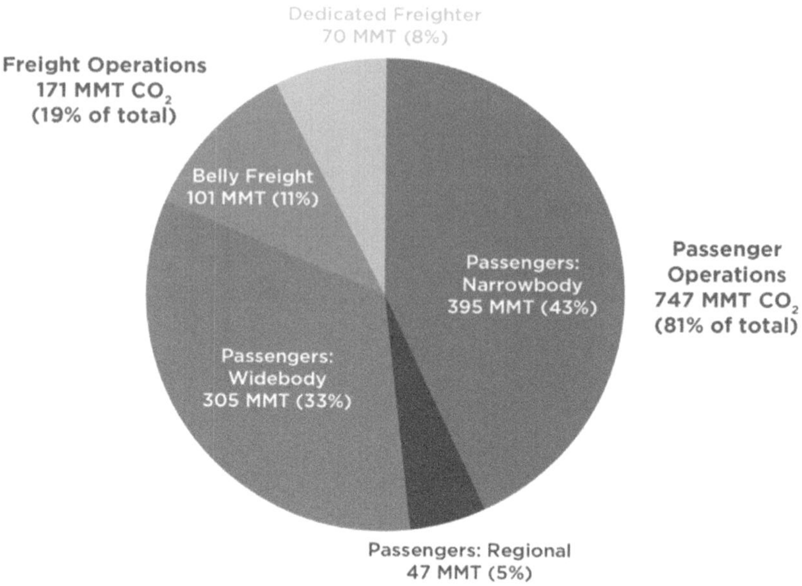

FIGURE 1.4
Emission from commercial aviation [11].

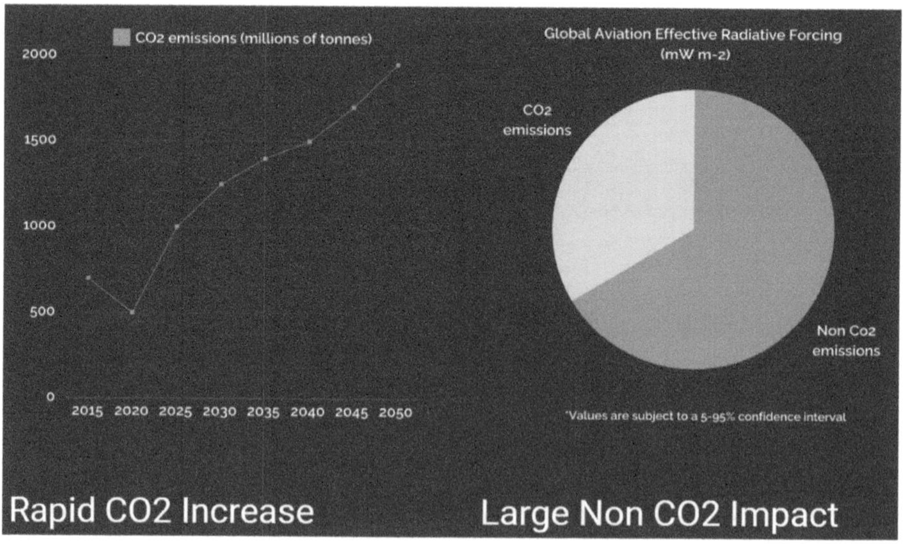

FIGURE 1.5
Projected global aviation CO_2 emissions [13].

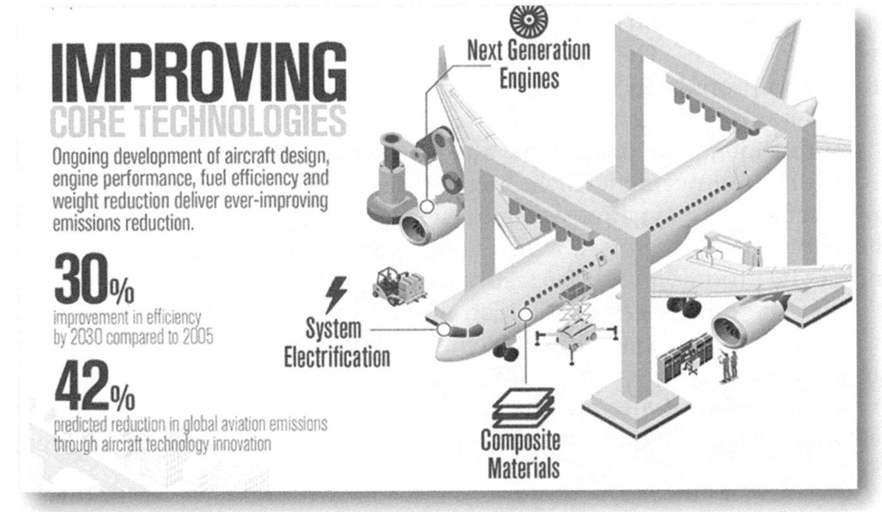

FIGURE 1.6
Innovation with technology on existing aircraft [10]

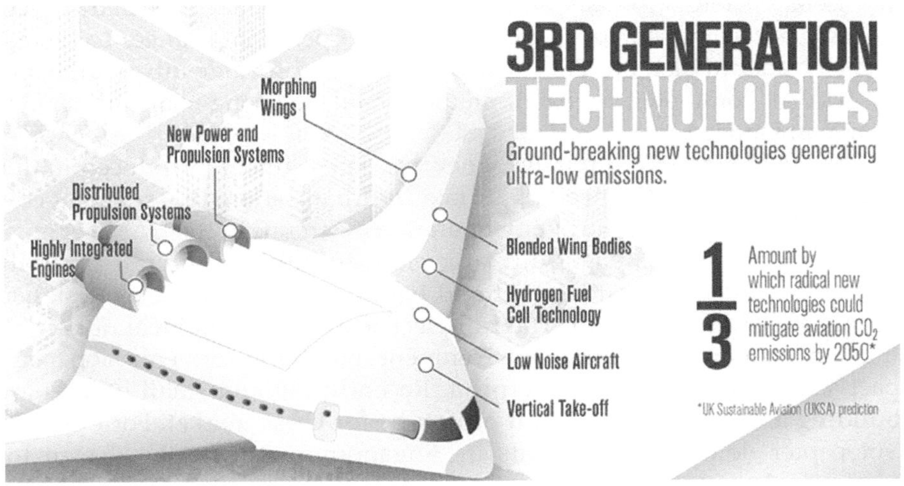

FIGURE 1.7
Innovation with technology on next-generation aircraft [10].

years, for instance, the overall fuel efficiency of the aircraft engine would improve 30% by 2030 when compared to 2005. Furthermore, the innovative aircraft design for better aerodynamics, new materials, new engine design, and advanced manufacturing technologies improves the efficiency of next-generation aircraft further, as shown in Figures 1.6 and 1.7. Every 1% increase in energy efficiency could result in saving US$75,000 per annum per aircraft, hence any incremental in energy efficiency of aircraft bring millions of dollar savings to airlines. The innovation also focuses on electrification and novel power and propulsion systems for aircraft, which includes battery-powered propeller and hydrogen fuel cell.

Improvements in operations and infrastructure focus on operational procedure improvement, including airline load factor improvements and improvements for aviation systems. The sustainable aviation fuel focuses on alternative fuel for aircraft, which would significantly reduce the emissions and can be produced with the feedstock or through carbon capture and electrolysis. The out-of-sector reduction market focuses on the aviation industry investment on accelerating development of carbon capture and direct air capture efficiency.

Electrification of aircraft power and propulsion systems offers a viable pathway for reducing the emissions particularly in the short haul segments.

1.5 Aviation Industry Market Segments

The aviation industry caters different market segments. The market segments are categorised based on the aircraft operational range, passenger capacity, and mission profile. The market segments include urban air mobility (UAM), commuter, regional, medium-haul and long-haul airliners, as shown in Figure 1.8.

The UAM market is a relatively new market segment introduced in this decade considering the need of air taxi in urbanisation; this segment covers the short-range and intra-city operations. Aircraft in the UAM segment cover two- to six-passenger aircraft, while full-electric aircraft typically aircraft includes vertical take-off and landing (eVTOL) vehicles designed for missions under 100 km. The full-electric aircraft is an ideal option for the densely populated urban environment due to the zero emission, compact size, and low acoustic footprint. Recently, big and small players are building and testing the aircraft prototype for UAM market. For example, Volocopter demonstrated flights in Singapore, maturing the design for public deployment [14].

Commuter market segment refers to the short-haul flights served by aircraft with capacity of 9–50 passengers. These flights typically connect smaller communities or serve as feeders for larger airlines. The typical operating range of the commuter aircraft is below 500 km. These aircraft often perform multiple short legs per day, which makes them suitable for full-electric or hybrid-electric propulsion systems. Their relatively modest energy requirements and predictable flight patterns allow for efficient integration of electric powertrains. Notable developments in this segment include Heart

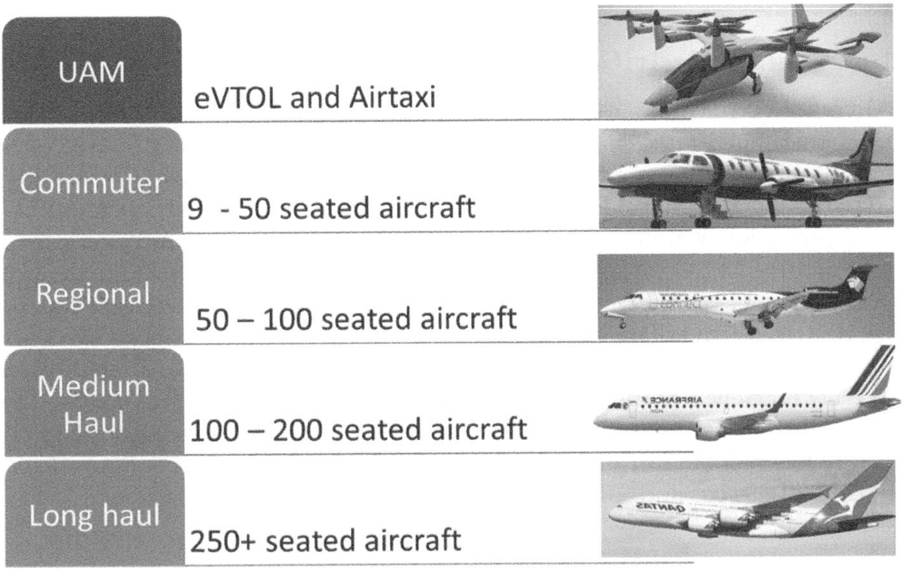

FIGURE 1.8
Aviation market segments.

Aerospace's ES-30 and Eviation's Alice, both of which aim to deliver zero-emission regional connectivity.

Regional market segment serves routes extending up to 1,000 km, regional aircraft accommodates 50–100 passengers. Electrification of this segment presents greater challenges due to increased energy demand and longer flight durations. The power density of battery poses technical challenge in offering full-electric aircraft for the market segment. However, the hybrid-electric configurations and hydrogen fuel cell systems offer promising alternatives. ZeroAvia's hydrogen-electric demonstrators illustrate the feasibility of clean propulsion technologies for regional applications.

The medium- and long-haul market segments cater flight services for intercontinental travel and are capable of carrying over 100 passengers. Electrification of this market segment with the current technologies pose significant barrier in terms of operating cost and safety. The sustainable aviation fuels (SAFs) and hydrogen combustion are considered more viable decarbonisation pathway for the market segment in the near term. Nevertheless, partial electrification of on-board systems such as environmental control, actuation, de-icing, auxiliary power, etc. offers improvements in energy efficiency and in turn reduced emissions [15].

In conclusion, the potential for electrification varies considerably across market segments. UAM and commuter aircraft are positioned well for early adoption, while regional and long-haul platforms likely depend on hybrid systems or alternative fuels until energy storage technologies advance further.

1.6 Aircraft Electrification

Aircraft electrification refers to the systematic replacement of conventional mechanical, hydraulic, and pneumatic systems with electrically powered alternatives. Electrification of the aircraft has three principal approaches: more-electric aircraft (MEA), hybrid-electric aircraft (HEA), and full-electric aircraft (FEA). The application of aircraft-electrification approaches depends mainly on aircraft size, and the MEA concept is widely applied for bigger aircraft where the propulsion power comes from the gas turbine engine; the MEA concept focuses on electrifying auxiliary systems for better energy efficiency and control, the hybrid-electric aircraft concept is applicable for commuter class of aircraft where the fuel efficiency of the conventional propulsion system can be increased by integrating electrical technology, and the full-electric aircraft concept caters primarily to small aircraft in the UAM market. The architectural configuration of the hybrid-electric and full-electric aircraft is shown in Figure 1.9.

1.6.1 More-Electric Aircraft

In MEA platforms, the propulsion systems remain conventional, but subsystems driven by pneumatic, hydraulic, and mechanical systems such as flight control actuators, gas turbine engine starting, landing gear, environmental control systems, etc. are replaced with electrical systems, Figure 1.10

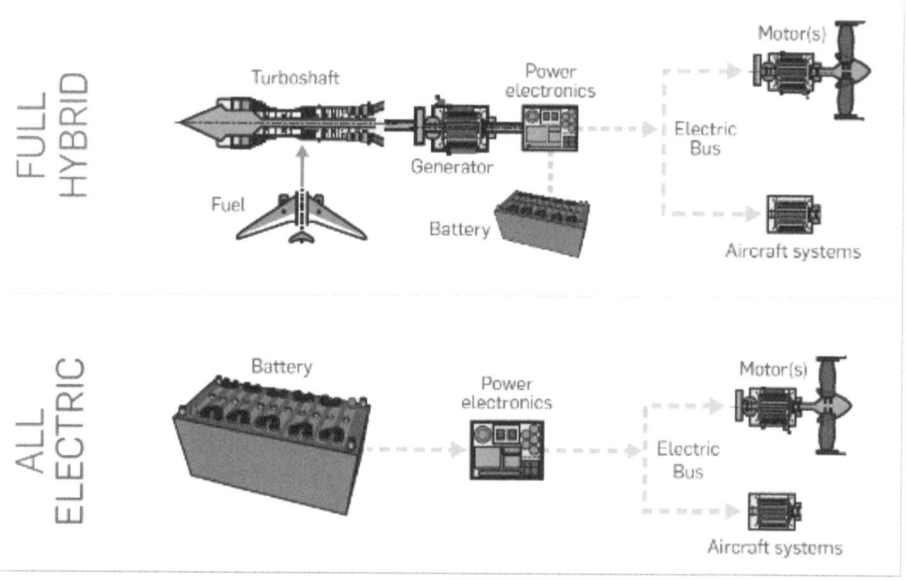

FIGURE 1.9
Full-electric and hybrid-electric architectural configuration.

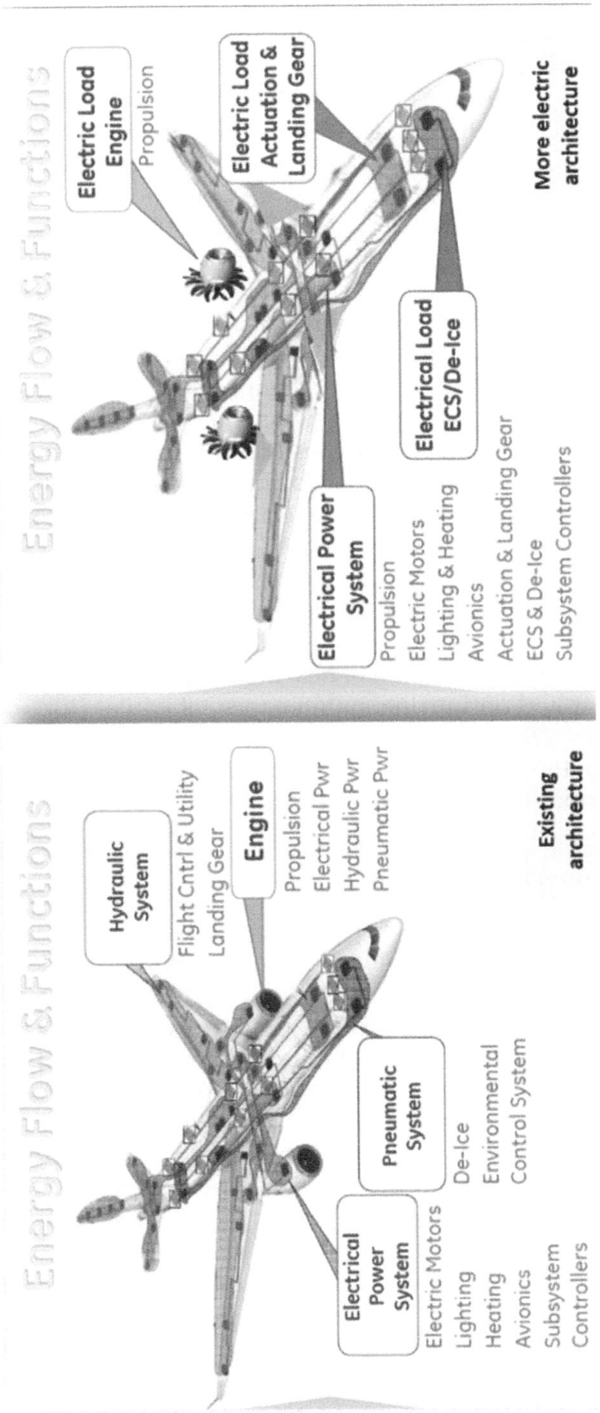

FIGURE 1.10
Comparison of conventional and MEA system configuration [17].

shows the potential MEA configuration of an aircraft compared to conventional aircraft. For instance, replacing the pneumatic system, which supplies compressed air from the gas turbine engine into the auxiliary power unit (APU) to start the gas turbine engine, with the electrical starter generator for starting the engine, which provides significant savings on the weight, space, and better control, increases the overall fuel efficiency. Similarly, replacing engine bleed air system with electrical air-conditioning compressors for environmental control system provides significant fuel savings. In addition, replacing the heavy and high maintenance hydraulic actuators for controlling stator vanes and other actuators with the electrical motors provides significant weight savings and reduces maintenance. These approaches have already been implemented in commercial aircraft such as the Boeing 787 and Airbus A380, which utilise electric compressors and actuators to reduce reliance on bleed air and hydraulic systems [16].

MEA offers superior performance and fault tolerance compared to their mechanical counterparts. MEA also facilitates the integration of health-monitoring systems, allowing for real-time diagnostics and predictive maintenance.

Projects such as iBearing and iGear, developed under the Clean Sky 2 initiative, exemplify the application of intelligent health monitoring in MEA platforms. These systems use embedded sensors and machine learning algorithms to assess the condition of bearings and gearboxes, enabling early detection of wear and optimising maintenance schedules [18].

MEA serves as a foundational step towards full electrification, preparing the industry for the integration of electric propulsion and supporting the transition to sustainable aviation.

1.6.2 Hybrid-Electric Propulsion

The hybrid-electric propulsion approach combines the advantages of the combustion engine and electrical drives, which allows flexible power management across different phases of flight driving towards better energy efficiency. The hybrid-electric propulsion system provides up to 5% reduction in fuel consumption compared to a standard flight [19], and this figure could be as high as 10% for helicopter. The electricity for the hybrid-electric system could come from batteries or fuel cell that converts hydrogen into electricity. The hybrid-electric propulsion system could be a serial hybrid or parallel hybrid, as shown in Figure 1.11. In the series hybrid configuration, the engine generates electricity that powers the motor; the energy storage system in the series hybrid system like battery operates as either source or sink, when there is an increased propulsion power demand like during take-off, the battery operates as a source supply of additional power to the motor along with the generated power from the engine. During the lesser propulsion power demand, for example, during crusing, the battery operates as a power sink to charge or store energy in the battery. In parallel configurations, the propeller

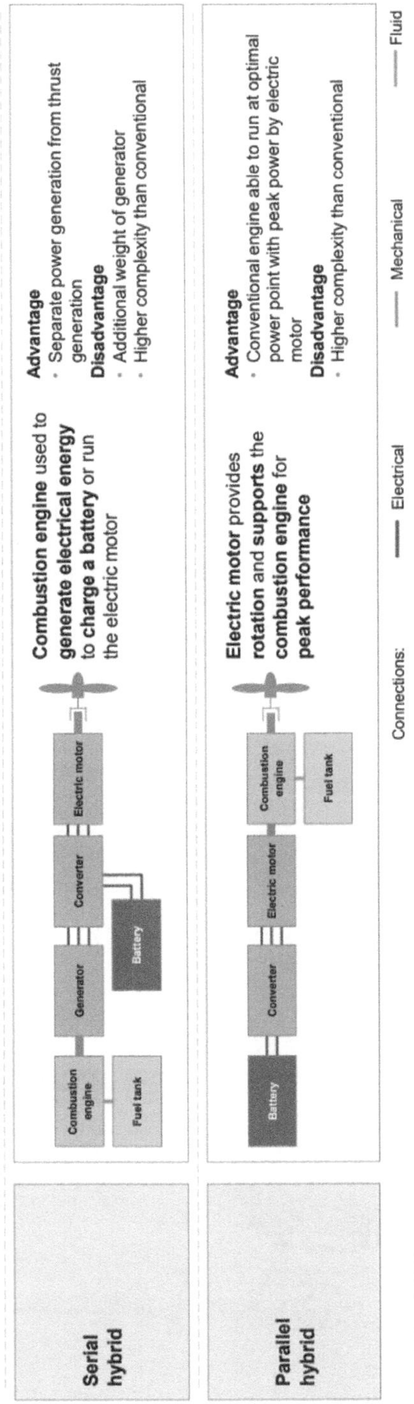

FIGURE 1.11
Hybrid-electric aircraft configurations.

is directly coupled to the engine and the engine is also coupled to an electric motor. At high propulsion power demand, the electrical motor complements the torque produced by the engine; and at lesser power demand, the electrical motor is driven by engine, so it operates as a generator and charges the battery.

Key aviation industry players are demonstrating the hybrid-electric propulsion system for aircraft, the demonstration programmes such as Airbus's E-Fan X and Avio Aero's RACER rotorcraft have explored hybrid-electric propulsion in various configurations. These platforms validate the integration of electric motors, power electronics, and energy management systems under real-world conditions [20]

Hybrid-electric aircraft offer a practical pathway to decarbonisation, leveraging existing technologies while enabling incremental improvements in performance and sustainability. They also provide a testbed for developing certification standards and operational procedures for electric propulsion [21]

E-FanX is a demonstrator where the key industry players like Airbus, Siemens, and Rolls-Royce worked together to demonstrate the series hybrid concept for a regional jet. The E-FanX series hybrid architecture is shown in Figure 1.12. Airbus developed energy storage and controller for the hybrid system, Rolls-Royce developed 2-MW generator coupled to one of their small gas turbine engines, and Siemens developed propulsion motor and drive for the demonstrator [22].

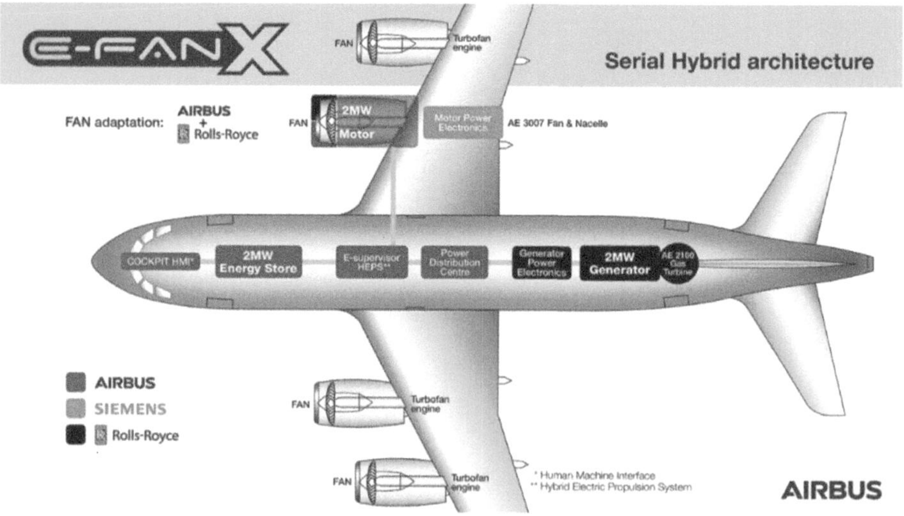

FIGURE 1.12
E-FanX series hybrid architecture [22].

1.6.3 Full-Electric Aircraft

Full-electric aircraft relies exclusively on batteries or fuel cells to power electric motors which are coupled to the propellers, as shown in Figure 1.13. This eliminates the use of fossil fuels entirely. In addition to the elimination of emissions, it offers significant advantages in terms of noise and maintenance. Hence, full-electric aircraft is an attractive option for aviation decarbonisation, but the size of full-electric aircraft is limited by the power density of the energy storage system on board, that is, battery and fuel cell. The current battery technologies limit the electric aircraft range and payload capacity. Hence the full-electric aircraft is limited to short-range missions, specifically UAM market such as urban air mobility, commuter flights, and pilot training. Hydrogen fuel cells are also being explored as a means to extend range while maintaining zero-emission operation [23].

As aerospace application is a safety and mission-critical application, it is important to ensure the safety of the on-board systems. Hence, the full-electric aircraft architecture considers a multi-lane fault-tolerant architecture to increase reliability and availability of the on-board system to ensure aircraft safety. The multilane propulsion system basically refers to a single-propulsion motor which is driven by multiple power electronics converters powered by independent energy sources such as battery. The propulsion motor for the multilane system is designed to have multiple sets of three-phase windings such as two or four sets of three-phase windings in the stator. Each set of three-phase winding is powered by a three-phase inverter which converts DC supply from a dedicated battery to a controlled three-phase supply for the motor, this is called single lane. Multiple lanes are integrated within a single propulsion motor, as shown in Figures 1.14 and 1.15 for driving propeller of the aircraft. If the propulsion motor encounters a winding fault such as inter-turn or other short circuits in a phase winding, or the inverter powering a set of three-phase motor windings has fault, then that particular lane is disconnected from the propulsion motor, and the motor propels the aircraft with rest of the lanes.

Urban air mobility platforms, including Volocopter's VoloCity and Joby Aviation's S4, are designed for short-range missions in urban environments. These aircraft prioritise safety, noise reduction, and operational efficiency, and have demonstrated successful flight trials in cities such as Singapore [26, 27].

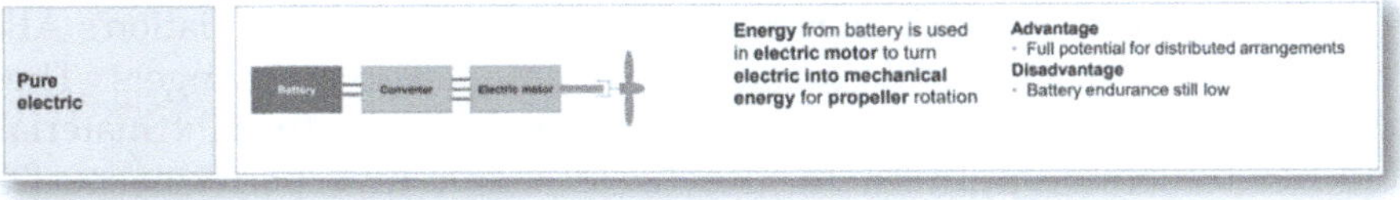

FIGURE 1.13
Full-electric aircraft configuration.

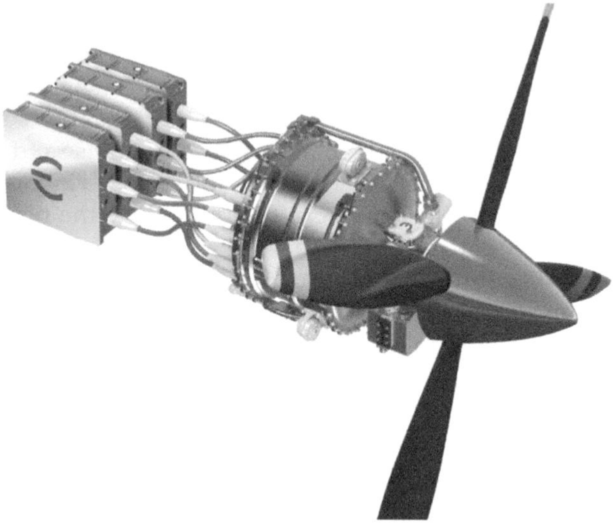

FIGURE 1.14
Emobility: multilane electrical propulsion unit [24].

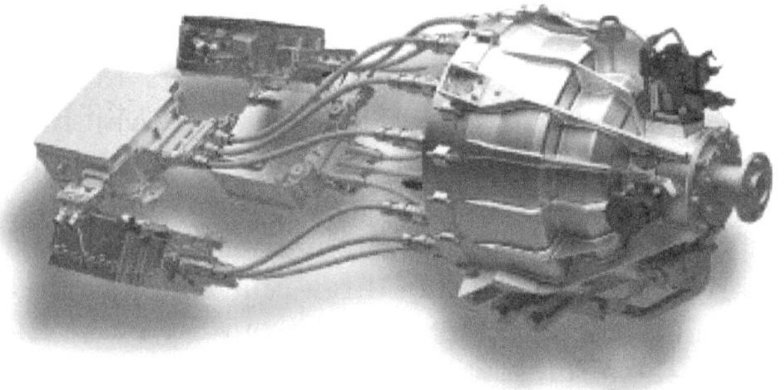

FIGURE 1.15
ZeroAvia: multilane electrical propulsion unit [25].

Figure 1.16 shows various models of the full-electric aircraft system from Volocopter.

Commuter aircraft such as Heart Aerospace's ES-30 and Eviation's Alice target regional routes with seating capacities of 9–19 passengers. These platforms leverage advances in battery technology, lightweight materials, and aerodynamic optimisation to achieve competitive performance [23]. Figure 1.17 shows the Eviation commuter aircraft.

Hydrogen fuel cells offer an alternative to battery-electric systems, particularly for the longer-range applications. Companies like ZeroAvia and

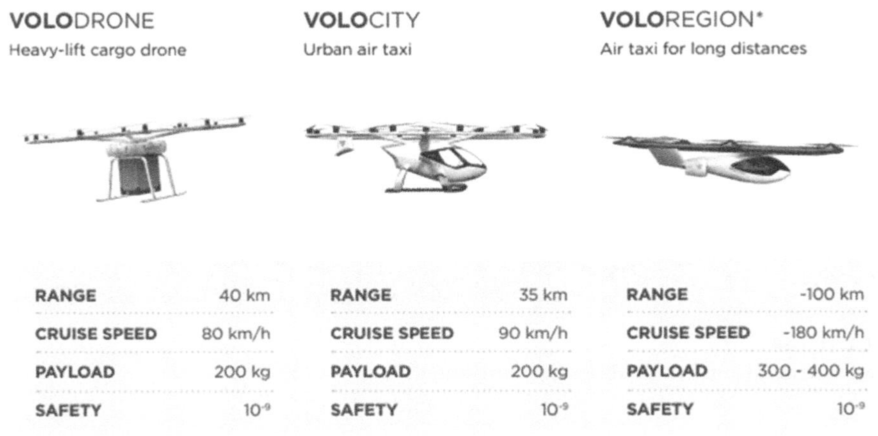

FIGURE 1.16
Volocopter full-electric aircraft.

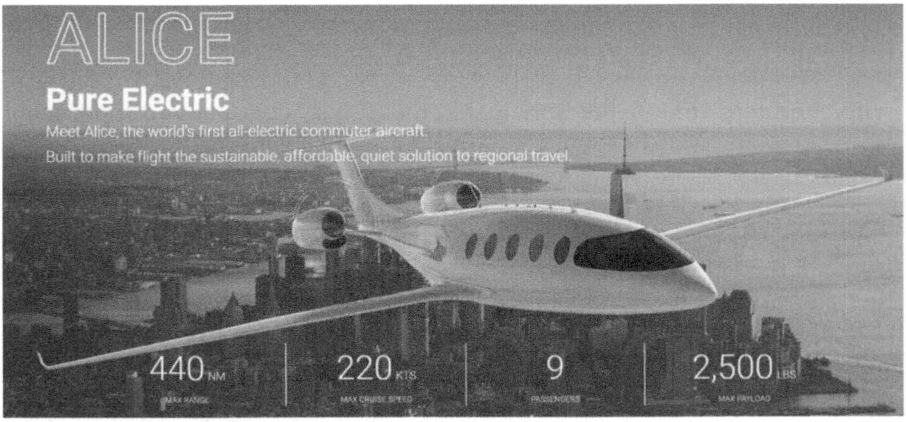

FIGURE 1.17
Eviation commuter electric plane [28].

Universal Hydrogen are developing fuel cell–powered aircraft for regional markets, as shown in Figure 1.18, with flight demonstrations validating the feasibility of hydrogen-electric propulsion [23].

The primary limitation of electric aircraft is energy density. Lithium-ion batteries currently offer 200–300 Wh/kg, which constrains range and payload. Research into solid-state batteries and lithium–metal chemistries aims to overcome these limitations, although challenges in safety, manufacturability, and cycle life remain [29]

FIGURE 1.18
ZeroAvia and Universal Hydrogen demonstrator flight.

Electric aircraft represent a paradigm shift in aviation, offering a sustainable, efficient, and scalable solution for specific market segments. As technology matures and regulatory frameworks evolve, electric propulsion is expected to become a cornerstone of future air transport systems.

The development of electric aircraft has gained momentum, with several prototypes undergoing flight testing and certification processes. Companies such as Joby Aviation, Archer Aviation, and Volocopter have made significant strides in advancing electric vertical take-off and landing (eVTOL) platforms, with commercial launches anticipated by 2025 [27, 29]. These developments signal growing industry readiness and regulatory alignment, particularly in the Urban Air Mobility segment.

Regulatory bodies have begun to adapt certification frameworks to accommodate electric propulsion systems. The Federal Aviation Administration (FAA) in the United States has introduced new categories such as "powered-lift" aircraft and has established pilot training standards specific to eVTOL operations [29]. Similarly, the European Union Aviation Safety Agency (EASA) has revised its Certification Specifications (CS-23) to include provisions for electric and hybrid-electric aircraft. These regulatory evolutions are essential for enabling the safe and efficient integration of electric aircraft into the existing airspace.

Despite these advancements, significant challenges remain. The technology readiness levels of key components such as high-energy-density batteries, power electronics, and electric motors must be elevated to meet the demands of commercial aviation. Infrastructure development, including charging stations, battery swapping facilities, and vertiports, is also critical. Furthermore, public acceptance, economic viability, and supply chain resilience must be addressed to ensure the successful deployment of electric aircraft at scale.

Electrically powered systems offer several operational benefits. They provide precise control, support modular design, and are inherently compatible with health-monitoring technologies. By reducing the use of mechanical,

	2020	2025	2030	2035	2040	2045	2050
Commuter » 9-19 seats » < 60 minute flights » <1% of industry CO_2	SAF	Electric or Hydrogen fuel cell and/or SAF	Electric or Hydrogen fuel cell and/or SAF	Electric or Hydrogen fuel cell and/or SAF	Electric or Hydrogen fuel cell and/or SAF	Electric or Hydrogen fuel cell and/or SAF	Electric or Hydrogen fuel cell and/or SAF
Regional » 50-100 seats » 30-90 minute flights » ~3% of industry CO_2	SAF	SAF	Electric or Hydrogen fuel cell and/or SAF	Electric or Hydrogen fuel cell and/or SAF	Electric or Hydrogen fuel cell and/or SAF	Electric or Hydrogen fuel cell and/or SAF	Electric or Hydrogen fuel cell and/or SAF
Short haul » 100-150 seats » 45-120 minute flights » ~24% of industry CO_2	SAF	SAF	SAF	SAF potentially some Hydrogen	Hydrogen and/or SAF	Hydrogen and/or SAF	Hydrogen and/or SAF
Medium haul » 100-250 seats » 60-150 minute flights » ~43% of industry CO_2	SAF	SAF	SAF	SAF	SAF potentially some Hydrogen	SAF potentially some Hydrogen	SAF potentially some Hydrogen
Long haul » 250+ seats » 150 minute + flights » ~30% of industry CO_2	SAF	SAF	SAF	SAF	SAF	SAF	SAF

FIGURE 1.19
Electrification journey in aviation industry [15].

hydraulic, and pneumatic systems, the on-board electrical systems enhance safety and performance. Moreover, electrification aligns with the broader industry goal of achieving net-zero emissions by mid-century. As the technology continues to evolve and matures over the year by addressing the current challenges on power density and reliability, commuter and regional aircraft market segments are expected to adopt electric propulsion systems, as shown in Figure 1.19. The shift towards electric aviation represents a critical step in the journey towards sustainable flight.

1.7 Electrical Equipment and Systems

The electric aircraft relies on electrical energy for propulsion and critical subsystems. This paradigm shift imposes new demands on the design, integration, and certification of electrical equipment. Innovation in power generation, distribution, conversion, and energy storage is required to meet the demands.

The electrical machines are the key components in electrical propulsion system, which drives the propeller blades to convert electrical energy into mechanical thrust. The state-of-the-art electrical machines typically offer the

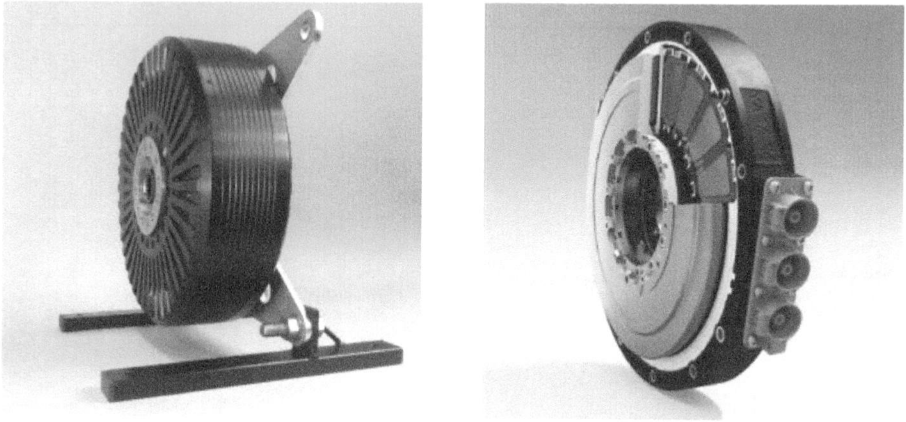

FIGURE 1.20
Axial flux motor developed by the magniX for electric propulsion unit (EPU) [30].

specific power density, that is, power-to-weight ratio of 8–10 kW/kg; but for aerospace application, specifically for the full-electric aircraft, the required power density is >13 kW/kg [30]. To achieve such high-power density as well as better performance, manufacturers are exploring advanced topologies such as axial flux, transverse flux, and radial flux machines. Axial flux motors, for instance, offer compact form factors and high torque density, making them suitable for distributed propulsion architectures in eVTOL and regional aircraft. MagniX is a leading developer of propulsion systems for electric aircraft, including motors, inverters, and motor controllers. The company manufactures a range of electric propulsion units (EPU), including the magni650 EPU rated at 640 kW peak power and 560 kW continuous power, as shown in Figure 1.20. The axial length of the motor P400 model is significantly as low as 80.4 mm.

Transverse flux machines, under development by companies like Rolls-Royce, promise even higher torque densities and simplified cooling requirements, although they present challenges in magnetic design and manufacturing [9]. The liquid cooled electrical machines are the key focus in aerospace industry for better power density; however, the liquid cooled machines require auxiliary cooling system which adds weight to the aircraft. The transverse flux machine has the stator windings placed on the surface of the stator unlike the radial or axial flux machine where the stator windings are placed inside the slots; as the transverse flux machines windings are placed on the stator, it can be cooled effectively by airflow generated by the propeller – so heavy liquid cooled system is not required for the transverse flux machine. Figure 1.21 shows the transverse flux machine prototype developed by Rolls-Royce for UAM market.

FIGURE 1.21
Transverse flux machine architecture and prototype developed by Rolls-Royce for UAM market [31].

Power electronics play a critical role in managing the flow of electricity between the energy source and the propulsion system. Inverters, converters, and rectifiers must operate efficiently under high-voltage and current conditions, often exceeding 800 VDC and hundreds of kilowatts. Silicon carbide (SiC) semiconductors have emerged as the preferred technology for these applications due to their superior thermal conductivity, high breakdown voltage, and switching efficiency. SiC-based inverters developed by GE Aerospace and ZeroAvia have demonstrated power densities exceeding 19 kW/kg and efficiencies approaching 99%, enabling compact and lightweight power conversion systems [23, 29]. Figure 1.22 shows the SiC converters key industrial players developed for aerospace application.

Energy storage remains one of the most critical components in electric aircraft. Lithium-ion batteries are well known for automotive use: the same battery chemistry is predominantly used in electrical aircraft, which offers the energy densities in the range of 200–300 Wh/kg. However, these values are insufficient for medium- and long-haul flights, prompting research into next-generation chemistries such as lithium–sulfur, lithium–metal, and solid-state batteries. These technologies promise energy densities exceeding 500 Wh/kg, but face challenges in cycle life, thermal stability, and manufacturability [30]. In parallel, hydrogen fuel cells are being explored as an alternative energy source, particularly for regional aircraft. Fuel cells offer higher specific energy and zero-emission operation but require complex storage and thermal management systems [23, 30].

Thermal management is a pervasive challenge in electric aircraft systems. High-power electrical components generate significant heat, which must be

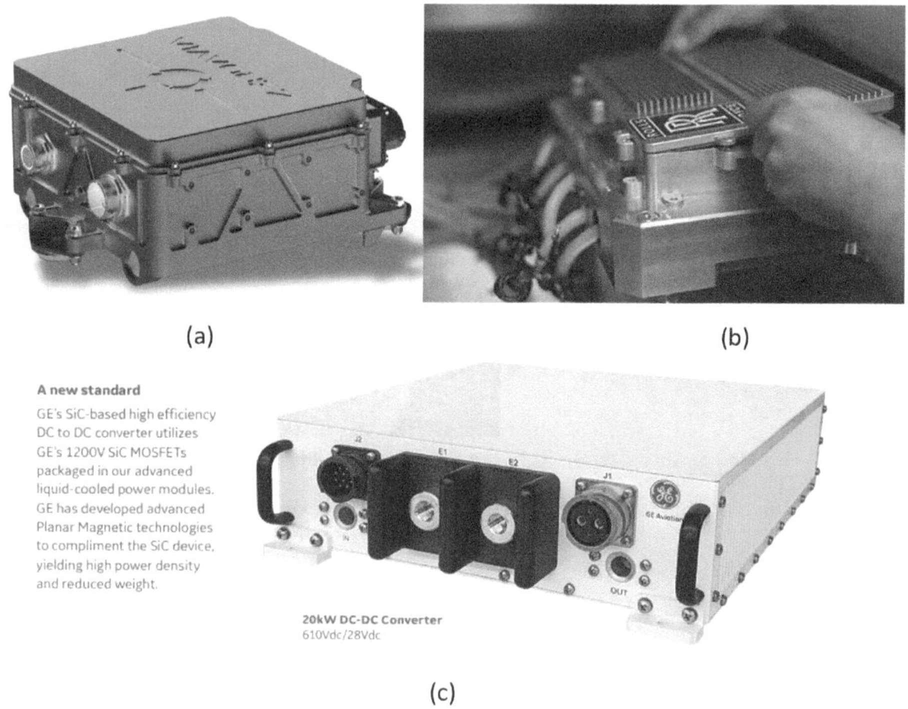

FIGURE 1.22
SiC aerospace power converters from (a) Zeroavia, (b) Rolls-Royce, and (c) GE Aviation.

dissipated to maintain performance and reliability. Advanced cooling techniques, including liquid cooling, phase-change materials, and heat pipes, are being integrated into motor housings, power electronics enclosures, and battery packs. The design of these systems must account for altitude-induced pressure variations, vibration, and electromagnetic interference.

Electrical distribution systems in electric aircraft must also evolve to handle higher voltages and power levels. Traditional 28 VDC and 115 VAC systems are inadequate for propulsion-grade power. As a result, new architectures operating at 540 VDC, 800 VDC, or even 1,000 VDC are being proposed. These systems require novel insulation materials, fault detection mechanisms, and arc suppression technologies to ensure safety and compliance with airworthiness standards [32].

Finally, the integration of digital control and monitoring systems is essential for managing the complexity of electric aircraft. Embedded sensors, real-time diagnostics, and predictive maintenance algorithms enable condition-based monitoring of motors, batteries, and power electronics. Projects such as iBearing and iGear exemplify the use of intelligent health monitoring to extend component life, reduce maintenance costs, and enhance operational safety [18].

In summary, the electrification of aircraft systems demands a holistic reengineering of electrical equipment, encompassing motors, power electronics, energy storage, thermal management, and digital control. These systems must be optimised for weight, efficiency, reliability, and safety, while meeting the stringent requirements of aerospace certification.

1.8 Challenges

The path towards net-zero aviation through electrification requires to addressing technical, regulatory, economic, and operational challenges. Though significant progress has been made in recent years, several critical barriers must be addressed to enable the widespread adoption of electric and hybrid-electric aircraft.

One of the foremost technical challenges is the limitation of current energy storage technologies. The energy density of lithium-ion batteries, while sufficient for small eVTOLs and short-range commuter aircraft, falls short of the requirements for regional and long-haul flights. To match the energy content of jet fuel, batteries would need to achieve energy densities exceeding 1,000 Wh/kg, a target that remains elusive with existing chemistries. Moreover, battery systems add significant weight to the aircraft, reducing payload capacity and range. The development of advanced battery technologies, such as solid-state and lithium–metal batteries, is essential but requires breakthroughs in materials science, manufacturing, and safety validation.

Thermal management is another critical issue. High-power electrical systems generate substantial heat, which must be effectively dissipated to prevent thermal runaway, degradation, and failure. The challenge is compounded by the low ambient temperatures and reduced convective cooling at high altitudes. Innovative cooling solutions, including liquid-cooled jackets, embedded heat exchangers, and active thermal control systems, are being explored, but add complexity and weight to the aircraft.

From a regulatory perspective, the certification of electric aircraft presents a significant hurdle. Existing airworthiness standards were developed for conventional aircraft and do not fully address the unique characteristics of electric propulsion, high-voltage systems, and energy storage. Regulatory agencies such as the FAA and EASA are working to update certification frameworks, but the process is time-consuming and resource-intensive. The lack of standardised testing protocols, failure modes, and safety margins for electric systems further complicates certification efforts.

Infrastructure development is another major challenge. The deployment of electric aircraft requires a supporting ecosystem of charging stations, battery swapping facilities, hydrogen refuelling stations, and vertiports. These infrastructures must be integrated with existing airport operations, comply with safety regulations, and be economically viable. The coordination between aircraft manufacturers, airport authorities, utility providers, and regulatory bodies is essential to ensure seamless infrastructure deployment.

Economic viability remains a concern, particularly in the early stages of adoption. Electric aircraft currently have higher upfront costs due to the expense of batteries, power electronics, and lightweight materials. While operational costs may be lower due to reduced fuel and maintenance expenses, the total cost of ownership must be competitive with conventional aircraft to attract operators. Additionally, the limited range and payload of electric aircraft constrain their revenue-generating potential, especially in commercial aviation.

Operational integration poses further challenges. Electric aircraft must be incorporated into existing air traffic management systems, flight planning tools, and maintenance procedures. Pilots, technicians, and ground crew require specialised training to handle high-voltage systems, battery management, and emergency protocols. The development of simulation tools, training curricula, and certification programmes is necessary to build the human capital required for electric aviation.

Finally, public perception and acceptance play a crucial role in the success of electric aircraft. Concerns about safety, reliability, and noise must be addressed through transparent communication, demonstration flights, and community engagement. Building trust among passengers, regulators, and stakeholders is essential to foster a supportive environment for innovation.

In conclusion, while the electrification of aviation offers transformative potential, it also presents a complex array of challenges that must be systematically addressed. Multidisciplinary collaboration, sustained investment, and regulatory innovation are key to overcoming these barriers and realising the vision of net-zero flight.

1.9 Summary

The pursuit of net-zero aviation represents one of the most ambitious and consequential endeavours in the history of aerospace engineering. As the global community grapples with the realities of climate change, the aviation sector must undergo a profound transformation to align with environmental sustainability goals. Electrification, in its various forms, stands at the forefront of this transformation, offering a pathway to reduce emissions, enhance efficiency, and enable new modes of air transport.

This chapter has provided a comprehensive overview of the net-zero targets set by international agreements and aviation stakeholders, highlighting the urgency and scope of the decarbonisation challenge. It has examined the current status of progress, noting both the advancements in electric aircraft development and the gaps that remain in technology readiness, regulatory frameworks, and infrastructure.

The chapter has also explored the segmentation of the aviation market, identifying the segments most amenable to electrification and the specific requirements of each. It has delved into the technical foundations of aircraft electrification, including the more-electric aircraft concept, hybrid-electric propulsion, and fully electric flight. The discussion of electrical equipment and systems has underscored the complexity and innovation required to support electric propulsion, while the analysis of challenges has illuminated the multifaceted barriers that must be overcome.

As the industry moves forward, the integration of electrification with other sustainability strategies such as sustainable aviation fuels, hydrogen propulsion, and carbon offsetting will be essential. The path to net-zero aviation is not linear, but it is navigable through coordinated action, technological innovation, and a shared commitment to a cleaner, quieter, and more equitable future for air travel.

References

[1] "https://berkeleyearth.org/global-temperature-report-for-2023/," [Online]. Available: https://berkeleyearth.org/global-temperature-report-for-2023/.

[2] A. T. A. G. (ATAG), "Waypoint 2050: Balancing growth in connectivity with a comprehensive global air transport response to the climate emergency," Sept. 2021.

[3] [Online]. Available: https://sciencebasedtargets.org/resources/files/SBTi-Net-Zero-Standard-Event-slides.pdf.

[4] "https://www.wri.org/insights/climate-action-progress-indicators-2030-2050-targets," [Online]. Available: https://www.wri.org/insights/climate-action-progress-indicators-2030-2050-targets.

[5] "https://www.wri.org/insights/climate-action-progress-1-5-degrees-c," [Online]. Available: https://www.wri.org/insights/climate-action-progress-1-5-degrees-c.

[6] "https://www.statista.com/chart/33334/total-greenhouse-gas-emissions-by-sector/," [Online]. Available: https://www.statista.com/chart/33334/total-greenhouse-gas-emissions-by-sector/.

[7] "https://ourworldindata.org/co2-emissions-from-transport," [Online]. Available: https://ourworldindata.org/co2-emissions-from-transport.

[8] "https://aviationbenefits.org/environmental-efficiency/climate-impact/," [Online]. Available: https://aviationbenefits.org/environmental-efficiency/climate-impact/.

[9] "https://www.icao.int/Meetings/FutureOfAviation/Pages/default.aspx," [Online]. Available: https://www.icao.int/Meetings/FutureOfAviation/Pages/default.aspx.
[10] "https://www.adsgroup.org.uk/knowledge/the-route-to-achieving-net-zero-aviation-emissions-by-2050/," [Online]. Available: https://www.adsgroup.org.uk/knowledge/the-route-to-achieving-net-zero-aviation-emissions-by-2050/.
[11] "https://theicct.org/publication/co2-emissions-from-commercial-aviation-2018/," [Online]. Available: https://theicct.org/publication/co2-emissions-from-commercial-aviation-2018/.
[12] "https://www.icao.int/Meetings/FutureOfAviation/Pages/default.aspx," [Online]. Available: https://www.icao.int/Meetings/FutureOfAviation/Pages/default.aspx.
[13] "https://zeroavia.com/," [Online].
[14] "https://cdn.volocopter.com/assets/vnrac6vfvrab/DcsQRmxTSPhAcK1xmKnS4/5290e554030c7b2a093e14d1eafbe7e5/Volocopter_Whitepaper_Singapore-Roadmap_web.pdf," [Online].
[15] "https://aviationbenefits.org/media/167417/w2050_v2021_27sept_full.pdf," [Online].
[16] G. K. a. N. M. H. Oyori, "A study on the More Electric Architecture for Aircraft and Propulsion (MEAAP) concept," *IHI Engineering Review*, 2019.
[17] P. Patel, "GE aviation systems – Electric power technologies for all electric aircraft," July 2010. [Online].
[18] A. Starr, "More electric aircraft – health monitoring, Impact.pub, 2022," [Online]. Available: https://www.impact.pub.
[19] "https://www.airbus.com/en/innovation/energy-transition/hybrid-and-electric-flight," [Online].
[20] Airbus, "E-fan X: Hybrid-electric propulsion demonstrator," Airbus Innovation, 2023. [Online]. Available: https://www.airbus.com," [Online].
[21] International Civil Aviation Organization (ICAO), "Environmental Report 2022," ICAO, Montreal, QC, Canada, 2022. [Online].
[22] "https://www.aerosociety.com/news/how-e-fan-x-will-jump-start-a-new-era-in-hybrid-electric-flight/," [Online].
[23] ZeroAvia, "Hydrogen-electric propulsion system: Technical overview," ZeroAvia, 2024. [Online]. Available: https://zeroavia.com," [Online].
[24] "https://www.emobility-engineering.com/electric-motors-for-aircraft/," [Online].
[25] "https://zeroavia.com/zeroavia-receives-signed-p-1-issue-paper-from-federal-aviation-administration-for-600kw-electric-propulsion-system/," [Online].
[26] V. GmbH, "The launch of urban air mobility in Singapore – a roadmap," 2022, [Online]. Available: https://www.volocopter.com.
[27] F. A. A. (FAA), "FAA clears the skies for eVTOL startups, paving the way for Air Taxi networks by 2025," TechCrunch, Oct. 2025, [Online]. Available: https://www.faa.gov.
[28] "https://flyone.com.au/eviation-alice/?srsltid=AfmBOopJPRwe3Y-x4eT1wJ-3TeUDbAJlwm4vPGRGbEOunIZU1UTnEOEno," [Online].
[29] GE Aerospace, "Power electronics for advanced air mobility," GE Aerospace, 2024, [Online]. Available: https://www.geaerospace.com.

[30] "https://www.hdi.global/globalassets/_local/international/downloads/specialty/aviation/e-flight_whitepaper.pdf," [Online].
[31] "https://www.rolls-royce.com/media/our-stories/discover/2024/developing-a-transverse-flux-machine-for-urban-air-mobility.aspx," [Online].
[32] Institute of Electrical and Electronics Engineers (IEEE), "IEEE Guide for Aircraft Electric Systems," IEEE Std., 2022, [Online].

2

Innovative Microgrid Solutions for Sustainable Electric Aircraft: A Pathway to Green Aviation

Pampa Sinha, Snehalika, and Kaushik Paul

This study proposes a comprehensive integrated framework for renewable energy–based airport microgrids designed specifically to support electric aircraft charging infrastructure. The proposed framework integrates distributed renewable energy sources (DERs) such as solar photovoltaic systems and wind turbines with hybrid energy storage systems (HESS) consisting of lithium-ion batteries and supercapacitors. Lithium-ion batteries provide medium- to long-duration energy storage, while supercapacitors enable rapid power delivery during sudden high-demand charging events. This hybrid architecture improves system flexibility, mitigates power fluctuations from renewable generation, and ensures stable energy supply for critical airport operations. To enhance operational efficiency and grid reliability, the framework incorporates advanced energy management strategies, including AI-based predictive scheduling, demand response mechanisms, peak shaving algorithms, and time-of-use optimisation. These intelligent control strategies dynamically coordinate energy generation, storage, and consumption within the microgrid, thereby reducing grid stress and improving overall energy utilisation. The system is capable of operating in both grid-connected and islanded modes, allowing airports to maintain uninterrupted charging services during external grid disturbances or outages. Furthermore, the proposed architecture integrates IoT-enabled monitoring systems and AI-driven microgrid controllers that enable real-time energy flow monitoring, predictive maintenance, and autonomous decision-making. The framework also considers critical aspects such as regulatory compliance, safety constraints, cybersecurity risks, and emerging digital technologies associated with next-generation smart airport infrastructure. Case studies of existing airport microgrid implementations demonstrate that renewable-powered microgrids can significantly reduce carbon emissions, enhance energy reliability, and lower long-term operational costs. By combining renewable energy integration, intelligent energy management, and hybrid energy

DOI: 10.1201/9781003747994-2

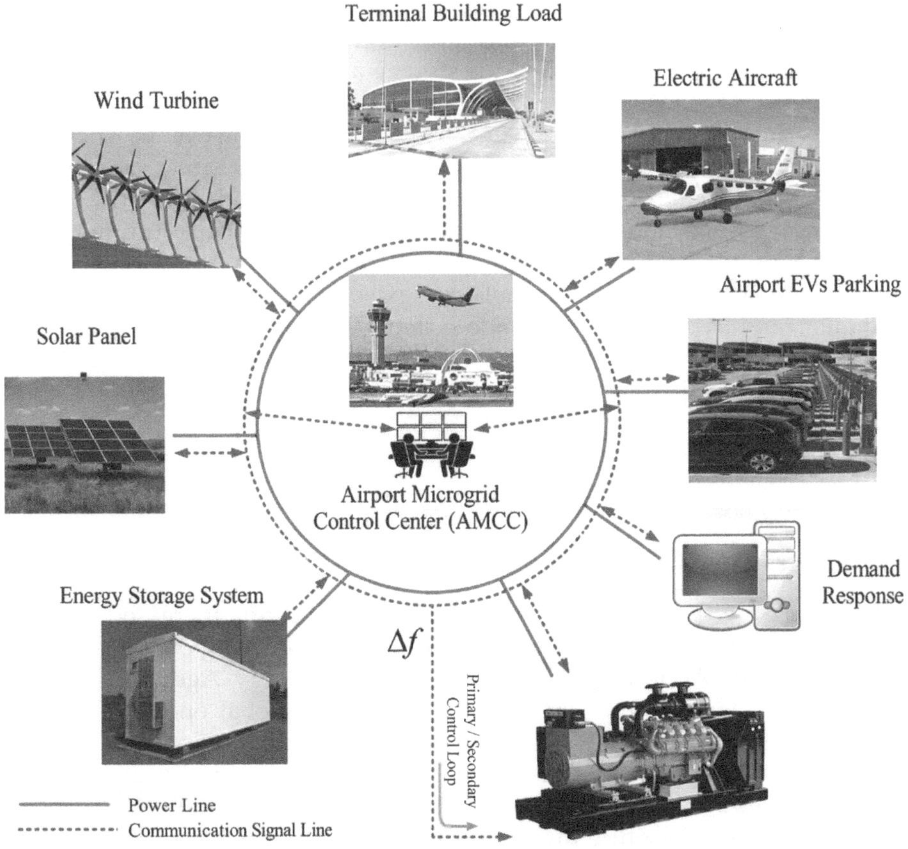

FIGURE 2.1
Conceptual framework of a smart airport microgrid integrating renewable energy, hybrid storage, and AI-based energy management.

storage technologies, the proposed framework provides a scalable and sustainable pathway towards carbon-neutral airport operations and future electric aviation ecosystems.

Figure 2.1 illustrates the conceptual architecture of a renewable energy–driven airport microgrid designed to support electric aircraft charging infrastructure. The system begins with distributed renewable energy generation units, including solar photovoltaic arrays, wind turbines, and hydrogen-based fuel cells, which provide clean and sustainable electricity to the airport energy network. The generated energy is coordinated through a smart airport microgrid platform equipped with AI-based control algorithms and IoT-enabled monitoring systems. These intelligent systems continuously monitor generation levels, storage conditions, grid interactions, and aircraft charging requirements, enabling optimal energy dispatch and predictive load scheduling. To maintain grid stability and handle the high-power charging demands of electric aircraft, the microgrid integrates a hybrid energy storage system (HESS). This storage architecture combines lithium-ion batteries for long-duration energy storage with supercapacitors capable of delivering rapid bursts of power during peak charging events. The hybrid configuration enhances system flexibility and reduces stress on the main grid. The stored and optimised energy is then supplied to electric aircraft charging stations and airport operational loads, ensuring uninterrupted airport activities and efficient aircraft turnaround operations. Through the coordinated operation of renewable energy sources, intelligent energy management, and hybrid energy storage systems, the proposed microgrid architecture significantly reduces carbon emissions, improves energy resilience, and enables sustainable airport electrification.

2.1 Literature Survey

The rapid growth of the aviation sector has significantly increased concerns regarding its environmental impact and contribution to global greenhouse gas emissions. According to the International Energy Agency, the transportation sector accounts for a major portion of global energy consumption and carbon emissions, with aviation contributing approximately 2–3% of global CO_2 emissions [1]. As a result, researchers and policymakers have intensified efforts to explore sustainable aviation technologies capable of reducing emissions while maintaining operational efficiency. These efforts include the development of electrified aircraft propulsion systems, renewable energy–powered airport infrastructure, and intelligent energy management frameworks. Early studies focused on operational strategies to reduce aviation emissions without major technological changes to aircraft systems. Wells et al. [2] investigated fuel-optimised routing strategies for transatlantic flights and demonstrated that optimised flight trajectories can significantly reduce fuel consumption and associated emissions. Similarly, Chiaramonti [3] examined the role of sustainable aviation fuels (SAF) in decarbonising aviation, highlighting that biofuels and synthetic fuels can reduce life cycle emissions by a substantial margin. However, the study also emphasised challenges related to production scalability and supply-chain infrastructure. In parallel, Dhara and Lal [4] reviewed sustainable aircraft design approaches and identified electrified propulsion, lightweight materials, and advanced aerodynamics as key technologies for improving aircraft efficiency. Electric propulsion systems have emerged as one of the most promising solutions for achieving low-emission aviation. Ansell and Haran [5] explored electrified airplane concepts and highlighted the potential of electric propulsion to reduce both greenhouse gas emissions and noise pollution around airports. Furthermore, Afonso et al. [6] conducted a comprehensive systematic review of sustainable aviation strategies and emphasised the importance of integrating electrification, renewable energy, and advanced propulsion technologies to achieve long-term environmental sustainability. Brelje and Martins [7] provided a detailed overview of electric, hybrid-electric, and turboelectric aircraft architectures, demonstrating how distributed electric propulsion systems can enhance propulsion efficiency and reduce overall fuel consumption.

In recent years, the development of electric aircraft prototypes has further accelerated research in this domain. For instance, the fully electric aircraft Alice developed by Eviation represents one of the most advanced electric regional aircraft concepts capable of carrying multiple passengers over short distances [8]. Similarly, Pipistrel has introduced electric aircraft technologies that demonstrate the feasibility of electric propulsion in aviation training and short-haul operations [9]. These technological advancements highlight the growing maturity of electric aviation systems and their potential role in future transportation networks. A critical component enabling electric aviation is energy storage technology. Mitali et al. [10] provided a comprehensive review of energy storage systems, emphasising the importance of high-energy-density batteries in supporting electric propulsion systems. Lithium-ion batteries have emerged as the dominant energy storage technology due to their high efficiency, energy density, and operational reliability. However, further advancements in battery technologies are required to extend aircraft range and improve charging efficiency. Several studies have also focused on the conceptual design and optimisation of hybrid-electric aircraft. Pornet [11] proposed conceptual design methods for hybrid-electric transport aircraft and demonstrated how advanced aircraft sizing techniques can improve system performance and efficiency. Similarly, Isikveren et al. [12] investigated pre-design strategies for dual-energy aircraft systems and highlighted the potential of hybrid propulsion architectures to enhance fuel efficiency. Brdnik et al. [13] further examined market and technological perspectives for next-generation regional passenger aircraft, emphasising electrification as a critical enabler for sustainable aviation. Another promising approach to aircraft electrification involves turboelectric propulsion systems. Alrashed et al. [14] analysed NASA's turboelectric distributed propulsion concept and demonstrated that such architectures can significantly improve propulsion efficiency and reduce emissions. Likewise, Marciello et al. [15] conducted a design exploration study of sustainable regional hybrid-electric aircraft and highlighted the importance of integrating energy storage technologies with advanced propulsion systems. Kiesewetter et al. [16] also provided a holistic review of advanced aircraft design concepts for future air mobility applications, identifying electric propulsion and distributed power architectures as key technological drivers for next-generation aviation systems. Further investigations have examined the performance characteristics of hybrid and turboelectric aircraft. Duffy and Jansen [17] analysed key performance parameters of turboelectric and hybrid-electric aircraft systems and demonstrated that optimised power distribution networks can significantly enhance aircraft efficiency. Thalin [18] presented fundamental concepts of electric aircraft systems, including propulsion architectures, power electronics, and energy storage requirements. Additionally, Sahoo et al. [19] reviewed electrical propulsion-based aircraft technologies and identified battery energy density and charging infrastructure as major challenges that must be addressed for large-scale adoption.

Research efforts have also explored broader strategies for achieving environmentally sustainable aviation. Kuśmierek et al. [20] discussed several green aviation technologies aimed at reducing the environmental footprint of air transport. Bradley and Droney [21] investigated hybrid-electric aircraft design within NASA's Subsonic Ultra Green Aircraft Research programme and demonstrated that hybrid-electric propulsion systems can significantly reduce fuel consumption and emissions compared to conventional aircraft technologies. While aircraft propulsion technologies are advancing rapidly, supporting energy infrastructure at airports is equally critical for enabling electric aviation. Guo et al. [22] proposed infrastructure planning strategies for airport microgrids integrating electric aircraft charging systems and electric vehicles. Their research demonstrated that coordinated planning of renewable energy generation, energy storage, and charging infrastructure is essential to handle the high power demands associated with electric aircraft charging. Babuder et al. [23] further examined the impact of emerging sustainable aircraft technologies on airport operational ecosystems, highlighting the need for infrastructure modernisation to accommodate electrified aviation systems. In addition, Trainelli et al. [24] investigated optimal sizing and operation of airport infrastructures supporting electric aircraft operations. Their findings showed that advanced energy management systems and optimised infrastructure planning are necessary to ensure reliable and efficient airport electrification. Research on charging infrastructure planning has also been influenced by developments in electric vehicle technologies. Zheng et al. [25] analysed battery charging and swapping stations in distribution systems and proposed optimisation techniques for infrastructure planning, providing valuable insights applicable to electric aircraft charging networks.

Energy management strategies are another crucial aspect of airport electrification. Wang et al. [26] investigated mobile energy storage scheduling strategies for AC–DC microgrids designed to support low-carbon airport operations. Their study demonstrated that mobile energy storage can improve energy flexibility and enhance the reliability of microgrid systems. Similarly, Buvarp and Leijon [27] proposed a mobile charging system for electric aircraft, which can provide flexible and scalable charging solutions at airports. High-power charging technologies also require advanced power electronics and converter systems. Kilicoglu and Tricoli [28] conducted a comprehensive review of power converters used in fast-charging stations and highlighted the importance of efficient converter architectures for enabling high-power electric aircraft charging infrastructure. Environmental impacts of emerging aviation technologies have also been analysed using life cycle assessment approaches. Thonemann et al. [29] developed prospective life cycle inventory datasets for conventional and hybrid-electric aircraft technologies, enabling a more comprehensive evaluation of environmental benefits associated with electrified aviation systems. Finally, Meindl et al. [30] investigated ground-based power supply systems designed to operate hybrid-electric aircraft at regional airports. Their study highlighted the importance of dedicated ground infrastructure and advanced power supply systems for supporting future electric aviation operations. Despite the substantial progress achieved in electric aircraft propulsion systems, energy storage technologies, and airport charging infrastructure, several challenges remain. Existing studies primarily focus on individual components of the electric aviation ecosystem, such as propulsion systems, charging infrastructure, or renewable energy integration. However, comprehensive frameworks that integrate renewable energy sources, hybrid energy storage systems, and intelligent energy management strategies within airport microgrids remain limited. Therefore, there is a critical need for integrated energy management frameworks capable of supporting high-power electric aircraft charging while ensuring energy reliability, operational efficiency, and environmental sustainability.

2.2 Overview of Electric Aircraft Technology and Its Potential Impact on Aviation

Electric aircraft technology is still in its developmental phase but holds immense promise for transforming the aviation sector. The core of this technology revolves around replacing the conventional fuel-burning engines with electric propulsion systems that rely on batteries, fuel cells, or hybrid solutions. A key advancement in this field is the development of high-energy density batteries and fuel cells that make electric flight more feasible. Currently, electric aircraft are mostly being designed for regional and short-haul flights, where battery limitations (such as lower energy density compared to jet fuel) are less of a constraint. For instance, the Alice aircraft by Aviation Aircraft, a fully electric aircraft, has a range of approximately 440 nautical miles and can carry up to nine passengers. However, as battery technology improves, such as through developments in solid-state batteries and lithium–sulfur batteries, the range is expected to increase substantially. According to a report by the International Air Transport Association (IATA), aviation accounts for approximately 2–3% of global CO_2 emissions, highlighting the potential for electric aircraft to reduce carbon emissions significantly. By transitioning to electric propulsion, carbon emissions per flight could be reduced by up to 70–90% depending on the source of electricity and operational efficiencies. Moreover,

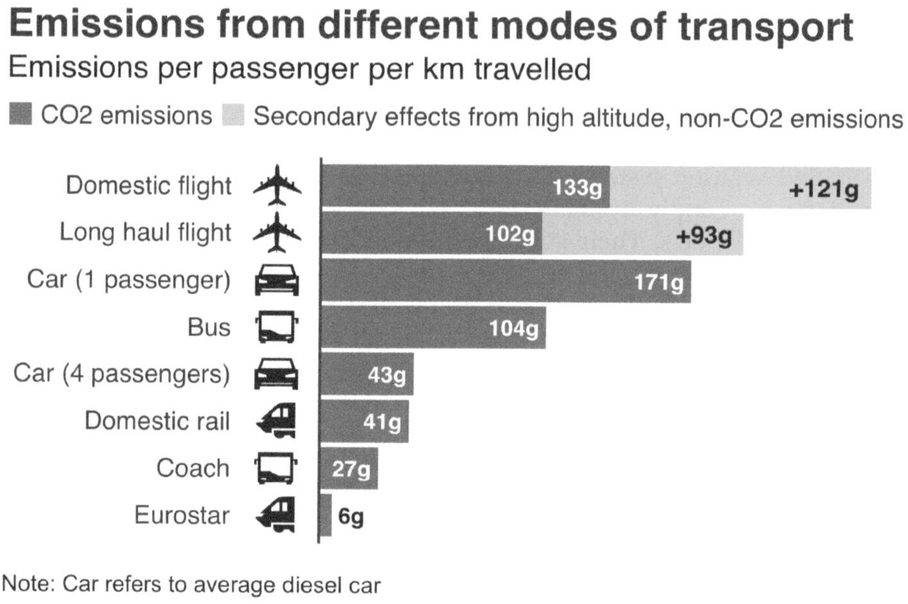

FIGURE 2.2
Comparison of carbon emissions: conventional aircraft versus electric aircraft.

electric aircraft are expected to offer a quieter and smoother flying experience due to the reduced noise from electric engines, a significant advantage in reducing noise pollution around airports. In addition, the operational costs of electric aircraft could be significantly lower. As per a study by PwC, it is estimated that the total operating costs for electric aircraft could be reduced by up to 50–70% per flight compared to traditional jet engines, mainly due to lower fuel costs, fewer moving parts, and reduced maintenance requirements. The growing support for electric aircraft development comes from major companies such as Airbus and Boeing, which are working on hybrid-electric and all-electric aircraft, respectively. The potential long-term benefits of electric aviation could redefine the aviation industry by making air travel more sustainable, affordable, and accessible (Figure 2.2).

2.3 Importance of Sustainability in the Aviation Industry

Sustainability in the aviation industry is a priority driven by the need to reduce the sector's environmental footprint and align with global climate targets. Aviation is a significant contributor to global greenhouse gas (GHG) emissions. According to the International Civil Aviation Organization

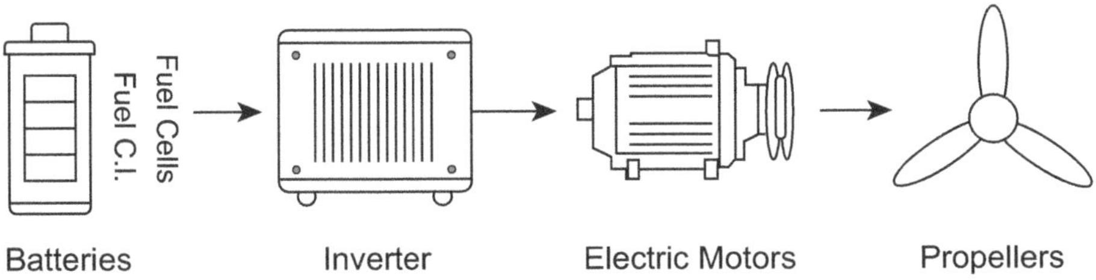

FIGURE 2.3
Electric aircraft propulsion system.

(ICAO), global aviation emissions in 2019 were estimated to be around 915 million tonnes of CO_2, with projections showing this could more than double by 2050 if no changes are made. This situation is prompting the industry to seek cleaner, more sustainable solutions. The industry has responded to this challenge through initiatives like sustainable aviation fuel (SAF), which reduces the carbon emissions of conventional aircraft. SAF can reduce CO_2 emissions by up to 80% over its life cycle compared to conventional jet fuel. Currently, the use of SAF is still in its early stages but is gaining momentum. In 2021, United Airlines became the first US airline to operate a commercial flight powered by SAF, signalling the shift towards greener aviation practices. Additionally, hydrogen fuel cell technology is also being explored as an alternative to jet fuel, with companies like ZeroAvia conducting test flights of hydrogen-powered aircraft.

Electric aviation adds another layer to these efforts. According to the Electric Aircraft Report (2023), electric aircraft have the potential to reduce the total global aviation emissions by up to 10% by 2050. This is due to their energy efficiency and the fact that they rely on electricity, which can be sourced from renewable resources such as wind, solar, or hydroelectric power, further decreasing their carbon footprint. In terms of fuel consumption, electric aircraft consume about 1–2 kWh per mile, which is much lower than the fuel consumption of traditional jet engines that consume about 5–8 gallons of fuel per mile for regional aircraft. The cumulative impact of these innovations can help the aviation sector meet its commitment to net-zero emissions by 2050, a goal supported by the Paris Agreement and many national and regional governments (Figure 2.3).

2.4 Definition and Role of Microgrids in the Context of Electric Aircraft

Microgrids play an essential role in the charging and energy supply infrastructure needed for the widespread adoption of electric aircraft. These localised energy systems are capable of generating, storing, and distributing electricity to support both critical airport operations and electric aircraft charging. Microgrids are especially beneficial in regions where the central grid may not be able to handle the increased load from electric aircraft charging without risking grid instability or increased carbon emissions. A typical electric aircraft requires substantial power for charging – ranging from 100 kW to several MW depending on the size and model. For example, the electric regional aircraft Alice by Aviation Aircraft is estimated to require up to 2.5 MW of power to charge. To meet these needs, microgrids can integrate renewable energy sources such as solar panels, wind turbines, and battery storage systems to provide a sustainable and reliable power supply for charging stations at airports. By using renewable energy, microgrids can help reduce the reliance on fossil fuels and support electric aviation's sustainability goals. A study by Rocky Mountain Institute found that airports can reduce their operational costs by up to 30% by integrating microgrids and renewable energy sources, while also improving energy reliability and resilience. Moreover, microgrids can operate in islanding mode – isolating from the central grid during power outages or disruptions. This capability is crucial for airport operations where consistent power supply is vital for aircraft charging and other critical functions. Microgrids can also optimise energy flow, reduce energy costs, and ensure that excess energy generated by solar or wind systems is stored for later use, rather than being lost to the grid. An example of a microgrid in action is the San Diego International Airport, which has already implemented a microgrid system that includes solar power and battery storage. This system has improved the airport's energy resilience and helped reduce energy costs by approximately 30%. As electric aviation grows, microgrids will be integral to ensuring that airports can meet the rising demand for charging infrastructure in a sustainable, cost-effective, and resilient manner.

2.5 Components of Microgrid Systems Relevant to Electric Aircraft

The microgrid system for electric aircraft incorporates several essential components designed to support both the generation and distribution of energy. These components are crucial for ensuring reliability, sustainability, and efficiency in electric aircraft operations, particularly at airports. Microgrids rely on renewable energy sources to minimise the carbon footprint and ensure sustainable operations. Key sources include solar panels, wind turbines, fuel cells, and sometimes biomass. Solar energy is becoming one of the most popular energy sources for microgrids. According to the International Renewable Energy Agency (IRENA), solar capacity grew by 18% globally in 2020, with solar contributing to 60% of renewable energy capacity growth. Solar panels, often installed on airport rooftops or adjacent lands, can produce clean, reliable electricity during daylight hours. For instance, the Kona International Airport in Hawaii has implemented solar power arrays to provide a significant portion of its energy needs, which could similarly be adapted to charge electric aircraft. Airports situated in regions with favourable wind conditions could integrate wind turbines into their microgrid systems. Wind energy is efficient for providing supplemental power, especially in areas where solar power is insufficient. As per Global Wind Energy Council (GWEC), the total global wind energy capacity in 2021 was around 850 GW, indicating a growing trend of wind energy usage worldwide. Wind turbines could serve as an additional energy resource, providing electricity to microgrids in coastal or high-altitude regions where wind speeds are consistently high. Fuel cells, particularly hydrogen fuel cells, are increasingly considered as backup power solutions for microgrids. These cells generate electricity through electrochemical reactions and are highly efficient. Hydrogen fuel cells could be a key player in providing backup power to electric aircraft charging stations, particularly in cases where renewable energy generation is low or storage is insufficient. The European Commission in 2020 projected that the aviation industry could reduce carbon emissions by up to 50% by 2050 through the adoption of hydrogen-powered aircraft.

1. Energy storage systems are critical in managing energy demand and ensuring stable power availability. Storage systems help balance intermittent renewable energy generation (e.g., solar and wind) with the continuous power demands of electric aircraft charging stations.

Lithium-ion batteries are the most commonly used energy storage technology in microgrids. These batteries are relatively lightweight, offer high energy density, and are rechargeable. For example, Tesla Powerwall and Powerpack systems are used in various microgrid projects to store energy. These batteries provide crucial backup power and facilitate the integration of renewable sources. On a smaller scale, airports could use lithium-ion batteries to store excess solar or wind energy, which could then be deployed during peak demand times for electric aircraft charging.

Flow batteries, which use liquid electrolytes, are also a promising option for long-term energy storage. Unlike lithium-ion batteries, flow batteries can scale more easily, making them suitable for large airports with substantial energy demands. Flow batteries are capable of storing large quantities of energy and discharging it over extended periods. Vanadium flow batteries are already being deployed in several large-scale energy storage projects globally, with capacities exceeding 1 MW, offering insights into their potential application in the aviation sector.

2. The EMS is the brain of the microgrid, responsible for real-time monitoring and optimising the power flow. It uses data from energy sources, storage, and consumption points to balance supply and demand efficiently.

These systems decentralise control across the microgrid, enabling local decision-making and enhancing reliability. A smart grid approach using predictive analytics and machine learning algorithms can help anticipate energy generation patterns and consumption needs. According to a 2021 study by the International Smart Grid Action Network (ISGAN), smart grids are expected to improve energy efficiency by up to 15% globally.

Microgrid controllers dynamically adjust power flows between the renewable generation units (e.g., solar, wind), storage, and the electric aircraft charging station. These controllers ensure the aircraft receive optimal charging power based on battery state and grid conditions. The use of advanced predictive modelling allows airports to reduce operational costs and reduce reliance on the main power grid. The US Department of Energy reported that microgrids could reduce energy costs by up to 25% by optimising the use of renewable resources and storage.

2.6 Types of Microgrid Configurations: Grid-Connected versus Islanded Operation

The configuration of a microgrid determines how it interacts with the main power grid and provides energy to electric aircraft charging stations. The two primary configurations – grid-connected and islanded – offer distinct advantages depending on energy needs, geographical location, and reliability requirements. In grid-connected microgrids, the system is integrated into the central utility grid. These systems can exchange power between the grid and microgrid, enabling the sharing of energy resources to balance supply and demand.

2.6.1 Advantages

Grid-connected systems benefit from a larger pool of energy resources, which reduces costs for energy storage and backup generation. Microgrids can leverage the central grid during times of low renewable energy generation or high demand. Microgrids can feed excess energy back to the main grid during periods of high generation, helping to stabilise the grid and reduce transmission losses. This feature is particularly valuable in airport microgrids, where large-scale charging operations for electric aircraft may require substantial power at peak times.

2.6.2 Challenges

The main limitation of grid-connected systems is that they are dependent on the grid for backup during power outages. However, advancements in distributed generation technologies like solar and wind are making this model more robust. According to a 2020 report by the US Department of Energy, more than 75% of microgrid systems globally are grid-connected, indicating their popularity and effectiveness in balancing energy demand. Islanded microgrids operate independently of the main grid. This configuration is especially useful in remote locations or where grid reliability is low. Islanded microgrids are typically powered by a combination of renewable energy sources and storage systems.

2.6.3 Advantages

Islanded systems ensure energy supply continuity during grid outages or disruptions. For airports located in regions with unreliable grid access, an islanded microgrid could serve as a critical backup power system. Airports can fully control their energy supply, making them less vulnerable to external energy price fluctuations or grid instability. Because these systems cannot rely on the central grid for backup power, the initial setup cost for energy storage and backup generators is higher. However, advances in battery storage technology are making islanded systems more cost-effective. The Isolated Microgrid at the University of California, San Diego, is a prime example of an islanded system. The microgrid provides power to over 10,000 buildings, showcasing the feasibility of large-scale, islanded systems.

2.6.4 Case Studies of Existing Microgrid Systems in Aviation or Similar Sectors

1. SAN has integrated a microgrid system to support its operations with renewable energy. The microgrid features solar power, energy storage, and grid connectivity. This system provides nearly 30% of the airport's energy needs, including powering electric vehicle charging stations. The microgrid also helps reduce the airport's carbon footprint by around 5,000 tonnes of CO_2 annually. As the adoption of electric aircraft accelerates, this model can be adapted to support aircraft charging.
2. McMurdo Station operates an islanded microgrid with a mix of solar power, wind energy, and diesel backup generators. The islanded microgrid is capable of supporting the station's energy needs year-round, even during harsh winter months when solar generation is limited. The system helps maintain power for critical infrastructure, and while the station does not yet support electric aircraft, it provides an excellent example of self-sufficiency in extreme conditions.
3. The Brooklyn Navy Yard's microgrid is a grid-connected system that incorporates solar panels, energy storage, and backup generators.

The microgrid serves several industrial facilities, including an electric vehicle charging station. The flexibility of grid-connected microgrids enables airports and other transportation hubs to use surplus energy to charge electric aircraft while simultaneously selling excess power back to the grid. This model shows how urban microgrids can benefit aviation infrastructure by reducing costs and improving sustainability.

In all these cases, microgrids offer a pathway for airports to reduce their environmental impact, ensure energy reliability, and support the future of electric aircraft operations. Through the integration of renewable energy, energy storage, and smart grid technologies, airports can optimise their energy usage and provide reliable, clean energy for electric aircraft charging.

2.7 Energy Management Strategies

2.7.1 Energy Management Strategies for Electric Aircraft Microgrids

The proposed airport microgrid architecture integrates renewable energy generation, hybrid energy storage systems, and digital-twin-based monitoring to ensure reliable energy supply for electric aircraft charging infrastructure. In this framework, solar and wind energy systems provide clean power that feeds into the airport microgrid. Energy storage technologies such as lithium-ion batteries and supercapacitors stabilise the energy supply and manage fluctuations caused by renewable intermittency and aircraft charging demand. The digital twin serves as a virtual representation of the physical energy infrastructure, continuously collecting operational data from sensors and monitoring systems. Using real-time data analytics, the digital twin simulates system behaviour and enables predictive energy management. This architecture allows airports to operate either in grid-connected mode, where the microgrid exchanges power with the utility grid, or in islanded mode, where the system functions independently during grid disturbances. As a result, the digital-twin microgrid enhances energy reliability, improves system efficiency, and supports sustainable electric aviation infrastructure (Figure 2.4).

Figure 2.5 shows that energy storage technologies play a crucial role in supporting the high-power charging requirements of electric aircraft. Lithium-ion batteries are widely used in microgrid applications because of their high energy density and long operational life. These batteries provide sustained energy supply for prolonged charging processes and maintain stability in airport energy systems. However, lithium-ion batteries are not ideal for sudden high-power discharge. To address this limitation, supercapacitors

FIGURE 2.4
Digital twin–driven airport microgrid architecture illustrating renewable generation, grid interaction, hybrid energy storage, AI-based energy management, and electric aircraft charging infrastructure.

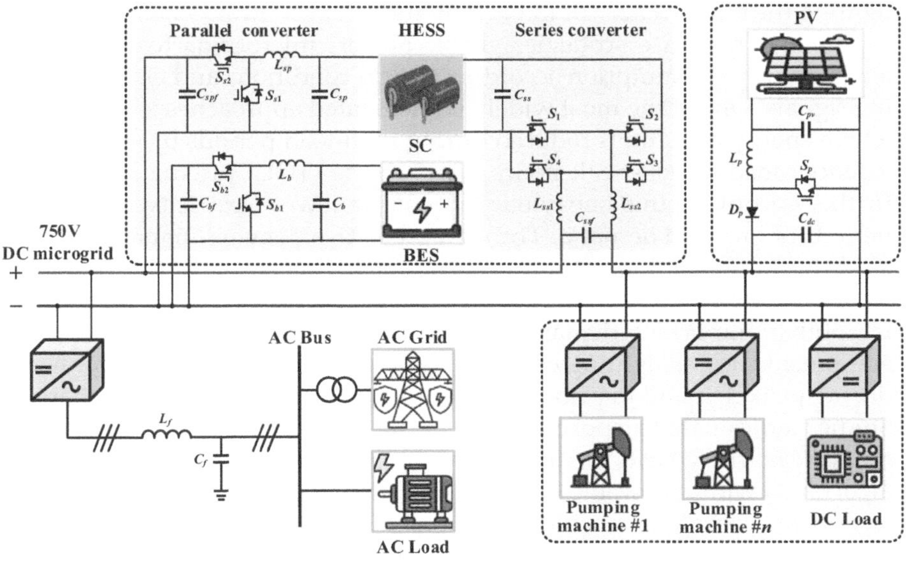

FIGURE 2.5
Hybrid energy storage system for electric aircraft charging.

are integrated into the system to deliver rapid bursts of energy during peak demand periods such as aircraft take-off preparation or fast charging operations. The hybrid energy storage system (HESS) coordinates both technologies, allowing batteries to supply steady base load energy while supercapacitors handle transient peak loads. Studies from the National Renewable Energy Laboratory indicate that hybrid storage architectures can improve overall energy efficiency by up to 25% in aviation microgrids. This coordinated storage strategy significantly enhances system flexibility, reduces stress on battery components, and ensures reliable energy delivery for electric aircraft charging.

Efficient energy management is (Figure 2.6) essential for maintaining reliable and cost-effective operation of airport microgrids. Advanced energy management systems combine digital twin simulation, real-time monitoring, and artificial intelligence algorithms to optimise energy flow within the microgrid. Supervisory control and data acquisition (SCADA) systems continuously monitor energy generation, consumption, and storage conditions across the airport infrastructure. The collected data is transmitted to the digital twin platform, which simulates system behaviour and predicts future energy demand. AI-based scheduling algorithms then determine the optimal distribution of power resources, ensuring that electric aircraft charging operations receive priority while maintaining grid stability. For instance, predictive scheduling models can analyse flight schedules and forecast charging demand to prevent sudden load spikes. Airports such as Los Angeles International Airport have demonstrated that real-time energy scheduling strategies can improve energy efficiency by nearly 18%, highlighting the effectiveness of intelligent microgrid control systems.

Demand response strategies enable airport microgrids to dynamically adjust energy consumption according to grid conditions and electricity pricing signals. One of the most widely implemented approaches is *peak shaving*, where energy demand is reduced during high-load periods by shifting nonessential loads to off-peak hours. Time-of-use (TOU) pricing mechanisms further encourage this behaviour by offering lower electricity rates during periods of low grid demand. Through predictive demand management, aircraft charging operations can be scheduled during cost-effective time intervals, reducing both operational expenses and grid congestion. Load shedding is another important demand response technique in which non-critical equipment such as lighting or HVAC systems are temporarily disconnected during peak demand to prioritise essential operations like aircraft charging. Studies indicate that airports implementing demand-response strategies can reduce electricity costs by up to 18–20%. By combining demand-response algorithms with renewable energy integration and digital twin monitoring, airport microgrids can achieve optimal energy utilisation while ensuring reliable operation of electric aircraft infrastructure (Figure 2.7).

Figure 2.8 shows that microgrids can operate in either grid-connected mode or islanded mode depending on the reliability and availability of the

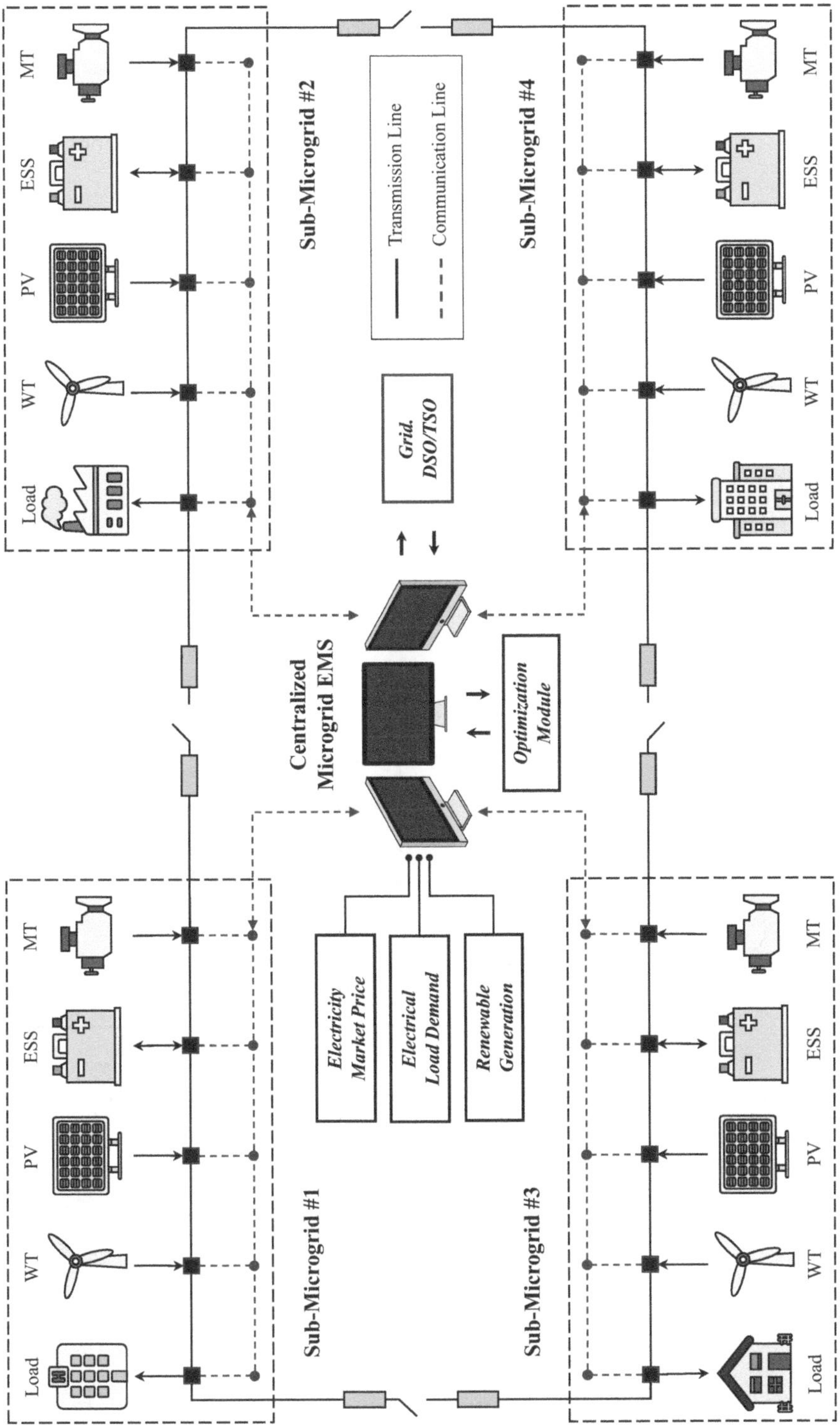

FIGURE 2.6
Energy management and digital twin control framework.

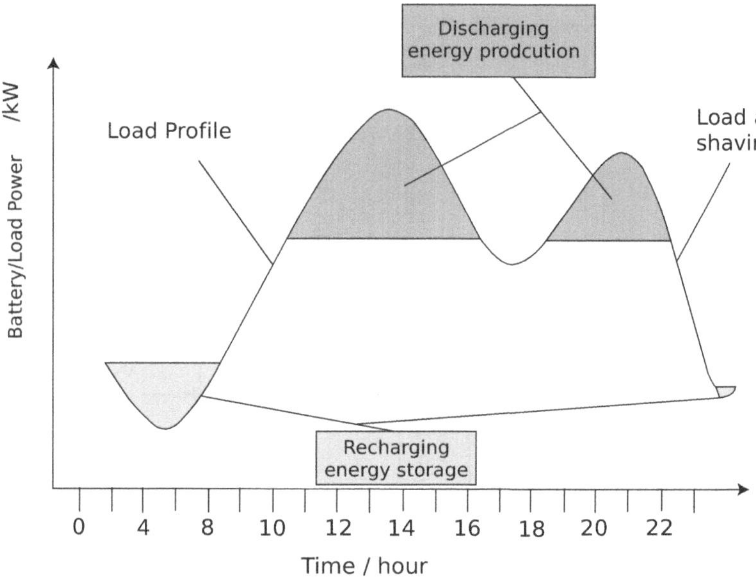

FIGURE 2.7
Demand response and smart grid optimisation.

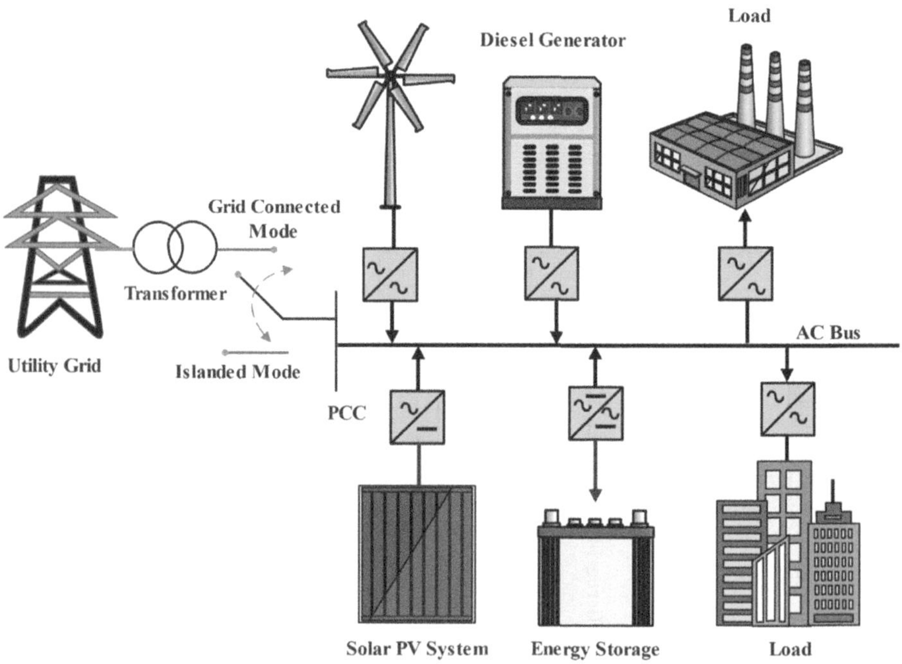

FIGURE 2.8
Grid-connected versus islanded microgrid operation.

main power grid. Grid-connected microgrids remain linked to the central utility network and can exchange electricity with the grid when required. This configuration is widely used because it provides backup power during periods of insufficient local generation. According to the US Department of Energy, more than 75% of microgrid systems worldwide operate in grid-connected mode due to their ability to balance supply and demand efficiently. In contrast, islanded microgrids function independent of the main grid and rely on local energy resources such as renewable generation and energy storage systems. Islanded operation is particularly valuable in remote areas or regions where grid reliability is limited. Airports can utilise islanded microgrids as critical backup energy systems during grid failures, ensuring uninterrupted operation of aircraft charging infrastructure and other essential airport services. Although the initial investment cost for storage systems is higher in islanded microgrids, advancements in battery technology are steadily improving their economic feasibility.

Several real-world microgrid implementations demonstrate the feasibility of integrating renewable energy and energy storage technologies into large infrastructure systems. San Diego International Airport has implemented a renewable-powered microgrid that supplies nearly 30% of the airport's electricity demand while significantly reducing carbon emissions. Similarly, McMurdo Station in Antarctica operates an islanded microgrid powered by wind, solar energy, and diesel backup generators, ensuring year-round energy availability under extreme environmental conditions. Another example is the Brooklyn Navy Yard microgrid in New York, which combines solar generation, energy storage, and grid connectivity to support industrial facilities and electric vehicle charging infrastructure. These case studies illustrate how microgrid systems can improve energy resilience, reduce environmental impact, and support emerging electrified transportation technologies such as electric aircraft. By integrating renewable energy sources, hybrid storage technologies, and intelligent energy management systems, airports can transition towards sustainable and reliable energy ecosystems capable of supporting future electric aviation operations (Figure 2.9).

2.8 Visual Data Insights for Intelligent Electric Aircraft Microgrids

The successful deployment of electric aircraft charging infrastructure requires intelligent energy management within airport microgrids. Advanced visualisation of energy consumption patterns, storage behaviour, and pricing dynamics provides valuable insights into how smart control strategies can improve efficiency and grid stability. By integrating demand side management (DSM), hybrid energy storage systems (HESS), and

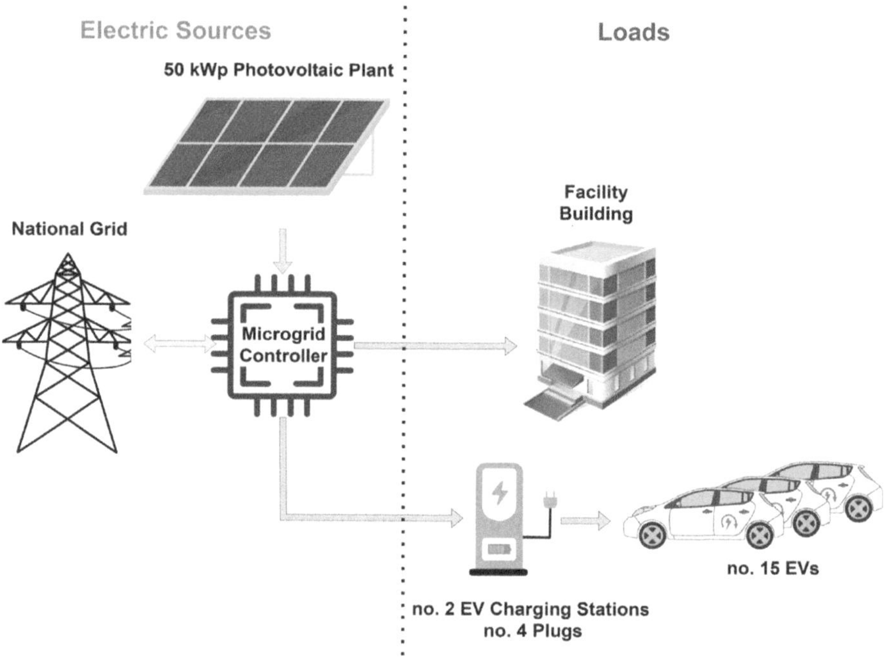

FIGURE 2.9
Case studies of microgrid applications in aviation and infrastructure.

dynamic pricing mechanisms, airport microgrids can optimise power usage while minimising stress on the electrical grid. To better understand these operational dynamics, several visual data analyses are presented that illustrate how energy demand shifts, how different storage technologies behave, and how economic incentives influence charging behaviour.

2.9 Energy Consumption Profile with DSM and Load Shifting

Electric aircraft charging can introduce large and sudden power demands at airports. Without intelligent control mechanisms, these charging loads may coincide with peak grid demand, resulting in grid congestion and increased operational costs. Demand side management (DSM) provides an effective strategy for redistributing energy consumption by shifting charging activities towards off-peak periods.

In traditional energy consumption scenarios, charging activities tend to occur during operational peaks, creating high demand spikes. DSM-enabled

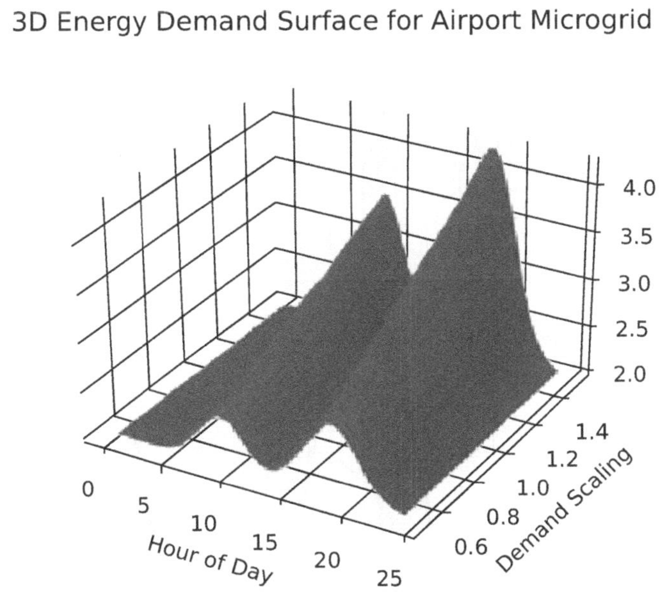

FIGURE 2.10
Three-dimensional visualisation of airport microgrid energy demand across a 24-hour cycle under varying demand conditions. The surface illustrates how demand side management (DSM) smooths peak loads by redistributing aircraft charging to off-peak hours.

systems, however, utilise predictive scheduling algorithms and real-time monitoring to redistribute charging loads throughout the day.

Through intelligent load shifting, aircraft charging is moved to periods when renewable energy generation is abundant or when grid demand is low. As a result, the overall energy consumption profile becomes smoother and more balanced, reducing the need for additional grid capacity and minimising peak demand stress.

2.9.1 Key Observations

Figure 2.11 illustrates the impact of demand side management (DSM) on the energy consumption profile of an airport microgrid over a 24-hour operational period. The curve with higher peaks represents the traditional load profile without DSM, where energy demand exhibits sharp peaks during active airport operational hours, particularly when multiple aircraft charging events occur simultaneously. These peaks increase stress on the electrical infrastructure and lead to higher electricity costs due to peak-period pricing. In contrast, the curve with lower peaks shows the optimised load profile when DSM strategies are applied. By intelligently scheduling aircraft charging during off-peak hours and distributing energy demand more

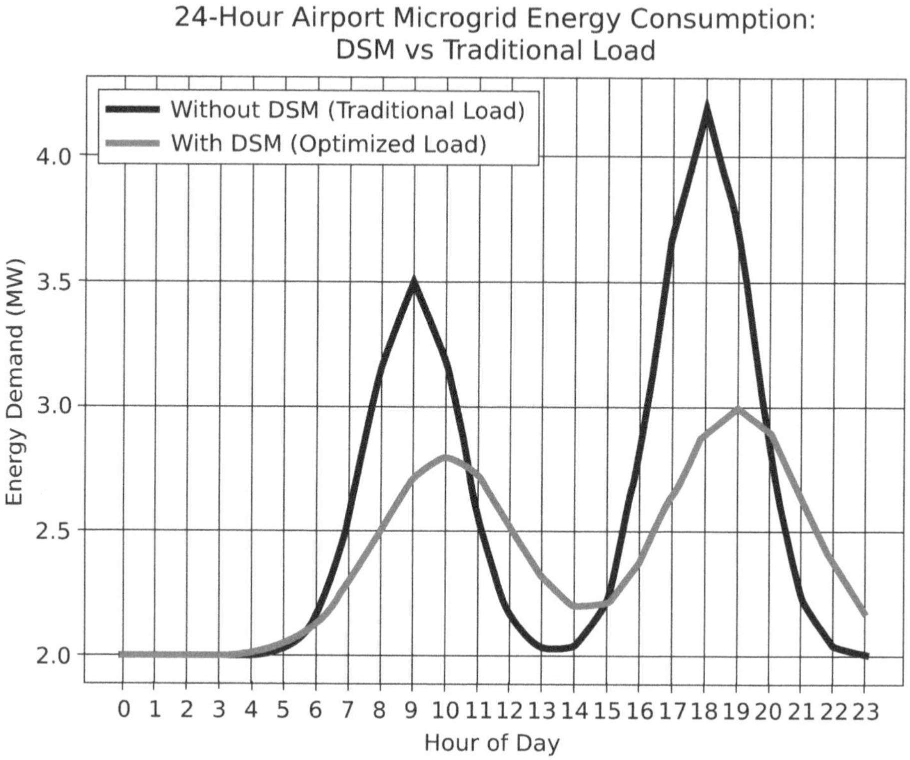

FIGURE 2.11
Twenty-four-hour airport microgrid energy consumption profile comparing traditional load demand without demand side management (DSM) and the optimised load profile with DSM-based load shifting, illustrating peak reduction and improved grid stability.

evenly throughout the day, DSM significantly smooths the demand curve. This load flattening reduces peak power requirements, improves grid stability, and enables more efficient utilisation of available energy resources, including renewable power sources integrated within the airport microgrid. As a result, DSM plays a critical role in enhancing both the economic and operational performance of electric aircraft charging systems.

2.10 Comparative Analysis of Energy Storage Technologies

Energy storage systems are essential components of airport microgrids supporting electric aircraft operations. Two major technologies dominate

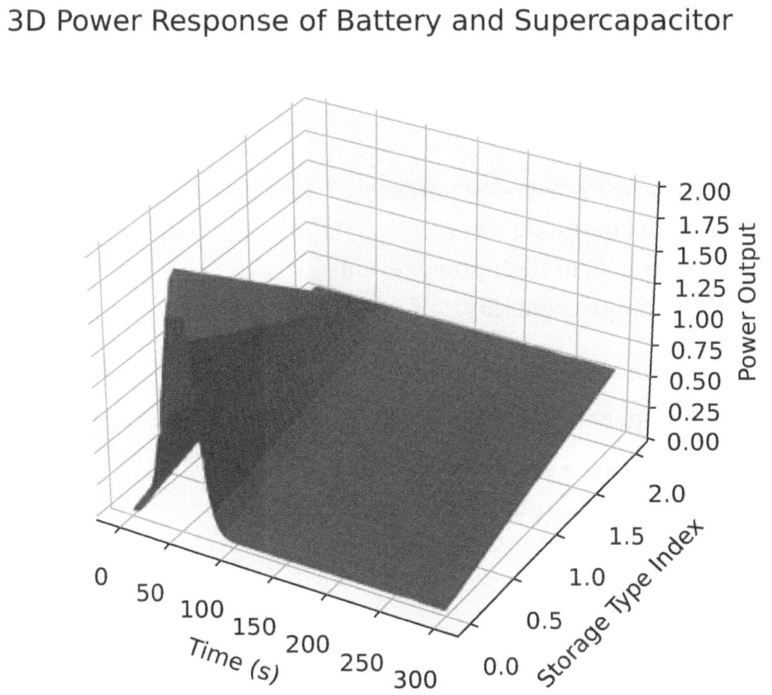

FIGURE 2.12
Three-dimensional power response characteristics of battery and supercapacitor storage technologies under dynamic charging conditions. Supercapacitors provide rapid high-power bursts, while batteries supply steady long-duration energy.

modern storage solutions: lithium-ion batteries and supercapacitors. Each technology offers distinct advantages in terms of energy density, discharge rate, and operational response time. Lithium-ion batteries are particularly suitable for long-duration energy storage, enabling airports to store renewable energy generated during periods of surplus and release it gradually over extended periods. Their relatively stable discharge characteristics make them ideal for maintaining continuous energy supply for aircraft charging. Supercapacitors, on the other hand, are designed for rapid energy delivery. Their extremely fast charge and discharge capabilities allow them to supply short bursts of high power, making them ideal for handling sudden load spikes during aircraft take-off preparation or rapid charging scenarios.

Storage Characteristics

Lithium-ion batteries:

- High energy density
- Long-duration storage

- Stable discharge behaviour
- Suitable for sustained charging loads

Supercapacitors:

- Extremely fast power discharge
- Short-term energy delivery
- Ideal for transient high-power events
- Rapid recharge capability

AI Visualisation Suggestion

Create an AI-generated performance graph showing

- Battery output → steady power curve
- Supercapacitor output → short power spikes

2.11 Efficiency Enhancement through Hybrid Energy Storage Systems (HESS)

Although batteries and supercapacitors provide valuable individual capabilities, combining them into hybrid energy storage systems (HESS) produces significantly greater efficiency and flexibility.

Hybrid systems integrate the long-term energy capacity of batteries with the rapid response characteristics of supercapacitors. In this configuration, supercapacitors handle short-duration power spikes, while batteries supply sustained energy demands.

This coordinated operation prevents excessive stress on battery systems and enhances overall microgrid responsiveness.

Performance Comparison

Battery-only system

- Slower response to sudden power demand
- Higher degradation under rapid load changes

Supercapacitor-only system

- Excellent transient performance
- Limited energy capacity

Hybrid system

- Fast dynamic response
- Stable long-duration supply
- Improved overall efficiency

Studies indicate that hybrid energy storage architectures can increase system efficiency by 15–30% while extending battery lifetime. Figure 2.13 compares the energy-efficiency characteristics of battery-only, supercapacitor-only, and hybrid energy storage system (HESS) configurations over varying airport microgrid load levels.

2.12 Impact of Peak Shaving and Time-of-Use Pricing

Energy pricing models significantly influence how and when electric aircraft charging occurs. Time-of-Use (TOU) pricing provides lower electricity rates during off-peak periods and higher prices during peak demand hours. By

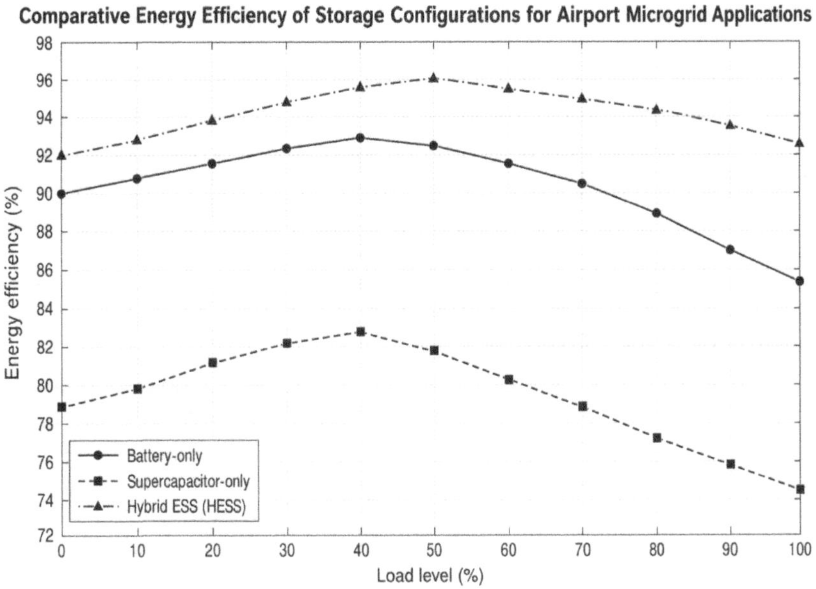

FIGURE 2.13
Comparative energy-efficiency characteristics of battery-only, supercapacitor-only, and hybrid energy storage system (HESS) configurations under varying load levels in an airport microgrid.

aligning charging schedules with these pricing signals, airports can reduce operational costs and improve grid efficiency.

Peak-shaving strategies further complement this approach by using stored energy during high-demand periods. When electricity prices rise during peak hours, energy stored in batteries or supercapacitors can be used instead of drawing power directly from the grid.

The combination of TOU pricing and peak shaving results in optimised charging patterns and significant cost savings.

Operational Benefits

- Reduced electricity expenditure
- Improved grid load balancing
- Increased renewable energy utilisation
- Reduced need for infrastructure expansion

2.13 Renewable Energy Integration in Airport Microgrids

The integration of renewable energy sources into airport microgrids represents a critical step towards achieving sustainable aviation. Electric aircraft charging requires substantial power capacity, and relying solely on conventional grid electricity would limit the environmental benefits of electrified aviation.

By incorporating renewable generation technologies such as solar photovoltaic systems, wind turbines, and geothermal energy, airports can significantly reduce their carbon footprint while enhancing energy independence.

2.14 Solar Energy Integration

Solar power is among the most practical renewable energy sources for airports due to the availability of large rooftop areas, parking structures, and unused land around runways.

Solar photovoltaic panels can generate significant electricity during daytime hours when aircraft charging demand is often high.

Advantages include the following:

- Low operating costs
- Abundant availability
- Scalable installation
- Reduced greenhouse gas emissions

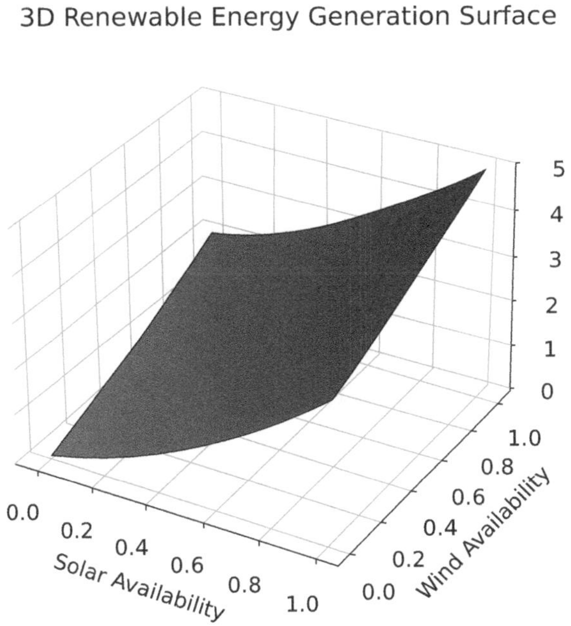

FIGURE 2.14
Three-dimensional surface showing combined renewable power generation from solar and wind resources within airport microgrids. The model illustrates how hybrid renewable resources can support electric aircraft charging infrastructure.

For example, Denver International Airport has implemented large-scale solar installations capable of generating over 10 MW of electricity, supporting sustainable airport operations. "Airport solar farm integrated with electric aircraft charging infrastructure and smart microgrid control, futuristic sustainable aviation illustration."

2.15 Wind Energy Integration

Wind energy provides a valuable complement to solar power by generating electricity during nighttime or cloudy conditions. Airports located in coastal or high-wind regions can deploy wind turbines to supply additional renewable power to microgrids. For instance, Copenhagen Airport utilises wind energy systems that produce approximately 2 MW of renewable electricity. "Airport wind turbine microgrid powering electric aircraft charging stations with renewable energy integration, futuristic smart airport concept."

Figure 2.15 presents a three-dimensional cost optimisation surface that illustrates the relationship between electricity pricing and aircraft charging

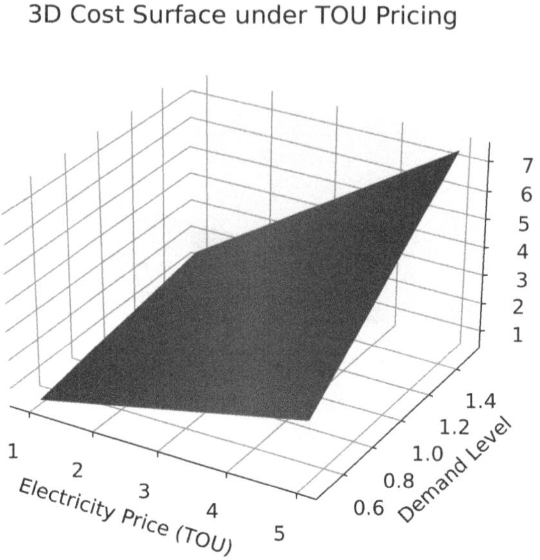

FIGURE 2.15
Cost surface under time-of-use pricing.

demand within an airport microgrid. The graph demonstrates how time-of-use (TOU) pricing influences energy consumption behaviour by introducing variable electricity prices throughout the day. In this visualisation, the electricity price represents one dimension, the charging demand level forms another dimension, and the resulting operational cost is shown along the vertical axis. As electricity prices increase during peak hours, the operational cost of charging aircraft also rises significantly, encouraging the system to reduce energy consumption during those periods. Conversely, during off-peak hours when electricity prices are lower, the charging demand shifts upwards as the system schedules energy-intensive charging operations at these cheaper time intervals. This dynamic interaction allows the microgrid to perform *peak-shaving*, where high-cost peak demand is minimised by redistributing loads to low-cost periods. Consequently, TOU-based scheduling improves economic efficiency, reduces grid stress, and enhances the overall sustainability of electric aircraft charging infrastructure.

2.16 Technological Innovations in Smart Microgrids

Modern airport microgrids are evolving into highly intelligent systems through the integration of artificial intelligence, Internet of Things (IoT), and advanced energy management systems (EMS). These technologies enable real-time monitoring, predictive maintenance, and autonomous energy optimisation. AI algorithms analyse weather conditions, charging schedules, renewable generation, and storage capacity to determine the most efficient energy distribution strategy. IoT sensors continuously collect operational data from microgrid components, allowing operators to detect anomalies and maintain system reliability. "AI-powered airport microgrid architecture diagram showing renewable energy, battery storage, electric aircraft charging, and intelligent energy management system."

2.17 Future Research Directions

Although substantial progress has been made in electric aviation and microgrid technology, several research challenges remain.

Future research should focus on the following:

1. *Advanced Energy Storage*

   Development of solid-state and lithium–sulfur batteries with higher energy density.
2. *Smart Microgrid Optimisation*

   Advanced AI-based energy forecasting and predictive load management.
3. *Lightweight Power Systems*

   Design of compact and lightweight power electronics for aviation applications.
4. *Cyber-Physical Security*

   Protection of microgrid communication networks from cyber threats.

2.18 Final Perspective

The integration of *electric aircraft technologies, renewable energy systems, hybrid energy storage architectures, and intelligent microgrid control frameworks* marks a transformative milestone in the evolution of modern aviation infrastructure. As airports transition towards electrified aircraft operations, the demand for resilient, sustainable, and intelligent energy ecosystems is rapidly increasing. The deployment of renewable-powered microgrids, supported by advanced storage solutions and *AI-driven energy management platforms*, enables airports to efficiently balance energy generation, storage, and consumption while maintaining operational reliability. By leveraging predictive analytics, real-time monitoring, and adaptive energy scheduling, intelligent microgrid systems can optimise aircraft charging processes, minimise peak demand stress, and significantly reduce carbon emissions associated with airport operations. Furthermore, the integration of solar and wind generation with hybrid energy storage systems enhances grid flexibility and energy resilience, allowing airports to operate effectively in both grid-connected and islanded modes. These technological advancements position airports as *next-generation intelligent energy hubs*, capable of supporting not only electric aircraft charging but also broader smart city energy networks. In this context, airports will evolve from traditional transportation facilities into *sustainable energy ecosystems*, where renewable energy generation, digital infrastructure, and intelligent energy management converge to support low-carbon aviation and environmentally responsible urban development. Ultimately, the convergence of these technologies will accelerate the global transition towards *clean, efficient, and sustainable aviation systems for the future.*

References

[1] International Energy Agency, "Transport," 2023, [Online]. Available: https://www.iea.org/energy-system/transport. Accessed: Feb. 7, 2024.

[2] C. A. Wells, P. D. Williams, N. K. Nichols, D. Kalise, and I. Poll, "Reducing transatlantic flight emissions by fuel-optimised routing," *Environmental Research Letters*, vol. 16, no. 2, 2021, https://doi.org/10.1088/1748-9326/abce82.

[3] D. Chiaramonti, "Sustainable aviation fuels: The challenge of decarbonization," *Energy Procedia*, vol. 158, pp. 1202–1207, 2019, https://doi.org/10.1016/j.egypro.2019.01.308.

[4] A. Dhara and J. M. Lal, "Sustainable technology on aircraft design: A review," *IOP Conference Series: Earth and Environmental Science*, vol. 889, 2021, https://doi.org/10.1088/1755-1315/889/1/012068.

[5] P. J. Ansell and K. S. Haran, "Electrified airplanes: A path to zero-emission air travel," *IEEE Electrification Magazine*, vol. 8, no. 2, 2020, https://doi.org/10.1109/MELE.2020.2985482.

[6] F. Afonso *et al.*, "Strategies towards a more sustainable aviation: A systematic review," *Progress in Aerospace Sciences*, vol. 137, Art. no. 100878, 2023, https://doi.org/10.1016/j.paerosci.2022.100878.

[7] B. J. Brelje and J. R. Martins, "Electric, hybrid, and turboelectric fixed-wing aircraft: A review of concepts, models, and design approaches," *Progress in Aerospace Sciences*, vol. 104, pp. 1–19, 2019, https://doi.org/10.1016/j.paerosci.2018.06.004.

[8] Eviation, "Alice aircraft," 2024. [Online]. Available: https://www.eviation.com/aircraft/. Accessed: Feb. 2, 2024.

[9] Pipistrel, "Systems and equipment," 2024. [Online]. Available: https://cloud.pipistrel.si/share/index.php/apps/files/?dir=/Systems%20and%20equipment&fileid=116315. Accessed: Feb. 2, 2024.
[10] J. Mitali, S. Dhinakaran, and A. A. Mohamad, "Energy storage systems: A review," *Energy Storage and Saving*, vol. 1, no. 3, pp. 166–216, 2022, https://doi.org/10.1016/j.enss.2022.07.002.
[11] C. Pornet, "Conceptual design methods for sizing and performance of hybrid-electric transport aircraft," Ph.D. dissertation, Technische Universität München, Munich, Germany, 2018.
[12] A. T. Isikveren, S. Kaiser, C. Pornet, and P. C. Vratny, "Pre-design strategies and sizing techniques for dual-energy aircraft," *Aircraft Engineering and Aerospace Technology*, vol. 86, no. 2, pp. 525–542, 2014.
[13] A. P. Brdnik, R. Kamnik, M. Marksel, and S. Božičnik, "Market and technological perspectives for the new generation of regional passenger aircraft," *Energies*, vol. 12, no. 10, p. 1864, 2019.
[14] M. Alrashed, T. Nikolaidis, P. Pilidis, and S. Jafari, "Utilisation of turboelectric distribution propulsion in commercial aviation: A review on NASA's TeDP concept," *Chinese Journal of Aeronautics*, vol. 34, no. 11, pp. 48–65, 2021.
[15] V. Marciello *et al.*, "Design exploration for sustainable regional hybrid-electric aircraft: A study based on technology forecasts," *Aerospace*, vol. 10, no. 2, p. 165, 2023.
[16] L. Kiesewetter *et al.*, "A holistic review of the current state of research on aircraft design concepts and considerations for advanced air mobility applications," *Progress in Aerospace Sciences*, vol. 142, Art. no. 100949, 2023.
[17] K. P. Duffy and R. H. Jansen, "Turboelectric and hybrid electric aircraft drive key performance parameters," in Proc. AIAA/IEEE Electric Aircraft Technologies Symposium (EATS), 2018, pp. 1–19.
[18] P. Thalin, *Fundamentals of Electric Aircraft*, SAE International, 2019.
[19] S. Sahoo, X. Zhao, and K. Kyprianidis, "A review of concepts, benefits, and challenges for future electrical propulsion-based aircraft," *Aerospace*, vol. 7, no. 4, 2020.
[20] A. Kuśmierek, C. Galiński, and W. Stalewski, "Special issue on green aviation," *Progress in Aerospace Sciences*, vol. 141, 2023.
[21] M. K. Bradley and C. Droney, *Subsonic Ultra Green Aircraft Research: Phase II Hybrid Electric Design Exploration*, NASA CR-218704, 2015.
[22] Z. Guo, B. Li, G. Taylor, and X. Zhang, "Infrastructure planning for airport microgrid integrated with electric aircraft and parking lot electric vehicles," *eTransportation*, vol. 17, Art. no. 100257, 2023.
[23] D. Babuder, Y. Lapko, P. Trucco, and R. Taghavi, "Impact of emerging sustainable aircraft technologies on the existing operating ecosystem," *Journal of Air Transport Management*, vol. 115, Art. no. 102524, 2024.
[24] L. Trainelli *et al.*, "Optimal sizing and operation of airport infrastructures in support of electric-powered aviation," *Aerospace*, vol. 8, no. 2, p. 40, 2021.
[25] Y. Zheng *et al.*, "Electric vehicle battery charging/swap stations in distribution systems: Comparison study and optimal planning," *IEEE Transactions on Power Systems*, vol. 29, no. 1, pp. 221–229, 2014.
[26] M. Wang *et al.*, "Mobile energy storage scheduling for AC-DC microgrids enabling low-carbon airport," in Proc. IEEE APPEEC, 2019, pp. 1–5.

[27] D. Buvarp and J. Leijon, "Mobile charging system for flexible and convenient charging of electric aircraft," in Proc. Int. Conf. More Electric Aircraft, 2024.

[28] H. Kilicoglu and P. Tricoli, "Technical review and survey of future trends of power converters for fast-charging stations of electric vehicles," *Energies*, vol. 16, no. 13, p. 5204, 2023.

[29] N. Thonemann *et al.*, "Prospective life cycle inventory datasets for conventional and hybrid-electric aircraft technologies," *Journal of Cleaner Production*, vol. 434, Art. no. 140314, 2024.

[30] M. Meindl, M. März, and K. J. Weber, "Ground-based power supply system to operate hybrid-electric aircraft for future regional airports," in Proc. IEEE ESARS-ITEC, 2023.

3

Power Electronics in More-Electric Aircraft

Sheron Figarado, Kranthi Panuganti, and Akshay Kumar Rathore

3.1 Introduction

The requirement for aviation electrification is accelerated with the global consensus to achieve net-zero greenhouse gas emission by 2050 [1]. The advent of more-electric aircraft (MEA) architectures represents a paradigm shift in aviation, where traditional hydraulic, pneumatic, and mechanical systems are progressively being replaced by electrically driven equivalents to enhance fuel efficiency, reduce maintenance costs, and lower emissions. Current MEA architectures incorporate higher bus voltages to increase the power levels to cater to more subsystems. In addition, one of the key change involved replacement of the constant speed drive (CSD) for maintaining generator input speeds with variable frequency output static converters, thereby reducing the weight of the aircraft. Non-propulsive loads in MEA encompass a wide array of subsystems critical to aircraft operation, including environmental control systems (ECS) for cabin pressurisation and air-conditioning, electro-hydraulic actuators (EHAs) for flight surfaces, electromechanical actuators (EMAs) for landing gear and brakes, galley equipment, lighting, and on-board infotainment. These loads collectively demand significant electrical power, amounting up to 1 MW in large commercial aircraft such as Boeing 787, necessitating versatile power converters to interface with the aircraft's variable frequency AC buses (typically 115/230 V at 360–800 Hz).

Power converters, such as AC–DC rectifiers, DC–DC buck/boost stages, and DC–AC inverters, are pivotal in conditioning power for these loads, ensuring compatibility within the MEA's electrical power generation and distribution system (EPGDS). While propulsion system design prioritises high power density for thrust generation, non-propulsive converters must balance efficiency, compactness, and reliability under stringent aerospace constraints. This chapter gives a description of the MEA electrical power system (EPS), the key requirements, and design

DOI: 10.1201/9781003747994-3

considerations for the converters involved, while highlighting some of the recent advancements in wide-bandgap (WBG) semiconductors and circuit topologies.

3.2 Overview of MEA Electrical Power Systems

In conventional aircraft, power generation and distribution is on constant-frequency 400 Hz AC systems. The architecture of a conventional EPS is depicted in Figure 3.1 [2].

The choice of 400 Hz helps to reduce the weight of the magnetics such as transformers and pump motors. The EPS of a conventional aircraft mainly comprises two turbo generators, each connected to the main engines through constant speed drive mechanisms (together termed as integrated drive generator) to produce a 115 V/400 Hz fixed AC supply. The auxiliary power unit (APU) of the aircraft also produces 115 V/400 Hz

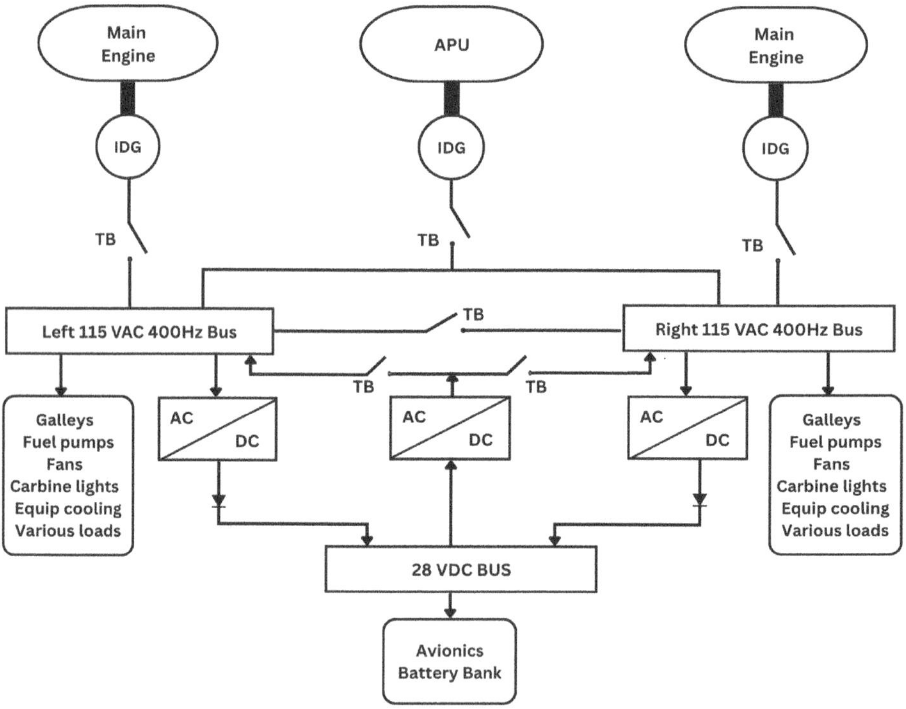

FIGURE 3.1
Electrical power system architecture in a conventional aircraft.

constant AC supply and is connected to the same grid. APU is used to power the essential electrical systems while the aircraft is on the ground before the main engines are engaged and is usually offline while in flight. A low voltage DC bus, typically of 28 V, is generated from the AC grid using transformer-rectifier units (TRU) and auto transformer rectifier units (ATRU), which may further have converted to 5 V and 3.3 V for avionics and other low voltage electronics.

The aircraft have a battery-based energy storage systems (ESS) and each subsystem of energy generation and storage is configured in such fashion to provide multiple redundancies for the power delivery to critical electrical systems during single-engine failure scenarios. Aircraft also have a ram air turbine (RAT), a wind turbine generator based on ram pressure, which supports critical systems such as cabin temperature, communication, and avionics in case of dual-engine failure during flight. Many of the systems in conventional aircraft rely on pneumatic and hydraulic systems instead of electrical based actuators. For instance, the wing bleed ice protection system of a conventional aircraft uses air from the pneumatic line to heat up the wings. The overall power rating of the electrical generation in conventional aircraft is 150–300kVA [4].

MEA architectures favour variable-frequency starter-generators (VFSGs) driven by gas turbines, producing 350–800 Hz AC and thereby eliminating heavy CSDs. The EPGDS typically features a high-voltage DC (HVDC) bus, typically rated at ±135 V or ±270 V, in parallel to AC buses, enabling efficient power sharing between propulsion and non-propulsive domains. Non-propulsive loads connect via remote power distribution units (RPDUs), where power converters perform rectification and regulation to supply DC intermediates (e.g., 28 V for avionics or 270 V for actuators). Electrical power system architecture of an MEA is illustrated in Figure 3.2. Apart from the removal of CSDs, other subsystems powered by pneumatics hydraulics are replaced by EMAs and EHAs. For example, in Boeing 787, WIPS uses an electric heater, thereby implementing a no-bleed architecture. There are two generators per either sides of the aircraft of 250 kVA each totalling to 1 MVA in Boeing 787 model. It also has 2 APUs of 225 kVA each with a total 450 kVA of power. Still, the percentage of electrification, $\frac{P_{el}}{P_{th}}$ given as the ratio of electrical power to the maximum thrust power during take-off, is still less than 0.2% [5].

The hybrid AC–DC architecture used in the MEAs reduces cabling weight by 20–30% compared to legacy systems but introduces challenges such as harmonic injection and voltage sags due to cable inductance, which scales with frequency. Recent studies highlight the need for converters that maintain unity power factor and low total harmonic distortion (THD <5%) to preserve system stability, particularly as non-propulsive demands fluctuate with flight phases [6].

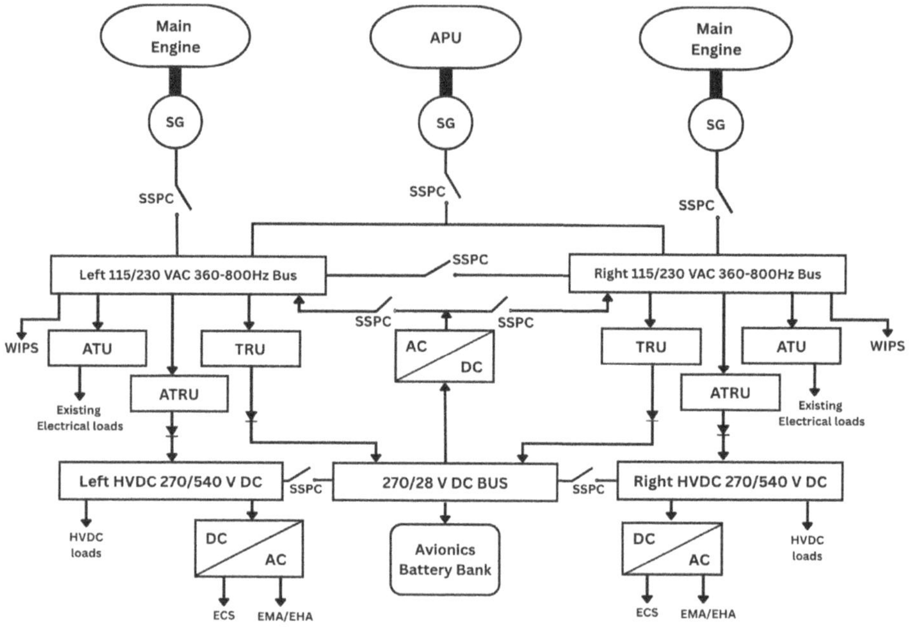

FIGURE 3.2
Electrical power system architecture in an MEA.

3.3 Other Non-Propulsive Subsystems in MEAs

Among the non-propulsive subsystems of aircraft, the following are the major subsystems that had undergone drastic electrification.

3.3.1 Main Engine Start

Traditionally, aircraft main engines are started pneumatically using compressed air from the APU-driven compressors, routed through ducts to air turbine starters (ATS) that initiate the spin of the engines. Ground cart-based compressors may also supply air, when the aircraft is on the ground. Electric aircraft architectures such as that in Boeing 787 replace this pneumatic start method with electric main engine start (MES) systems, where starter-generators powered by APU-generated electricity replace air ducts and ATS units. This shift demands AC–DC–AC conversion for variable voltage and frequency control, involving rectifiers and inverters. Proper sizing of generators, converters, and thermal components is critical to meet torque requirements without adding excess weight. Repurposing the same power converters for other functions after MES can help reduce the weight and improve the overall system efficiency, thereby reducing fuel consumption.

3.3.2 Environmental Control Systems

Modern aircraft have replaced traditional bleed-air ECS with electrically powered ones to improve efficiency and reduce pneumatic complexity. While Airbus A350 and A380 still use bleed-air ECS, the Boeing 787 employs electric compressors for cabin temperature and pressure control, eliminating engine bleed-air and ducts. Electric ECS design faces challenges in power sizing, as compressors and converters demand substantial energy, increasing main generator capacity. Reliability, redundancy, and thermal management are key considerations. ECS includes numerous fans, pumps, and compressors requiring AC–DC–AC power conversion, advanced motor controllers, and communication protocols like CAN bus for monitoring, EMI compliance, and system integration.

3.3.3 Electrification of Hydraulic Systems

In conventional aircraft, hydraulics power primary and secondary controls, brakes, and landing gear using mechanically driven actuators. In MEAs, these are replaced by EHAs and EMAs, both driven by electric motors and inverters. EHAs integrate a pump, cylinder, and fluid reservoir in a self-contained unit, eliminating external hydraulic lines and offering reduced weight and cost as well as reliability benefits. EMAs, using gear or screw mechanisms to convert rotary to linear motion, are more efficient and leak-free but risk mechanical jamming, limiting their use in safety-critical functions. Design focus is mainly on fault-tolerant PM motors, inverter design, regenerative power recovery, and thermal design to ensure better efficiency and reliability. Boeing 787 already employs EHAs for hydraulics and EMAs for braking, reflecting this transition towards electric actuation systems.

Table 3.1 summarises the comparison of important electrical subsystems of some of the commercially available more-electric aircraft.

TABLE 3.1

Comparison of the Electrical Subsystems of Large Commercial Transport Aircraft

Aircraft Model	Boeing 787	Airbus 380	Airbus 350
No. of engines	2	4	2
No. of generators per engine	2	1	2
Generator rating	250 kVA	150 kVA	100 kVA
Generator output voltage	235 V	115 V	230 V
No. of generators per APU	2	1	1
Generator rating per APU	225 kVA	120 kVA	150 kVA
RAT rating	Not available	150 kVA	100 kVA
Environmental control systems	Electric (4 × 100 kW compressors)	Bleed-air	Bleed-air
Brake system	Electric	Hydraulic	Hydraulic
Actuation system	EHA	Conventional + EHA	Conventional + EHA

3.4 Requirements for Power Converters

Designing power converters for non-propulsive loads in MEA mandates adherence to multifaceted requirements, governed by standards such as DO-160 for environmental conditions and MIL-STD-704 for power quality [3]. For more-electric aircraft and all-electric aircraft, new standards such as AIR7502 and AIR7506 define the higher voltage levels and impact of these elevated voltages on wiring insulation.

3.4.1 Electrical Requirements

Non-propulsive loads operate across diverse voltage levels: 28 V DC for low-power avionics, 270 V DC for actuators, and variable AC for motors in EHAs/EMAs. Converters must handle input frequencies from 360 Hz to 800 Hz with output regulation to ±1% accuracy under load steps up to 100% rated power. Power quality offered by the converter is critical in this scenario to ensure power system stability. The input current THD must remain below 5–8% to avoid generator overload, while output ripple should be <1% to prevent actuator jitter. For instance, boost-type AC–DC rectifiers are required to achieve power factors >0.99, mitigating reactive power demands that could destabilise the bus [6].

3.4.2 Performance Requirements

Efficiency targets for power converters in MEAs exceed 98% across a 10–100% load range to minimise thermal dissipation and fuel burn. Power density is another constraint, with gravimetric targets of >10 kW/kg and volumetric >20 kW/L, driven by space constraints in avionics bays. Recent investigations of interleaved Vienna rectifiers demonstrate 99.3% peak efficiency at 3 kW, underscoring the feasibility of these metrics for galley and lighting loads.

3.4.3 Environmental and EMI Requirements

Aircraft environments impose harsh conditions: ambient temperatures from 55° C to 70° C, vibration levels as per DO-160 Category R, and altitudes to 50,000 ft (~15 km) with low air density for cooling. Converters must operate without forced air cooling, relying on natural convection or heat pipes to redistribute the heat [7]. Electromagnetic interference (EMI) compliance as per RTCA/DO-160 standards have to be complied by the equipment for conducted and radiated emissions, necessitating shielded enclosures and high-frequency filtering. This is especially challenging due to higher probability of exposure to lightning. Effective electromagnetic compatibility design requires (a) optimised switching transitions and gate drive timing to reduce dv/dt and di/dt; (b) using common-mode chokes, isolation transformers, and differential mode filters sized for aircraft frequency spectra; (c) effective PCB/cable routing and ground plane management; and (d) enclosure shielding and filtered feedthroughs for I/O connectors. Cosmic radiation and humidity further accelerate component ageing, requiring derating by 20–50%.

3.5 Power Electronic Converters in MEAs

The previous section outlined power architecture in conventional and MEA, highlighting the migration from fixed-frequency AC systems to variable-frequency AC and high-voltage DC buses. This section focuses on the converter topologies that form the building blocks for these architectures to function cohesively. Power electronic converters perform the essential task of energy adaptation and conditioning power between the given sources, buses, and required loads across multiple voltage levels and frequencies. The requirements are diverse, starting from regulated low-voltage DC for avionics and actuation to variable-frequency AC for propulsion drives and environmental control systems. This section provides an integrative overview of the major converter categories:

(i) DC–DC converters (non-isolated and isolated converters)
(ii) AC–AC converter (cycloconverters and matrix converters)
(iii) DC–AC converters (inverters)
(iv) AC–DC converters (rectifiers)

3.5.1 DC–DC Converters in MEAs

DC–DC converters in MEA are central to on-board DC microgrids. They implement step-up/step-down and isolation, enable bidirectional energy flow (e.g., between buses and batteries), perform point-of-load regulation, and support high-voltage DC intermediate buses used for propulsion or high-power systems. Because MEA architectures increasingly favour higher DC link voltages (to reduce current and wiring mass), DC–DC designs emphasise high conversion ratio, galvanic isolation (for safety and fault containment), high efficiency across wide load ranges, and bidirectional capability for energy storage and regenerative flows. High-conversion interfacing between LVDC (28V) and HVDC buses as well as reducing to low-voltage power supplies for communication and avionic circuits call for DC–DC converters. High-voltage (HCHV) isolated topologies are commonly discussed for MEA because they allow high switching frequencies (reducing magnetics mass) while meeting isolation requirements, and will be suitable for the shift towards AEAs. A detailed review of these converters are discussed in Chapter 5.

3.5.2 AC–AC Converters in MEAs

AC–AC conversion covers direct frequency/voltage conversion without an intermediate DC link (cycloconverters, matrix converters), and indirect schemes using controlled rectifier plus inverter (i.e., active frequency converters). In aircraft, the most familiar AC–AC functions are frequency conversion (e.g., between 400 Hz and variable-frequency loads) and variable-frequency power supplied directly to advanced variable-speed generators or electric machines. Matrix and direct AC–AC converters offer potential benefits in reduced passive filtering and compactness by eliminating bulky DC link electrolytic capacitors and by providing direct frequency transformation; however, they impose higher switching-stress and more complex control. For propulsion and high-performance actuation, modular multilevel and matrix topologies are gaining attention because they can meet high voltage/power requirements with lower harmonic distortion and smaller filters. Detailed discussion on these converters are beyond the scope of this chapter.

3.5.3 DC–AC Converters in MEAs

Inverters in electric aircraft are employed in applications that require AC power, such as engine starter/generator (ES/G), propulsion drives, actuators, and various other applications as illustrated in Figure 3.3. Therefore, inverters are essential components of electric aircraft [8]. Inverters used in electric aircraft operate under harsh conditions such as low pressure, high altitude, cosmic radiation, and wide temperature variations. These environments demand exceptional reliability and performance from the power conversion system. Consequently, aircraft inverters are designed with a strong focus on achieving high power-to-weight ratios, efficiency, and robustness, while minimising electromagnetic interference (EMI) and total harmonic distortion (THD) in both input and output waveforms [9, 10].

The two-level voltage source inverter (2L-VSI), as shown in Figure 3.4, is widely used in aircraft applications such as ES/G [11], actuators [12], and propulsion [13, 14], owing to its uniform voltage-blocking switches, reduced component count, and high reliability. In this topology, each switch must block the full DC-link voltage. High switching frequencies used to reduce filter size increase EMI, while the two-level output causes higher common-mode voltage (CMV), THD, and greater stress on motor windings [15]. A detailed design methodology for a 2L-VSI suitable for a 50 kW motor in modern aviation applications was presented in Ref. [16]. The developed inverter, rated at 540 V/70 kVA, employs SiC MOSFETs to reduce weight and enhance efficiency. Furthermore, several switching frequencies and

PWM strategies were explored to improve the performance. The final 2L-VSI prototype achieved an efficiency of 99.4% and a power density of 15 kW/kg.

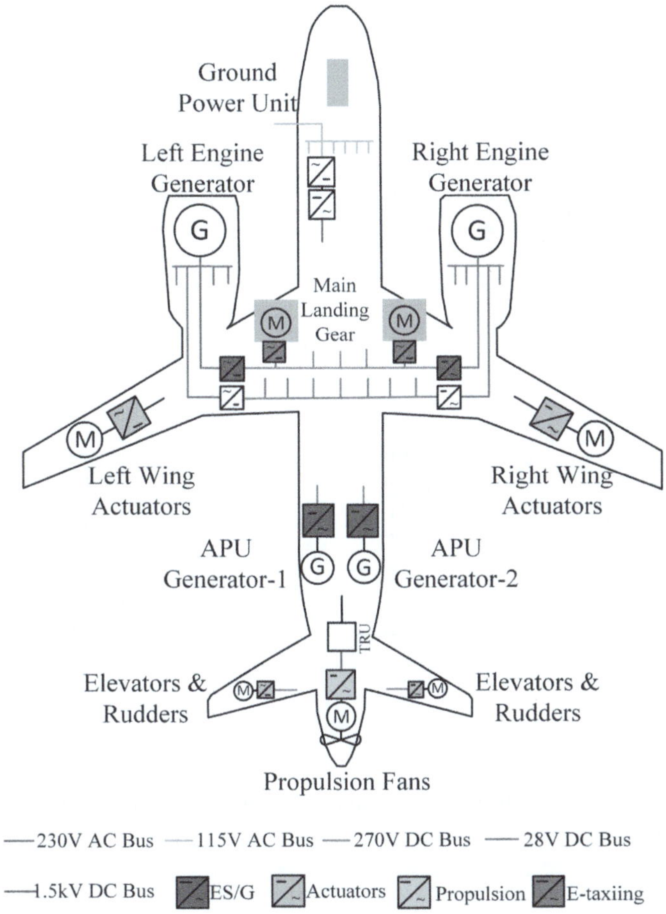

FIGURE 3.3
Subsystems of a turboelectric aircraft with inverters involved [8].

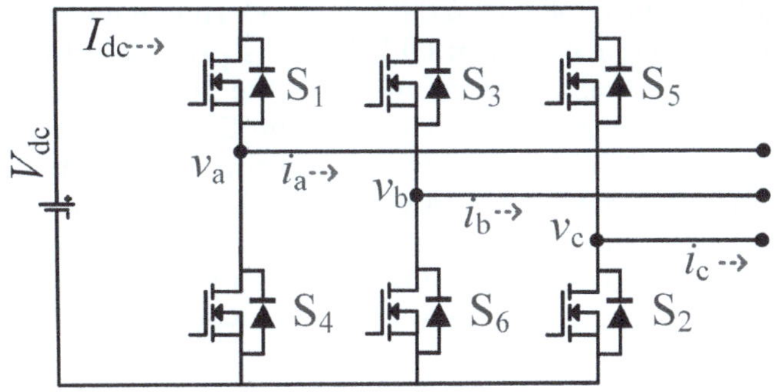

FIGURE 3.4
2L-VSI circuit topology.

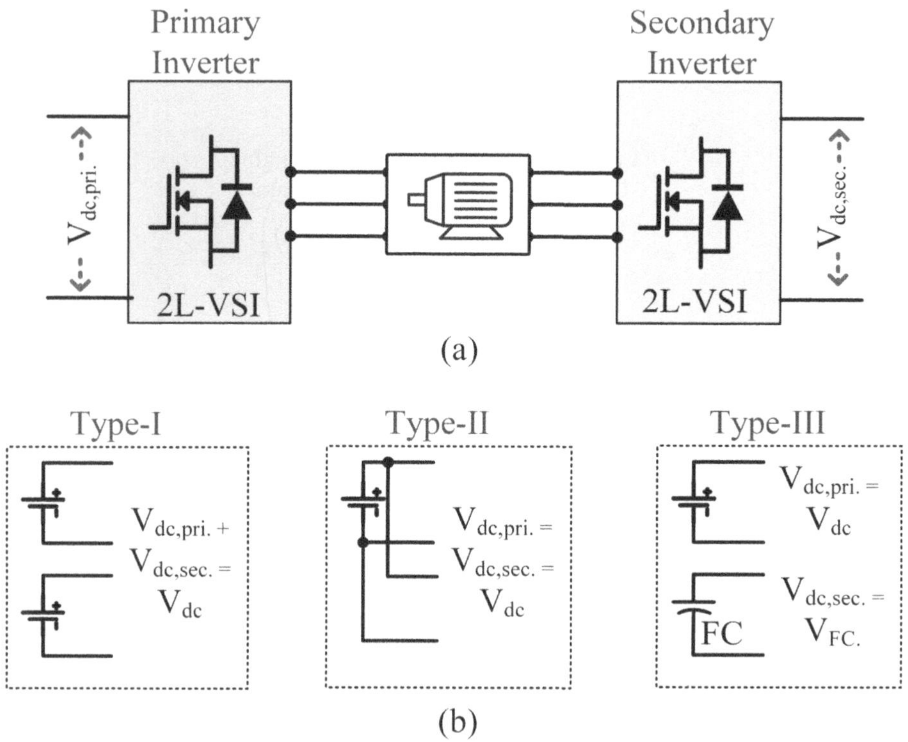

FIGURE 3.5
(a) Schematic of the motor drive with a dual inverter. (b) Three configurations of DC-links of dual-inverter topologies.

While achieving high power density and efficiency is crucial, these alone are not sufficient. Motor drive systems in aircraft must also ensure robust fault-tolerant operation to maintain reliability under abnormal conditions [17]. To address this need, advanced configurations such as open-end winding motor drives (OEWD) fed by dual inverters have been developed [17], as shown in Figure 3.5a. In OEWDs, it is presented that the voltage ratio between two dual inverters improves the output voltage levels and power quality [18, 19, 20]. Also, based on the DC-link voltage regulation on dual inverters, OEWDs are divided into three types, as illustrated in Figure 3.5b. Here, Type-I, has individual DC-links for both primary and secondary inverters and the sum of both DC-links equal to overall DC-link voltage [21]. In Type-II, same DC-link is utilised for both primary and secondary inverters [22]. In Type-III, secondary inverter uses flying capacitor as the DC-link [23].

On the other hand, as aircraft power demand continues to rise, high-voltage DC buses are increasingly adopted to reduce current levels, minimise

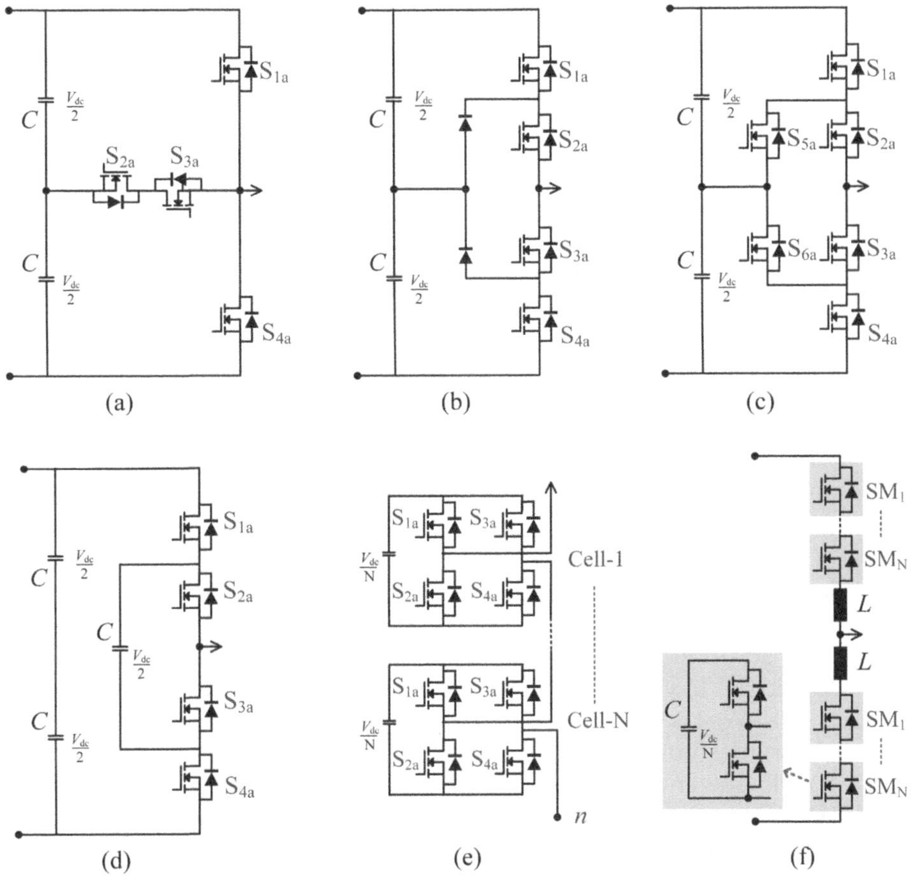

FIGURE 3.6
Single-phase configuration of various inverter topologies. (a) Three-level TNPC. (b) Three-level NPC. (c) Three-level ANPC. (d) Three-level FC. (e) (2*N*+1) level CHB. (f) (*N*+1)-level MMC.

cable weight, and improve system efficiency [24]. For instance, NASA's N3-X architecture operates at ±5 kV [25]. However, higher DC-link voltages bring new challenges, such as increased semiconductor blocking voltage, higher dielectric stress on passive components, and aggravated EMI at elevated switching frequencies. In addition, high-voltage devices and passives remain limited in availability and costlier than their low-voltage counterparts.

In two-level inverter configurations, these high-voltage requirements lead to greater switching losses, increased *dv*/*dt* stress, and larger filter needs, making the topology less suitable for next-generation aviation systems [24]. Consequently, multilevel inverter topologies have gained attention as promising alternatives, and these topologies are illustrated in Figure 3.6. The neutral point clamped (NPC)–type circuits, illustrated in Figure 3.6a–c, are the multilevel topologies that are being investigated frequently due to their high power density [15].

(a) **Neutral Point–Clamped Inverters**

A 2.4 kV, 1 MW three-level ANPC inverter developed by General Electric (GE) demonstrated the feasibility of high-frequency operation for aircraft motor drives [26] [27]. To meet the requirements of high-speed motors, the fundamental frequency was set to 1.4 kHz, while the switching frequency was maintained at 20 kHz. A hybrid configuration combining Si IGBTs and SiC MOSFETs was employed to achieve a balance between performance and cost, resulting in a peak efficiency of 99% and a gravimetric power density of 10.2 kW/kg [27]. In parallel, Boeing, in collaboration with the University of Tennessee under a NASA-sponsored programme, developed a fully SiC-based three-level ANPC inverter rated at 1 kV and 1 MW [28, 29]. The system comprised two interleaved 500 kW modules operating at a fundamental frequency of 3 kHz and a switching frequency of 60 kHz. This configuration achieved an efficiency of 99.2% at half load and 99% at full load, with a gravimetric power density of 18 kVA/kg.

In addition to these implementations, several other NPC-based topologies have been explored for aircraft applications. For instance, a 100 kW, 1 kV T-type NPC converter was developed as a power electronics building block, achieving 98.2% efficiency and a gravimetric power density of 27.7 kW/kg [30]. Furthermore, a MW-class ANPC topology has been proposed for propulsion applications [31], although its prototype is yet to be realised.

(b) **Flying Capacitor Inverter**

A flying capacitor (FC)–based inverter was developed at the University of Illinois as part of a NASA-sponsored project [32, 33]. The converter operates from a 1 kV DC link and delivers 15–20 kW per cell at a fundamental frequency of 3 kHz. Ten to twelve such cells are paralleled to achieve approximately 200 kW per module, and multiple modules are further connected in parallel to reach megawatt-level power ratings. Utilising GaN devices enabled switching frequencies as high as 120 kHz, resulting in a gravimetric power density of 17.3 kW/kg and a peak efficiency of 98.6%. In a related study, a 13-level FC inverter topology employing GaN switches was reported in Ref. [34]. Operating from an 800-V DC link, this design achieved a peak efficiency of 98.3% and a gravimetric power density of 12.24 kW/kg. Further advancements in FC-based inverter configurations have also been proposed in Ref. [35], highlighting their potential for compact and efficient power conversion in high-performance electric aircraft systems.

(c) **Modular Multilevel Inverters**

Modular multilevel converter topologies, including the cascaded H-Bridge (CHB) and modular multilevel converter (MMC), have emerged as promising candidates for medium- and high-voltage applications owing to their modular structure, scalability, and inherent fault tolerance. Recent investigations have extended their applicability to aircraft power conversion systems, with studies on CHB [36] and MMC [37] highlighting their potential benefits; however, comprehensive experimental validation is still lacking.

Although MMCs typically offer superior efficiency, their adoption in aircraft systems is constrained by a comparatively lower power density, primarily due to the large number of components required to synthesise the desired voltage levels and the significant volume of passive elements such as capacitors and inductors. The component-based weight comparison presented in Ref. [15] tends to underestimate MMC suitability, as it neglects the additional weight contributions from filters and circuit breakers required, as MMC has less filtering requirements. Also, an MMC with N submodules per arm can produce $(2N + 1)$ voltage levels. Considering these a more balanced assessment could reveal the MMC's potential as a viable candidate for electric aircraft applications, particularly when fault tolerance is taken into account [37].

From the above discussion, it is evident that substantial progress has been made by government agencies, industries, research laboratories, and academic institutions towards enhancing inverter efficiency and power density for aviation applications. Despite these advancements, future aircraft systems demand further improvements to satisfy the stringent performance benchmarks set by NASA, which specify efficiencies exceeding 99% and power densities of 19 kW/kg for non-cryogenic-cooled inverters, and over 99% efficiency with 26 kW/kg power density for cryogenic-cooled systems [38]. In parallel, achieving robust and fault-tolerant operation continues to be a fundamental design requirement to ensure operational reliability and safety in electric aircraft propulsion environments.

3.5.4 Future Trends of Inverters Used in Aviation

To meet NASA's ambitious targets for next-generation aircraft power systems, future inverter development is directed towards achieving higher efficiency, increased power density, lower harmonic distortion, and enhanced reliability. The most direct route to improved efficiency lies in minimising both conduction and switching losses. Conduction losses are influenced by the ON-state resistance of semiconductor devices, whereas switching losses depend on the device transition speed under given operating conditions. Wide bandgap (WBG) semiconductors, such as silicon carbide (SiC) and gallium nitride (GaN), have therefore emerged as key enablers due to their low ON-state resistance and fast-switching capability, nearly 20 times faster than conventional silicon-based IGBTs. Despite their commercial availability, GaN modules currently suffer from limited current ratings, rendering them less suitable for high-power applications such as aircraft propulsion. Efficiency can also be improved through advanced modulation strategies, such as bus-clamping techniques [39], which minimise switching transitions, and by replacing electrolytic capacitors with film capacitors to mitigate conduction losses associated with high ESR values.

Enhancing power density remains another critical objective. This can be achieved by reducing filter size while maintaining the required harmonic attenuation. Furthermore, advanced control and modulation techniques that effectively regulate DC-link voltage and current ripple can substantially reduce the required capacitance, often a major contributor to inverter volume. Integration of the inverter with the motor, known as an integrated motor drive (IMD), further improves overall system density, although it introduces challenges in thermal management and mechanical robustness.

Reliability improvement is equally vital, particularly under the harsh environmental and thermal conditions encountered in aircraft systems. Advances in packaging technology can enhance the thermal endurance and mechanical stability of semiconductor devices at high altitudes. From a control perspective, strategies that actively balance switching losses among devices can significantly extend system lifetime. Topologically, modular converter structures inherently enhance reliability through redundancy and fault isolation, although often at the expense of power density. Hence, future research should emphasise the development of hybrid or reconfigurable modular topologies that preserve the reliability advantages of modular architectures while achieving competitive power density [37].

3.5.5 AC–DC Converters in More-Electric Aircraft

Since the electric power generation in aircraft is directly from the AC generators connected to variable speed engines, rectifiers are needed in the aircraft power architecture to generate the low-voltage DC bus (28 V) and high-voltage DC bus (270 V or 540 V). In conventional aircraft architecture, TRU and controlled rectifiers are widely used, which are simple and more reliable, but has poor power quality. In MEA, however, AC–DC front ends must meet stricter EMI and power-quality standards while supporting bidirectional operation and higher efficiency. For power conversion from 115 V, 400 Hz AC to DC service buses, topologies range from simple diode/thyristor rectifiers (with heavy passive filtering) to actively controlled PWM rectifiers and multi-pulse autotransformer-based rectifier banks that reduce harmonic injection. Novel multi-pulse/phase-shifted rectifiers and active front-end PWM rectifiers are also explored in the literature to meet DO-160 EMI limits and to reduce bulky filtering that would otherwise penalise mass. A number of recent studies focus on avionics-grade rectifiers that combine low emissions, high efficiency, and reduced magnetics.

In modern MEA designs, the trend is towards active rectifiers with power-factor correction (PFC) and bidirectional capability so that the same converter can feed energy back to the AC source during regenerative events. Such active front ends also enable tighter DC bus regulation and better transient performance. As with other converter categories, use of SiC devices, optimised packaging, and improved EMI design are recurring themes to hit the aircraft targets for weight, reliability, and emissions.

MEA designs target to enhance fuel efficiency, reduce emissions, and improve reliability by electrifying aircraft subsystems. This shift demands robust power conversion systems, particularly AC/DC rectifiers, to convert variable-frequency AC power (115 V, 360–800 Hz) from on-board generators into regulated DC power for avionics, battery systems, and electric propulsion.

Key requirements for MEA rectifiers:

- High efficiency and power density
- Low input current harmonics
- Fault tolerance and EMI resilience
- Compact and lightweight design

3.5.5.1 AC–DC Converters Topologies

Unidirectional power flow rectifiers can be broadly classified into two categories:

- Passive rectifiers
- Active rectifiers

In this section, a few of the popular topologies for rectification will be discussed.

3.5.5.2 Passive Rectifiers

Passive rectifiers employ uncontrolled diode rectifiers along with passive filter components to generate DC supply. Power converters such as the 6-pulse and 12-pulse, diode bridge rectifiers, are widely used in high-power MEA applications due to their simplicity and reliability. The obvious advantages are the lack of need for active control, high reliability, and lower cost. However, the AC side current drawn from the 115V AC bus has poor harmonic characteristics and power factor and lack of output regulation. In addition, the use of power frequency transformers and huge filter elements makes the system bulky. However, it is noteworthy that since the power frequency is in the

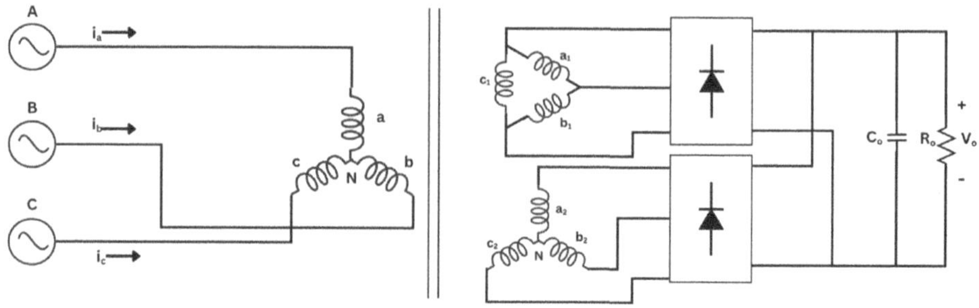

FIGURE 3.7
Twelve-pulse rectifier-based TRU.

range 360–800 Hz, the magnetics are still less bulky compared to the ground power system of the same rating.

TRU and ATRU are used in aircraft for generating the isolated DC link and non- isolated DC link, respectively. In order to meet the harmonic limits, set by the airborne standard DO-160G and the military standard MILSTD704F, multi-pulse rectifiers such as12-pulse diode-based passive ATRUs and TRUs were used commonly in conventional aircraft, which can provide a power factor of near 0.98. More advanced multi-pulse converters such as 24-pulse rectifiers are proposed to improve the harmonic profile in literature.

Figure 3.7 illustrates a standard 12-pulse diode rectifier–based TRU. A conventional three-phase rectifier generates a rectified DC voltage waveform with a six-pulse ripple profile. This is because the three-phase rectifier follows the envelope of three-phase line voltages (three positive peak envelope generated by positive group diodes and negative peak envelope generated by negative group diodes). A 12-pulse rectifier generates a DC output with a ripple with 12 times the fundamental line frequency. This is achieved by paralleling two three-phase rectifiers, one powered by the three-phase voltage in phase with the AC grid (from the Y-connected secondary) and the other rectifier excited by 30-degree phase-shifted three-phase voltages (from the delta- connected secondary). This parallel combination results in a DC voltage with ripple frequency of 12 times that of line frequency, and reduces the output filtering requirements while improving the input harmonic content due to multi-stepped input current waveform.

The operation of ATRU is similar in output waveforms except the transformer used is an autotransformer, with the secondary and tertiary windings connected in wye and delta configuration, respectively, as given in Figure 3.8. In both configurations, input current waveform is devoid of 3rd, 5th, 7th, and 9th harmonic and the dominant harmonics are 11th and 13th. While this is closer to sinusoidal current, a further improvement is desired.

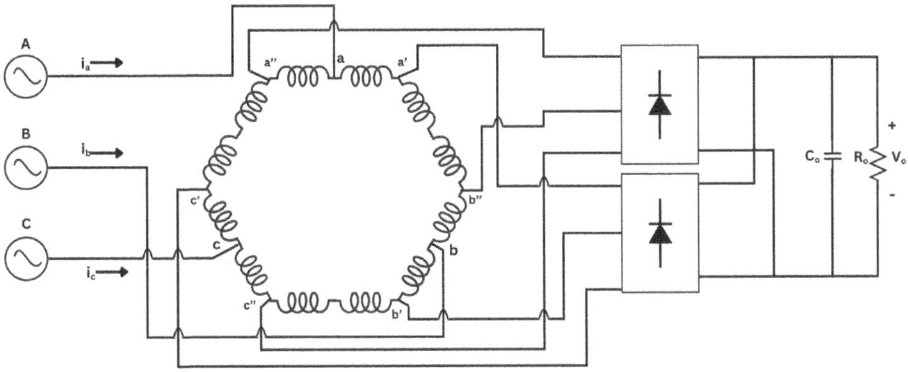

FIGURE 3.8
Twelve-pulse rectifier-based TRU.

3.5.5.3 Active Rectifiers

Active rectifiers use controlled switches such as IGBTs and MOSFETs, which allow control of output voltage and help improve power quality [41]. They typically allow power factor correction and a near-unity power factor operation while getting a regulated output voltage. The high-frequency switching operation allows reduced weight for magnetics for filtering and isolation.

Some of the popular topologies are as follows:

a) Boost, buck, or buck-boost-derived PFC converters
b) Vienna rectifier-based topologies
c) Multilevel converters and matrix converters

(a) **PFC Converters**
One of the classic PFC boost topologies is shown in Figure 3.9. The idea here is to use the boost converter operation to shape average of the line inductor current closer to a sinusoid instead of a quasi-square wave for a passive three-phase diode rectifier. This converter will produce a PWM modulation with a current control mode, which enables a simultaneous control of output voltage regulation and input current shaping. This three-phase, single switch, PFC boost converter can improve the power factor close to 0.99. Another aspect for this topology is that it can be operated in continuous conduction mode (CCM), critical conduction mode (CrCM), or discontinuous conduction mode (DCM). DCM operation offers a lot of advantages such as improved control speed, lower filter inductance, and reduced number of sensors. Further, it is possible to realise the zero-current switching (ZCS) of the converter switches and the zero reverse

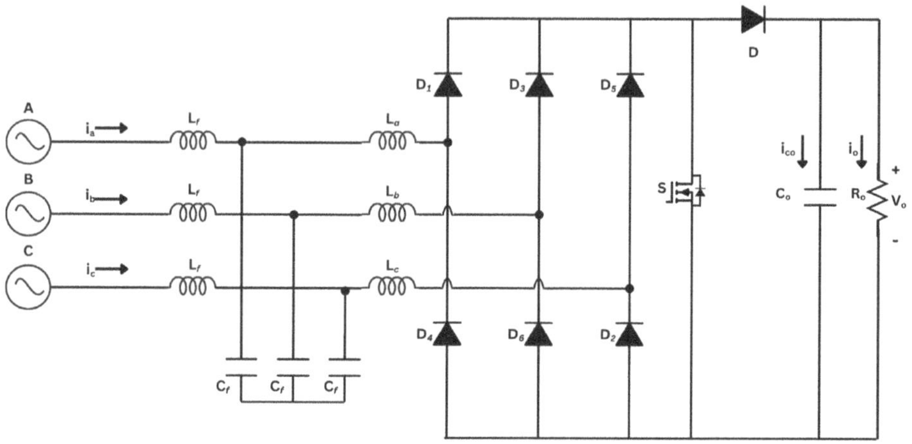

FIGURE 3.9
A three-phase, single-switch PFC boost converter.

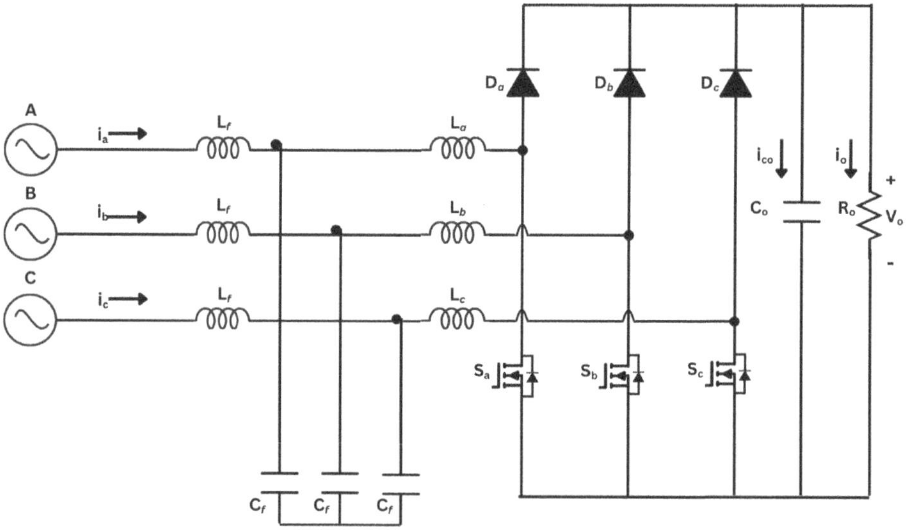

FIGURE 3.10
A three-phase semi-active bridge rectifier.

recovery losses in the converter diodes while operating in DCM [42]. However, the current stresses on the components are higher in DCM operation for a given power rating, due to increased peak current for the same average as compared to CCM operation.

Another widely used active boost-based converter is a three-phase, semiactive bridge rectifier that comprises three diodes and three controlled switches [43], as shown in Figure 3.10. This is a compromise between fully controlled PWM rectifiers and passive rectifiers. Another advantage compared to fully controlled PWM rectifiers and two- switch boost converter is that we need only low side drivers and single supply for the driver circuits. Since only negative group switches are controllable, sinusoidal current control is possible only in those sectors when two of the phase voltages are positive. The control can be simultaneous switching mode or turning on a single switch corresponding to highest negative phase voltage. Thus, converter also yields power factor near to 0.99.

(b) **Vienna Rectifier–Based Topologies**
Vienna rectifier–based converters are a class of rectifier topologies based on a 1993 patented three-switch converter, known as Vienna converter, which offers boost action, power factor correction, inherent multilevel operation, and getting reduced switch stress. It is also capable of operating under unbalanced conditions. Figure 3.11 shows a classical three-phase Vienna rectifier used for MEA application [44]. Depending on the states of the switches and current direction, each phase may be connected to $+V_O/2, 0, -V_O/2$.

In the modulation strategy for a Vienna rectifier, a phase-locked loop (PLL) is utilised to synchronise with the grid frequency and phase angle, which is then used to transform to the *dq* synchronous rotating reference frame. The three-phase input voltages and currents are thereby converted into DC quantities in the rotating *dq* frame and regulated via proportional-integral (PI) controllers to

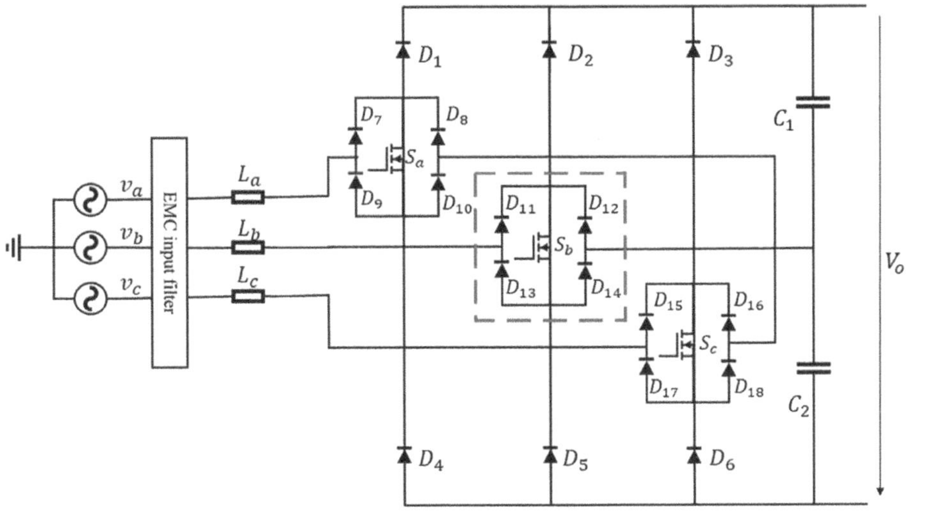

FIGURE 3.11
Classical Vienna rectifier topology in aircraft applications [44].

achieve precise current control. The resulting outputs from these *dq*-axis current regulators form the reference modulation signals for space vector pulse-width modulation (SVPWM), ensuring boost rectifier operation.

It can be observed that the four diode plus one switch combination (for example, D_{11}, D_{12}, D_{13}, D_{14}, and S_b) represent a four quadrant switch with bidirectional current conduction and unidirectional voltage blocking. Variations of the converter can be obtained once this four-quadrant switch is replaced with other four-quadrant switch options, as shown in Figure 3.12 [45].

A variant of the Vienna rectifier with switch realisation in Figure 3.12 is depicted in Figure 3.13. The 2-MOSFET switch based variant will have six controlled switches [46, 47]. However, the conduction losses in this variation will be lower since only one MOSFET and one diode will be involved in conduction, as compared to two diodes and one MOSFET in the topology in Figure 3.11. With appropriate control, the rectifier is demonstrated to have power factor >0.99 for both fixed frequencies input and variable frequency input.

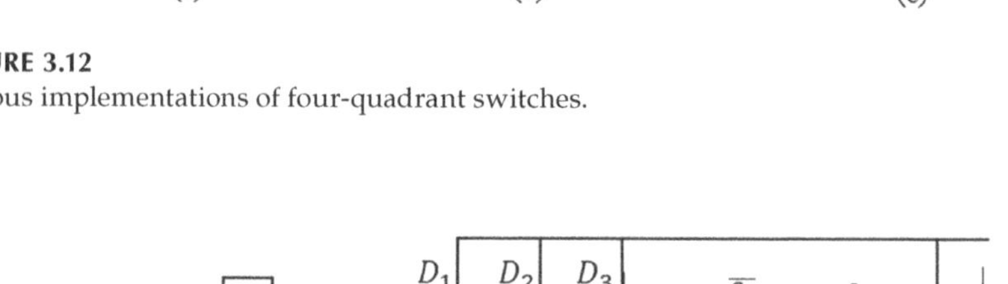

FIGURE 3.12
Various implementations of four-quadrant switches.

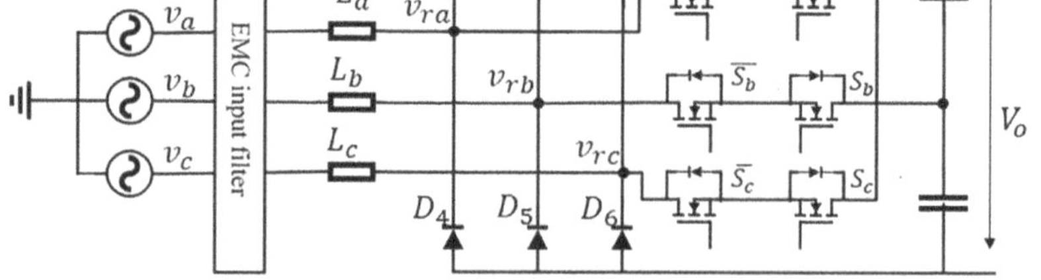

FIGURE 3.13
Vienna converter topology with star connected four-quadrant switches.

(c) **Multilevel and Matrix Converters as Rectifiers**
The inverter circuit topologies discussed in Section 3.5.3 are capable of bidirectional power flow and hence can be operated as PWM rectifiers with appropriate modifications. Multilevel converters (e.g., MMC, NPC) and matrix converters offer scalability and fault tolerance for medium-voltage MEA systems, and are considered as options for high power quality AC–DC conversion as well. The circuit complexity, however, will be the challenge with high number of controlled switches and the corresponding gate driver boards. In case of electric propulsion motor drives and high-voltage DC buses with regenerative capability, the topologies may still present itself as attractive options.

3.6 Design Challenges of WBG Converter in Aviation Electrification

In the preceding sections, the utilisation of WBG devices for converter development was highlighted as an effective means to enhance conversion efficiency. However, the adoption of WBG devices introduces several distinct design challenges. The foremost among them arises from their high-frequency switching characteristics. Although rapid switching facilitates reduced passive component size and increased power density, it simultaneously induces pronounced voltage overshoots, oscillations, and electromagnetic interference due to elevated dv/dt and di/dt transients. Consequently, the design of power electronic systems employing WBG devices necessitates careful consideration of device, parasitic effects, layout optimisation, and robust gate driving strategies to ensure reliable and effective operation in the stringent environment of aviation applications.

(a) *SiC versus GaN:* Between the two WBG semiconductor technologies, SiC and GaN, SiC has gained prominence as the preferred choice for electric aircraft power conversion systems. This preference primarily arises from its superior voltage-handling capability, with commercially available SiC MOSFETs rated up to 3.3 kV/700 A, in contrast to GaN devices that are typically limited to 650 V/120 A. Moreover, GaN transistors possess a lower gate threshold voltage, which increases the sensitivity of the device to parasitic transients and complicates the design of the gate driver circuit, particularly under high dv/dt conditions. Another limitation of GaN devices is their poor third-quadrant conduction behaviour; when operating with reverse current in the ON state, they exhibit significantly higher losses, up to four times greater than those observed in conventional

Si MOSFETs. Consequently, SiC technology offers a more robust and practical solution for high-power, high-voltage converter applications envisioned in next-generation electric aircraft systems.

(b) *False Turn-On:* False turn-on in SiC MOSFETs occurs due to crosstalk and the Miller effect. The Miller effect depends on gate resistance and gate-drain capacitance, while crosstalk arises when the rapid switching of the complementary device induces an overvoltage on the gate, momentarily turning it ON. To mitigate crosstalk, advanced gate-assist techniques have been introduced, incorporating adaptive control of gate voltage. Such approaches enable dynamic shaping of the gate signal, thereby moderating the drain-source voltage transition [40].

(c) *Overvoltage:* Overvoltage spikes result from the interaction between high *di/dt* currents and the parasitic inductances in the commutation loop. The problem becomes more pronounced in WBG-based converters due to their high switching speed and low on-state resistance. To mitigate overvoltage stress, it is essential to minimise loop inductance through careful device packaging and optimised layout design. Techniques such as laminated bus bars and vertical loop layouts are particularly effective, as they exploit magnetic field cancellation to reduce both short- and long-loop parasitics.

(d) *Inverter and Motor Interaction:* In long cable-fed motor drives, voltage reflections at the motor terminals can cause overvoltages that may reach twice the inverter output, stressing insulation. Fast switching transients further amplify this high-frequency stress, accelerating insulation ageing. Long cables also increase EMI emission and add conduction and reactive losses. For SiC devices, which operate at higher switching frequencies than Si, these phenomena are even more pronounced, further challenging motor and cable reliability. Commonly used solutions to mitigate this include the *dv/dt* filter at the AC terminals of the WBG converters to limit the *dv/dt* seen by the load or source. Another way is to utilise the multilevel converters, as they have lesser *dv/dt* stresses [36].

(e) *Radiated EMI:* Electric aircraft are highly sensitive to EMI due to the close integration of power electronics, avionics, and communication systems, while high-frequency switching in WBG converters can radiate EMI and disrupt critical signals. EMI filters are thus essential to meet standards like DO-160, but their passive components (capacitors and inductors) add significant weight and losses, often accounting most of the equipment weight. To further reduce EMI filter size, EMI should be mitigated at the source, which can be achieved through modified PWM techniques. Additional approaches include multistage filter topologies and hybrid filters that combine active and passive components.

References

1 H. Huang and K. Rajashekara, "Net-zero greenhouse gas emission electrified aircraft propulsion for large commercial transport," *World Electric Vehicle Journal*, vol. 15, no. 9, pp. 411–411, Sep. 2024.
2 J. Chen, C. Wang, and J. Chen, "Investigation on the selection of electric power system architecture for future more electric aircraft," *IEEE Transactions on Transportation Electrification*, vol. 4, no. 2, pp. 563–576, June 2018.
3 B. Sarlioglu and C. T. Morris, "More electric aircraft: Review, challenges, and opportunities for commercial transport aircraft," *IEEE Transactions on Transportation Electrification*, vol. 1, no. 1, pp. 54–64, June 2015.
4 L. Dorn-Gomba, J. Ramoul, J. Reimers, and A. Emadi, "Power electronic converters in electric aircraft: Current status, challenges, and emerging technologies," *IEEE Transactions on Transportation Electrification*, vol. 6, no. 4, pp. 1648–1664, Dec. 2020.
5 H. Schefer, L. Fauth, T. H. Kopp, R. Mallwitz, J. Friebe, and M. Kurrat, "Discussion on electric power supply systems for all electric aircraft," *IEEE Access*, vol. 8, pp. 84188–84216, 2020.
6 S. Gangavarapu, "Analysis and design of discontinuous conduction mode AC-DC power factor correction converters," *Concordia.ca*, Oct. 2025, https://spectrum.library.concordia.ca/id/eprint/986643/7/Gangavarapu_PhD_S2020.pdf.
7 J. Diebold, C. Tarau, K. -L. Lee, W. G. Anderson, and R. W. Dyson, "Electric aircraft thermal management using a two-phase heat transport system with solid-state thermal switching capability," 2021 AIAA/IEEE Electric Aircraft Technologies Symposium (EATS), Denver, CO, USA, 2021, pp. 1–11.
8 D. Wang et al., "Multilevel inverters for electric aircraft applications: Current status and future trends," *IEEE Transactions on Transportation Electrification*, vol. 10, no. 2, pp. 3258–3282, June 2024.
9 High Voltage Connection, "Paschen curve: Voltage breakdown vs pressure," 2010, [Online]. Available: https://www.highvoltageconnection.com/articles/paschen-curve.html. Accessed: Jun. 1, 2025.
10 H. You, Z. Wei, B. Hu, Z. Zhao, R. Na, and J. Wang, "Partial discharge behaviors in power modules under square pulses with ultrafast dv/dt," *IEEE Transactions on Power Electronics*, vol. 36, no. 3, pp. 2611–2620, Mar. 2021.
11 Securaplane, "Securaplane start power unit," [Online]. Available: https://www.securaplane.com/products/inverters/. Accessed: Dec. 9, 2022.
12 T. Liu and M. Fadel, "An efficiency-optimal control method for monoinverter dual-PMSM systems," *IEEE Transactions on Industry Applications*, vol. 54, no. 2, pp. 1737–1745, Mar. 2018.
13 Y. Cong et al., "Submodule design of a 2 kV 1 MW integrated modular motor drive for aviation applications," in Proc. WiPDA, Redondo Beach, CA, USA, Nov. 2021, pp. 345–350.
14 Siemens, "HYPSTAIR-system architecture, certifiability and safety aspects," [Online]. Available: http://www.hypstair.eu/wp-content/uploads/2013/10/E2-Fliegen-Symposium-Stuttgart_SAG.pdf. Accessed: Dec. 9, 2022.
15 J. Ebersberger, M. Hagedorn, M. Lorenz, and A. Mertens, "Potentials and comparison of inverter topologies for future all-electric aircraft propulsion," *IEEE Journal of Emerging and Selected Topics Power Electronics*, vol. 10, no. 5, pp. 5264–5279, Oct. 2022.

16 D. H. Tran, B. Cougo, G. Segond, and H. H. Sathler, "Optimal design of a three-phase 540 V/70 kVA SiC inverter for aircraft applications," in Proc. IEEE ITEC., 2023, pp. 1–6.
17 B. P. Reddy and S. Keerthipati, "A multilevel inverter configuration for an open-end-winding pole-phase-modulated-multiphase induction motor drive using dual inverter principle," *IEEE Transactions on Industrial Electronics*, vol. 65, no. 4, pp. 3035–3044, Apr. 2018.
18 M. G. Majumder, A. K. Yadav, K. Gopakumar, K. Raj, R. U. Loganathan, and L. G. Franquelo, "A 5-level inverter scheme using single DC link with reduced number of floating capacitors and switches for open-end IM drives," *IEEE Transactions on Industrial Electronics*, vol. 67, no. 2, pp. 960–968, Feb. 2020.
19 M. Ghosh Majumder, R. R, K. Gopakumar, L. Umanand, K. Al-Haddad, and W. Jarzyna, "A fault-tolerant five-level inverter topology with reduced component count for OEIM drives," *IEEE Journal of Emerging and Selected Topics in Power Electronics*, vol. 9, no. 1, pp. 961–969, Feb. 2021.
20 Z. Huang, T. Yang, P. Giangrande, S. Chowdhury, M. Galea and P. Wheeler, "An active modulation scheme to boost voltage utilization of the dual converter with a floating bridge," *IEEE Transactions on Industrial Electronics*, vol. 66, no. 7, pp. 5623–5633, July 2019.
21 S. Lakhimsetty and V. T. Somasekhar, "A four-level open-end winding induction motor drive with a nested rectifier–inverter combination with two DC power supplies," *IEEE Transactions on Power Electronics*, vol. 34, no. 9, pp. 8894–8904, Sep. 2019.
22 V. T. Somasekhar, K. Gopakumar, and M. R. Baiju, "Dual two-level inverter scheme for an open-end winding induction motor drive with a single DC power supply and improved DC bus utilisation," *IEE Proceedings - Electric Power Applications*, vol. 151, no. 2, pp. 230–238, Mar. 2004.
23 A. Amerise, M. Mengoni, L. Zarri, A. Tani, S. Rubino, and R. Bojoi, "Open-end windings induction motor drive with floating capacitor bridge at variable DC-link voltage," *IEEE Transactions on Industry Applications*, vol. 55, no. 3, pp. 2741–2749, May 2019.
24 A. Barzkar and M. Ghassemi, "Electric power systems in more and all electric aircraft: A review," *IEEE Access*, vol. 8, pp. 169314–169332, 2020.
25 M. Ghassemi, A. Barzkar, and M. Saghafi, "All-electric NASA N3-X aircraft electric power systems," *IEEE Transactions on Transportation Electrification*, vol. 8, no. 4, pp. 4091–4104, Dec. 2022.
26 D. Zhang, J. He, and S. Madhusoodhanan, "Three-level two-stage decoupled active NPC converter with Si IGBT and SiC MOSFET," *IEEE Transactions on Industry Applications*, vol. 54, no. 6, pp. 6169–6178, Nov. 2018.
27 D. Zhang, J. He, and D. Pan, "A megawatt-scale medium-voltage high-efficiency high power density 'SiC+Si' hybrid three-level ANPC inverter for aircraft hybrid-electric propulsion systems," *IEEE Transactions on Industry Applications*, vol. 55, no. 6, pp. 5971–5980, Nov. 2019,
28 H. Gui et al., "Modeling and mitigation of multiloops related device overvoltage in three-level active neutral point clamped converter," *IEEE Transactions on Power Electronics*, vol. 35, no. 8, pp. 7947–7959, Aug. 2020.
29 H. Gui et al., "Methodology of low inductance busbar design for three level converters," *IEEE Journal of Emerging and Selected Topics in Power Electronics*, vol. 9, no. 3, pp. 3468–3478, Jun. 2021.

30 A. Deshpande, Y. Chen, B. Narayanasamy, Z. Yuan, C. Chen, and F. Luo, "Design of a high-efficiency, high specific-power three-level T-type power electronics building block for aircraft electric-propulsion drives," *IEEE Journal of Emerging and Selected Topics in Power Electronics,* vol. 8, no. 1, pp. 407–416, Mar. 2020.
31 X. Liu, C. Bentheimer, F. Hilpert, M. Hofmann, B. Eckardt, and M. Maerz, "Digital twin for a 10 MW electrical drive system for future electric aircraft applications," in Proc. 11th Int. Electric Drives Prod.Conf. (EDPC), Erlangen, Germany, Dec. 2021.
32 T. Modeer, N. Pallo, T. Foulkes, C. B. Barth, and R. C. N. Pilawa-Podgurski, "Design of a GaN-based interleaved nine-level flying capacitor multilevel inverter for electric aircraft applications," *IEEE Transactions on Power Electronics,* vol. 35, no. 11, pp. 12153–12165, Nov. 2020.
33 C. B. Barth et al., "Design, operation, and loss characterization of a 1-kW GaN-based three-level converter at cryogenic temperatures," *IEEE Transactions on Power Electronics,* vol. 35, no. 11, pp. 12040–12052, Nov. 2020.
34 C. B. Barth et al., "Design and control of a GaN-based, 13-level, flying capacitor multilevel inverter," *IEEE Journal of Emerging and Selected Topics in Power Electronics,* vol. 8, no. 3, pp. 2179–2191, Sep. 2020.
35 C. C. O. Mendes, H. H. Sathler, B. Cougo, M. M. Stopa, F. Costa, and D. Labrousse, "Impact of PWM techniques in efficiency and power density of a 70 kVA multilevel inverter for more electric aircrafts," in Proc. 23rd Eur. Conf. Power Electron. Appl. (EPE ECCE Europe), Ghent, Belgium, Sep. 2021, pp. P.1–P.10.
36 C. Buccella et al., "A cascade multilevel configuration for commercial transport aircraft," in Proc. AEIT Int. Annu. Conf., Bari, Italy, Oct. 2018, pp. 1–5.
37 K. Panuganti, S. Figarado, and A. K. Rathore, "A fault tolerant modular multilevel converter topology for electric aircraft applications," in Proc. IEEE SEFET, 2025, pp. 1–6.
38 A. Misra, "Technical challenges and barriers affecting turbo-electric and hybrid electric aircraft propulsion," Presented at the ENERGYTECH Conf., Cleveland, OH, USA, 2017.
39 R. T. Arumalla, S. Figarado and N. Harischandrappa, "Dodecagonal voltage space vector based PWM techniques for switching loss reduction in a dual inverter fed induction motor drive," *IEEE Journal of Emerging and Selected Topics in Industrial Electronics,* vol. 1, no. 2, pp. 182–191, Oct. 2020.
40 F. F. Wang, R. Chen, and K. Rajashekara, "Wide bandgap semiconductor-based power electronics for aviation," *IEEE Power Electronics Magazine,* vol. 9, no. 3, pp. 26–36, Sept. 2022.
41 Michael Hartmann, "Ultra-compact and ultra-efficient three-phase PWM rectifier systems for more electric aircraft," ETH Zurich, [Online]. Available: https://www.ams-publications.ee.ethz.ch/uploads/tx_ethpublications/Diss_ETH19755_Hartmann_2011.pdf. Accessed: Oct. 14, 2025.
42 G. Sivanagaraju, A. K. Rathore, and D. M. Fulwani, "Discontinuous conduction mode three phase buck-boost derived PFC converter for more electric aircraft with reduced switching, sensing and control requirements," 2018 IEEE Applied Power Electronics Conference and Exposition (APEC), San Antonio, TX, USA, 2018, pp. 1467–1472.

43 J. Kikuchi, M. D. Manjrekar, and T. A. Lipo, "Performance improvement of half controlled three phase PWM boost rectifier," 30th Annual IEEE Power Electronics Specialists Conference. Record. (Cat. No.99CH36321), Charleston, SC, USA, 1999, pp. 319–324, vol. 1.
44 M. C. Alphan and G. Soykan, "Hybrid control strategy for VIENNA rectifiers in more electric aircraft electrical systems," *IEEE Access*, vol. 13, pp. 86686–86704, 2025, https://doi.org/10.1109/ACCESS.2025.3567113
45 M. Hartmann, J. Miniboeck, H. Ertl, and J. W. Kolar, "A three-phase delta switch rectifier for use in modern aircraft," *IEEE Transactions on Industrial Electronics*, vol. 59, no. 9, pp. 3635–3647, Sept. 2012.
46 Z. Xu, X. Ren, Z. Zheng, Z. Zhang, Q. Chen, and Z. Hao, "A quadrature signal-based control strategy for vienna rectifier under unbalanced aircraft grids," *IEEE Journal of Emerging and Selected Topics in Power Electronics*, vol. 10, no. 5, pp. 5280–5289, Oct. 2022.
47 TI Designs: TIDM-1000, "Vienna rectifier-based, three-phase Power Factor Correction (PFC) reference design using C2000™ MCU," [Online] June 2017.

4

Harmonic Mitigation in Multi-Pulse Rectifiers Using DC Injection Circuits for Electric Aircraft Applications

Saravana Prakash P. and Kalpana R.

4.1 Introduction

In the concept of more-electric aircraft (MEA), electrical systems replace conventional ones. Conventionally, mechanical, hydraulic, or pneumatic energy systems are also used with electrical energy systems to power aircraft. In MEA, a starter generator is used to initiate the engine; then, the generator begins generating electricity and supplying power to the electrical apparatus [1]. MEA eliminates the need to maintain the bleeding system and pneumatic ducts, valves, pre-coolers, and lubrication oil systems. Thus, the number of components, cost, and, most significantly, weight is reduced by MEA, making the system more straightforward, reliable, and efficient [2]. To store as much energy as possible in a given weight is the challenge for MEA, and another is to convert and control power using efficient power electronics [3].

Multi-pulse converters (MPCs) are an essential application of AC–DC conversion and are widely used in MEA. Harmonics cause the converter's power quality issues, manifesting as a low power factor, AC voltage distortion, and rippled DC outputs. There are a variety of solutions for these problems, including LC filters, power factor correction (PFC), pulse width modulation (PWM) rectifier technology, and MPCs. Further, these techniques can be categorised as active, passive, or hybrid PFC circuits. Active PFC consists of switches, PWM techniques, and other components. Further, multi-pulse transformers, diode bridge rectifiers (DBRs), interphase transformers (IPT) or reactors (IPR), and injection circuits make up the passive PFC harmonic reduction technique. Furthermore, varieties of isolated and non-isolated transformers are available [4, 5]. Transformers without isolation consume less power for the same rating and have their benefits and drawbacks.

DOI: 10.1201/9781003747994-4

In MEA, where high power quality with low input harmonic contents is needed, multi-pulse transformer and rectifier unit applications are prevalent. Rectifiers produce harmonic contents. For harmonics mitigation, multi-pulse transformers are the most reliable, robust, and economical components [6] and suitable for MEA applications. To cancel out certain harmonics in AC mains, the currents are drawn with phase variation through a transformer in MPCs. Specific low-order harmonics can be eliminated by selecting the correct phase angle shift. Phase-shifting transformers are a crucial part of MPRs and have three main functions: (a) providing the necessary phase shift between primary and secondary side line voltages for harmonic cancellation and high-power factor; (b) providing an appropriate secondary voltage; and (c) providing electrical isolation between the rectifier and the input supply.

Transformers are high-capacity magnetic devices that significantly affect the overall kVA rating of magnetic devices in rectification systems. In specific isolated topologies, the kVA rating of the transformer is exceedingly high and, in most cases, may exceed the output power. Certain varieties of autotransformers have a significantly lower kVA rating than isolated transformers. Compared with isolated transformers, autotransformers may reduce the system's size, overall cost, and power losses while providing the same load power. When electrical isolation is not required, autotransformers may be a cost-effective means of increasing power density.

In addition to mitigating lower-order harmonics, MPC reduces higher-order harmonics as well. Increasing the number of pulses can improve power quality indices but also increase the system's complexity and cost as the magnetic circuit size and number of bridges increase. Theoretically, 12-, 18-, 24-, and 30-pulse converters have THD values of 15.2%, 10.1%, 7.5%, and 6.2%, respectively, but their sizes, complexities, and costs have increased [7]. Thus, the technique of DC ripple reinjection is utilised for pulse multiplication. It reduces magnetic circuit cost, size, and ratings [8].

IPT interrupts the current flow between DBRs. Due to a variation in the properties of semiconductor devices made by the same model and manufacturer, in practice, the output voltages of parallel DBRs are not identical. The output voltage of bridges might drop due to losses caused by the circulatory current, which can flow between the bridges. To solve this issue, IPTs are employed [9]. IPTs are centre-tapped reactors that give the same DC voltage. The voltage across IPT is an AC voltage with a ripple frequency six times the source frequency [10].

There are three varieties of injection circuits used for harmonic reduction: passive, active, and hybrid. Passive devices are used in the passive injection techniques and have a distinct realisation on both sides of the IPT [11]. Active injection circuits include active devices and pulse width modulation (PWM) technology. A combination of passive and active techniques is called a hybrid injection circuit with both passive and active elements. Each type has distinctive characteristics. The pulse count of the input current is multiplied by these injection circuits, and thus the type of injection circuit determines the pulse number.

A passive injection circuit comprises passive components. So, passive harmonic reduction techniques do not need extra driving or control circuits. Depending on the number of passive circuits, this approach may be divided into two categories: the single passive injection circuit and the dual passive injection circuit. In a single passive injection circuit, harmonic injection circuits are often connected to the IPR's secondary side. A dual passive harmonic reduction strategy combines single passive harmonic reduction approaches. For example, on the primary side of the IPR, two tapings are installed, and either a single-phase full-wave rectifier or a single-phase diode-bridge rectifier is installed on the secondary side. Together, these elements can raise the MPR's pulse count. The diodes in the auxiliary circuit are conducted alternatively, creating a circulating current that controls the rectification system's input currents and adds additional DC voltages to the load side. In addition, IPT links the injection circuit with load [12].

To minimise the THD of the input line current, one of the most crucial design considerations is to calculate the tap ratio and turn ratio of the IPT [13]. The adaptability of the rectifiers has been investigated in earlier literature, and the findings imply that the THD value swings very little in response to changes in resistive load, fulfilling the power quality criteria of most applications. Even though the multi-winding IPR's kVA rating is higher than the single-winding IPR, it contributes approximately 3% of the load power. Additionally, these rectifiers utilise only uncontrolled devices, and the circuit characteristics of the auxiliary circuits determine their operational modes. Therefore, this method of harmonic reduction is simple and relatively economical to implement.

The proposed system consists of a zigzag transformer that can expel zero-sequence blocking transformer (ZSBT), decreasing the magnetic devices' total kVA rating and thereby consuming only 26–29% of the load power. The proposed injection circuit is a Dickson voltage multiplier, which is a passive injection circuit. This is widely used in integrated circuits to boost the supply voltage [14]. The combination of two capacitors and two diodes makes one stage. The input of the multiplier is in AC form, whereas the output is DC and connected to load [15]. The detailed description is presented in Ref. [16].

4.2 System Description

Figure 4.1A depicts the schematic diagram of the proposed 12-pulse DBR with injection circuit. It consists of 12-pulse DBR with Dickson current injection circuit at DC side. Two three-phase DBR are fed by a zigzag autoconnected transformer, which performs phase staggering.

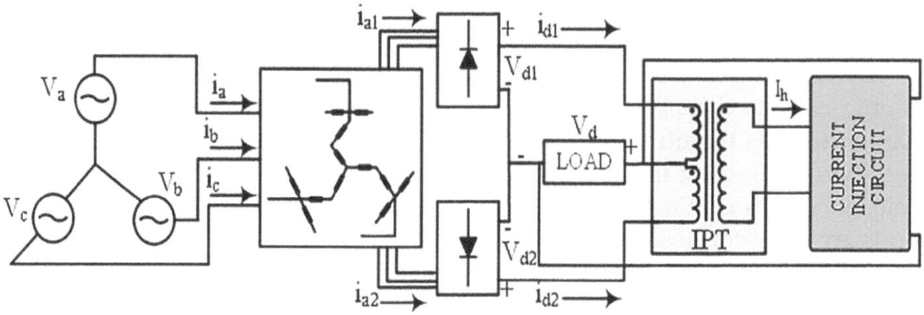

FIGURE 4.1 (A)
Schematic diagram of the proposed 12-pulse DBR with current injection circuit at DC.

Further, the ZSBT is not necessary because this transformer has an inherent capacity to block zero-sequence components. Furthermore, IPT ensures equal current distribution between the two DBRs, and it is connected to the proposed Dickson voltage multiplier circuit. The load is connected between the IPT and the negative bus of DBRs. The pulse multiplication is achieved by the injection circuit, and 24-pulse is obtained in the current input. Since the input line current approaches nearly sinewave, the THD is thereby significantly minimised.

4.2.1 Description about Zigzag Transformer, IPT, and Injection Circuit

4.2.1.1 Zigzag Transformer

The zigzag autoconnected transformer is a crucial element in the proposed 12-pulse rectifier system. It provides the necessary phase-shift of ±15° between its secondary windings, enabling 12-pulse operation by feeding two diode bridge rectifiers (DBRs). Unlike isolated transformers, the zigzag transformer offers reduced kVA rating, smaller size, and lower cost while inherently blocking zero-sequence components, thereby eliminating the need for an additional zero-sequence blocking transformer (ZSBT). This property makes it especially suitable for electric aircraft applications where weight, compactness, and efficiency are critical. By supplying phase-shifted voltages, the zigzag transformer ensures harmonic cancellation in input currents and supports high power factor operation.

4.2.1.2 Interphase Transformer (IPT)

The interphase transformer is placed between the two DBRs to ensure equal current sharing and balance in their DC outputs. Without the IPT, circulating currents may arise due to mismatched output voltages of the parallel DBRs. The IPT effectively mitigates this issue by enforcing equalised load distribution, while also providing the necessary AC ripple voltage at six times the supply frequency. This ripple serves as the input for the harmonic injection circuit, thereby linking the DC link and the injection stage. The IPT is designed with optimal turns ratio so that its kVA rating remains only a small fraction (around 1–3%) of the load power, ensuring compactness and efficiency.

4.2.1.3 Injection Circuit

To enhance harmonic mitigation, a *Dickson voltage multiplier injection circuit* is employed on the DC side. The Dickson circuit is a passive multiplier, composed of diodes and capacitors, that generates an auxiliary DC voltage from the IPT's AC ripple. This injected voltage modifies the DBR currents such that the effective pulse number of the system doubles – from 12-pulse to 24-pulse – leading to significant reduction in total harmonic distortion (THD) of the input current (as low as 4.42%). The injection current circulates only in the IPT primary, shaping the DBR currents without directly stressing the load. Since only small auxiliary power is drawn, the component ratings of the Dickson circuit remain minimal, making the solution simple, cost-effective, and well-suited for more-electric aircraft power systems.

4.3 Design of Zigzag Transformer

4.3.1 Winding Diagram of Zigzag Transformer

Figure 4.1B depicts the 12-pulse zigzag transformer's winding arrangement. The phase shift $\frac{360^\circ}{12} = 30^\circ$ is required for 12-pulse operation.

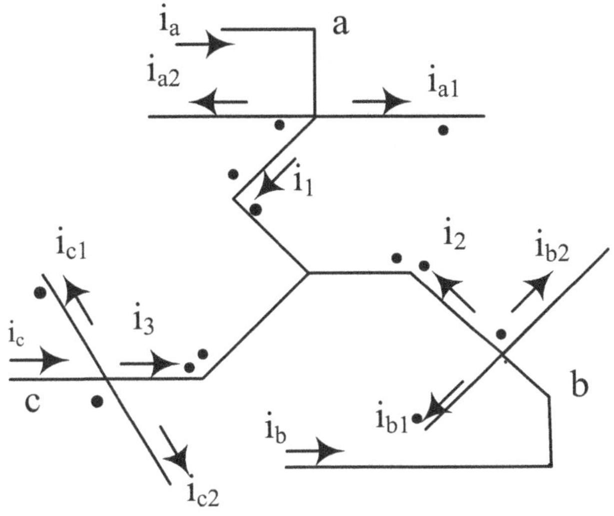

FIGURE 4.1B
Winding configuration.

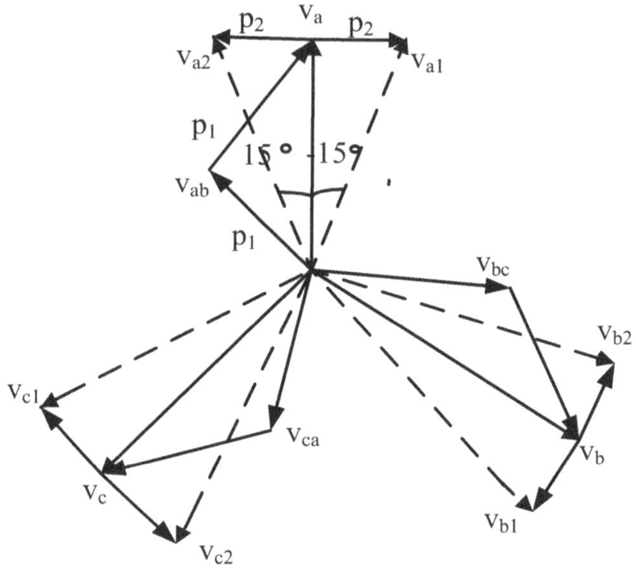

FIGURE 4.1C
Phasor diagram of zigzag autoconnected transformer.

4.3.2 Phasor Diagram of Zigzag Transformer

The zigzag autoconnected transformer produces two sets of three-phase voltages that are phase-shifted 30° with respect to one another and 15° and –15° with respect to the input voltage [17], as illustrated in Figure 4.1C.

4.3.3 Calculation of Turns Ratio of Zigzag Transformer

V_a, V_b, and V_c are three-phase input voltages whose phases shifted 120° with each other. V_{a1} and V_{a2} are expressed in terms of line voltages as follows:

$$V_{a1} = p_1(V_{ab} - V_{ca}) + p_2 V_{bc} \tag{4.1}$$

$$V_{a2} = p_1(V_{ab} - V_{ca}) - p_2 V_{bc} \tag{4.2}$$

The calculated values of $p1$ and $p2$ are 0.5573 and 0.2679, respectively.

4.3.4 Expression for Winding Currents of Zigzag Transformer

Using Kirchoff's current law and law of MMF, the following equations are derived:

$$i_a = i_{a1} + i_{a2} + i_1 \tag{4.3}$$

$$i_1 + i_2 + i_3 = 0 \tag{4.4}$$

$$p_1(i_1 - i_2) + p_2(i_{c2} - i_{c1}) = 0 \tag{4.5}$$

$$p_1(i_2 - i_3) + p_2(i_{a2} - i_{a1}) = 0 \tag{4.6}$$

$$p_1(i_3 - i_1) + p_2(i_{b2} - i_{b1}) = 0 \tag{4.7}$$

As shown in Figure 4.1B, i_{a1}, i_{b1}, and i_{c1} are the first set of three-phase currents fed to the input of the first DBR. Similarly, i_{a2}, i_{b2}, and i_{c2} are the second set of three-phase current feed to input of second DBR. i_1, i_2, and i_3 are zigzag transformer winding currents, which are expressed as follows:

$$i_1 = \frac{p_2}{3p_1}(i_{b2} - i_{b1} + i_{c1} - i_{c2}) \tag{4.8}$$

$$i_2 = \frac{p_2}{3p_1}(i_{c2} - i_{c1} + i_{a1} - i_{a2}) \tag{4.9}$$

$$i_3 = \frac{p_2}{3p_1}(i_{a2} - i_{a1} + i_{b1} - i_{b2}) \tag{4.10}$$

where p_1 and p_2 are constants. V_{a1}, V_{b1}, V_{c1}, and V_{a2}, V_{b2}, V_{c2} are the phase-to-phase voltages of the input of DBR1 and DBR2, respectively. V_{ab}, V_{bc}, and V_{ca} are the line-to-line voltages of the system. i_a, i_b, and i_c are the input currents of the zigzag autotransformer. i_1, i_2, and i_3 are the currents of internal windings of the zigzag autotransformer. i_{a1}, i_{b1}, i_{c1}, and i_{a2}, i_{b2}, i_{c2} are the input currents of the DBR1 and DBR2, respectively. V_d is the voltage across load terminals. V_{d1} and V_{d2} are the output voltages of DBR1 and DBR2, respectively.

4.4 Diode Bridge Rectifier

4.4.1 Expression for Input Voltages of Diode Bridge Rectifier

The equations associated with the input phase voltages of the system and the input voltages of the DBRs are as follows:

$$V_a = V_m \sin \omega t \tag{4.11}$$

$$V_b = V_m \sin\left(\omega t - \frac{2\pi}{3}\right) \tag{4.12}$$

$$V_c = V_m \sin\left(\omega t + \frac{2\pi}{3}\right) \tag{4.13}$$

$$V_{a1} = \sqrt{2}\left(\sqrt{3}-1\right)V_m \sin\left(\omega t + \frac{\pi}{12}\right) \tag{4.14}$$

$$V_{b1} = \sqrt{2}\left(\sqrt{3}-1\right)V_m \sin\left(\omega t - \frac{7\pi}{12}\right) \tag{4.15}$$

$$V_{c1} = \sqrt{2}\left(\sqrt{3}-1\right)V_m \sin\left(\omega t + \frac{3\pi}{4}\right) \tag{4.16}$$

$$V_{a2} = \sqrt{2}\left(\sqrt{3}-1\right)V_m \sin\left(\omega t - \frac{\pi}{12}\right) \tag{4.17}$$

$$V_{b2} = \sqrt{2}\left(\sqrt{3}-1\right)V_m \sin\left(\omega t - \frac{3\pi}{4}\right) \tag{4.18}$$

$$V_{c2} = \sqrt{2}\left(\sqrt{3}-1\right)V_m \sin\left(\omega t + \frac{7\pi}{12}\right) \tag{4.19}$$

4.4.2 Expression for Output Voltages of Diode Bridge Rectifier

The F_{a1}, F_{b1}, F_{c1}, F_{a2}, F_{b2}, and F_{c2} are the switching functions which are phase-shifted with each other and have quasi-square waveforms [18]. These functions represent the three phases of DBRs associated with the three legs of the DBR working with the phase shift.

$$V_{d1} = F_{a1}V_{a1} + F_{b1}V_{b1} + F_{c1}V_{c1} \tag{4.20}$$

$$V_{d2} = F_{a2}V_{a2} + F_{b2}V_{b2} + F_{c2}V_{c2} \tag{4.21}$$

where

$$F_{a1} = 0.5\left\{\text{sgn}\left[V_{a1} - V_{c1}\right] - \text{sgn}\left[V_{b1} - V_{a1}\right]\right\} \tag{4.22}$$

$$F_{b1} = 0.5\left\{\text{sgn}\left[V_{b1} - V_{a1}\right] - \text{sgn}\left[V_{c1} - V_{b1}\right]\right\} \tag{4.23}$$

$$F_{c1} = 0.5\left\{\text{sgn}\left[V_{c1} - V_{b1}\right] - \text{sgn}\left[V_{a1} - V_{c1}\right]\right\} \tag{4.24}$$

$$F_{a2} = 0.5\left\{\text{sgn}\left[V_{a2} - V_{c2}\right] - \text{sgn}\left[V_{b2} - V_{a2}\right]\right\} \tag{4.25}$$

$$F_{b2} = 0.5\left\{\text{sgn}\left[V_{b2} - V_{a2}\right] - \text{sgn}\left[V_{c2} - V_{b2}\right]\right\} \tag{4.26}$$

$$F_{c2} = 0.5\left\{\text{sgn}\left[V_{c2} - V_{b2}\right] - \text{sgn}\left[V_{a2} - V_{c2}\right]\right\} \tag{4.27}$$

V_{d1} and V_{d2} are the output voltage of six-pulse DBRs, which have six pulses in each cycle with a period 60°. V_d is the load voltage with 12-pulse in each cycle, and its period is 30°. Substituting the values from Equations (4.14)–(4.19) and

Equations (4.22)–(4.27) in Equations (4.20) and (4.21), the expression of V_{d1} and V_{d2} are obtained as follows:

$$V_{d1} = \begin{cases} \sqrt{3}V_m \sin\left(\omega t + \dfrac{\pi}{2}\right), & \omega t \in \left[60^\circ, 75^\circ\right) \\ \sqrt{3}V_m \sin\left(\omega t - \dfrac{\pi}{12}\right), & \omega t \in \left[75^\circ, 120^\circ\right] \end{cases} \tag{4.28}$$

$$V_{d2} = \begin{cases} \sqrt{3}V_m \sin\left(\omega t + \dfrac{\pi}{12}\right), & \omega t \in \left[60^\circ, 105^\circ\right) \\ \sqrt{3}V_m \sin\left(\omega t - \dfrac{\pi}{4}\right), & \omega t \in \left[105^\circ, 120^\circ\right] \end{cases} \tag{4.29}$$

In Figure 4.1a, using KVL at DBRs output and IPT primary, the following equation is obtained:

$$V_d = 0.5\left(V_{d1} + V_{d2}\right) \tag{4.30}$$

Substituting V_{d1} and V_{d2} from Equations (4.28) and (4.29) into Equation (4.30), the following equation is obtained:

$$V_d = \begin{cases} \sqrt{3}V_m \cos\left(\dfrac{\pi}{12}\right)\sin\left(\omega t + \dfrac{\pi}{3}\right), & \omega t \in \left[30^\circ, 45^\circ\right] \\ \sqrt{3}V_m \cos\left(\dfrac{\pi}{12}\right)\sin\left(\omega t + \dfrac{\pi}{6}\right), & \omega t \in \left[45^\circ, 60^\circ\right] \end{cases} \tag{4.31}$$

4.5 Injection Circuit and IPT

4.5.1 Modes of Operation of the Injection Circuit

As shown in Figure 4.2, a Dickson voltage multiplier is proposed as an injection circuit. The nature of the input to the injection circuit is AC, which is obtained from the secondary of the IPT. The DC output of the injection circuit is connected to the load. This circuit can multiply the number of input current pulses by modifying the input of the injection circuit. The Dickson injection circuit has a lower output voltage drop, but it also has an output capacitor at the very end. Therefore, a capacitor with a higher rating is required. However, the power absorbed by the proposed injection circuit is meagre; hence the components' rating is minimal. The optimised turns ratio results in pulse multiplication, and hence, 24-pulse is achieved in the input line current with less THD and a higher fundamental component [19]. Only the current properties are modified by this injection circuit, whereas

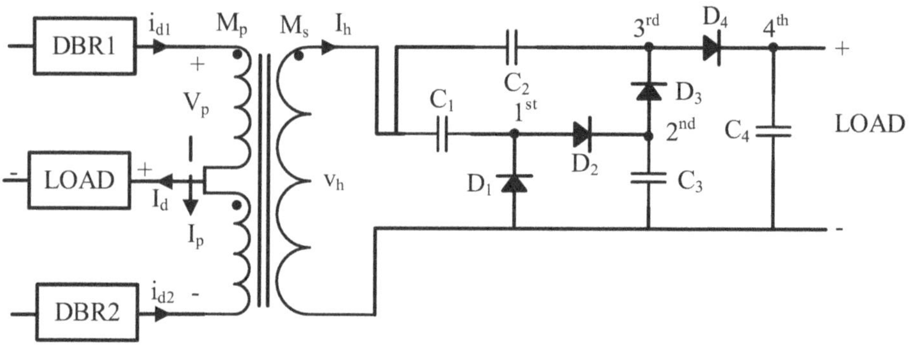

FIGURE 4.2
IPT with Dickson injection circuit.

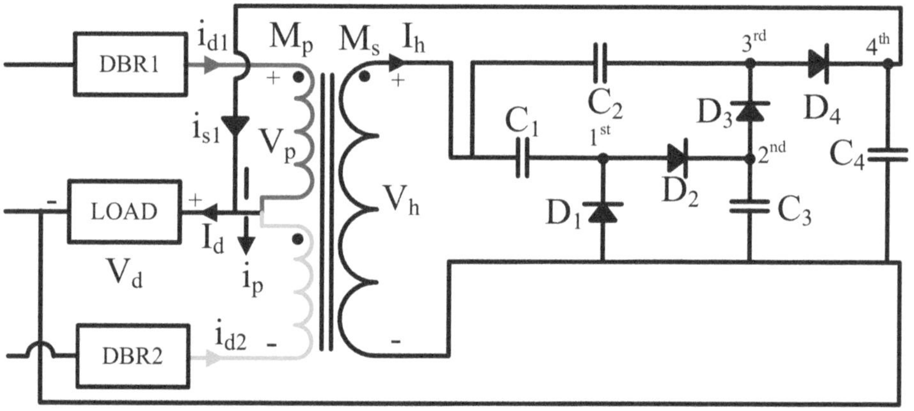

FIGURE 4.3
IPT primary conduction in mode II (red) and mode III (blue).

the voltage characteristics are the same. The symmetrical triangular voltage V_p at IPT primarily results from the phase shift between V_{d1} and V_{d2}, and V_d is the output load voltage. The injection current I_h causes the circulatory current I_p to exist in the IPT's primary winding. It only circulates through the IPT's primary winding and not the load. The harmonic reduction in input current depends on I_p, which also shapes the input current. It modifies the output current of DBR, followed by the input current of DBR, and ultimately the input current of the system to a 24-pulse waveform that approaches nearly a sine wave.

There are three operating modes of the proposed converter [20]. When $V_h < V_d$, it operates in mode I and in this mode, the injection circuit does not operate. In operation mode II, the primary winding's upper part (in red) conducts, and in mode III, the lower part conducts (in blue), as shown in Figure 4.3. Both contain $M_p/2$ turns. Figures 4.4 and 4.5 show the conduction

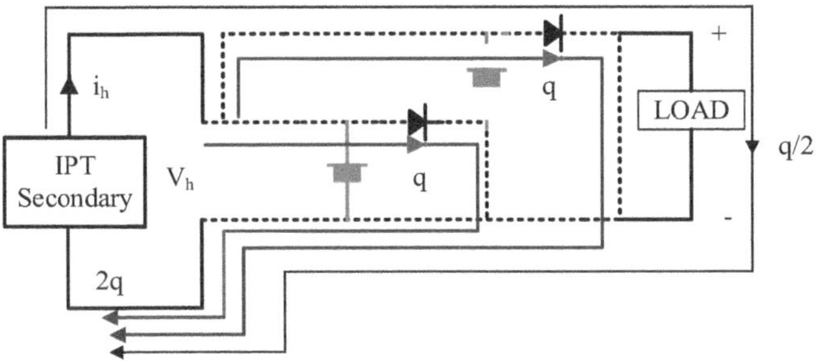

FIGURE 4.4
Charge-flow in injection circuit in mode II ($V_h > V_d$).

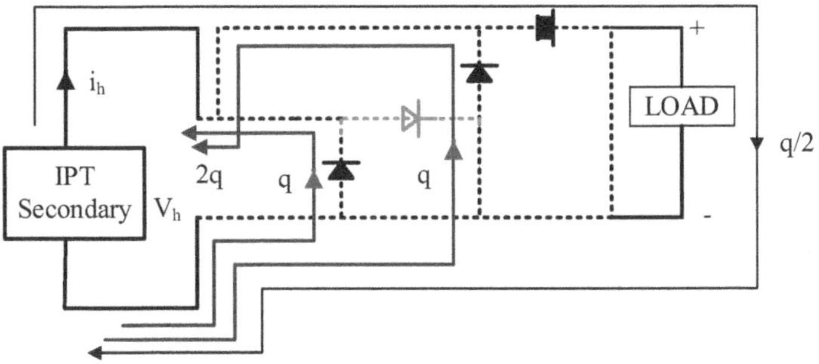

FIGURE 4.5
Charge-flow in injection circuit in mode III ($-V_h > V_d$).

of diodes and capacitors in different modes. Using MMF balance in operating mode II, the following equation can be obtained:

$$0.5i_{d1}M_p = i_{s1}M_s \tag{4.32}$$

$$r = \frac{V_h}{V_s} = \frac{M_s}{M_p} \tag{4.33}$$

Using the above two equations, i_{d1} is written as follows:

$$i_{d1} = 2ri_{s1} \tag{4.34}$$

The load current is the summation of the DBR current and injected current from the injection circuit. From mode II, the following equation can be obtained:

$$i_{d1} + i_{s1} = I_d \tag{4.35}$$

Substituting Equation (4.34) into Equation (4.35), the following equation is obtained for mode II:

$$i_{d1} = \frac{2r}{2r+1} I_d \tag{4.36}$$

Similarly for mode III,

$$i_{d2} = \frac{2r}{2r+1} I_d \tag{4.37}$$

4.5.2 Expression for DC Currents and Input AC Current

By using Equations (4.36) and (4.37), the following DBR output currents expression can be found:

$$i_{d1} = \begin{cases} \frac{I_d}{2}, & wt \in \left[60^\circ, \beta + 60^\circ\right) \\ 0, & wt \in \left[\beta + 60^\circ, 90^\circ - \beta\right) \\ \frac{I_d}{2}, & wt \in \left[90^\circ - \beta, 90^\circ + \beta\right) \\ \frac{2r}{2r+1} I_d, & wt \in \left[90^\circ + \beta, 120^\circ - \beta\right) \\ \frac{I_d}{2}, & wt \in \left[120^\circ - \beta, 120^\circ\right] \end{cases} \tag{4.38}$$

$$i_{d2} = \begin{cases} \frac{I_d}{2}, & wt \in \left[60^\circ, \beta + 60^\circ\right) \\ \frac{2r}{2r+1} I_d, & wt \in \left[\beta + 60^\circ, 90^\circ - \beta\right) \\ \frac{I_d}{2}, & wt \in \left[90^\circ - \beta, 90^\circ + \beta\right) \\ 0, & wt \in \left[90^\circ + \beta, 120^\circ - \beta\right) \\ \frac{I_d}{2}, & wt \in \left[120^\circ - \beta, 120^\circ\right] \end{cases} \tag{4.39}$$

Using switching functions and zigzag transformer equations, following equation can be written [19].

$$i_a = \left\{ \begin{array}{l} \left[\left[s_{a1} - \frac{p_2}{3p_1} s_{b1} + \frac{p_2}{3p_1} s_{c1} \right] i_{d1} + \right. \\ \left. \left[s_{a1} + \frac{p_2}{3p_1} s_{b1} - \frac{p_2}{3p_1} s_{c1} \right] i_{d2} \right] \end{array} \right. \tag{4.40}$$

Substituting i_{d1} and i_{d2} from Equations (4.38) and (4.39) into Equation (4.40), the following expression for system input current can be obtained:

$$i_a = \begin{cases} 0, & wt \in [0, \beta) \\ \left(\frac{\sqrt{2}-1}{2}\right)\frac{2r}{2r+1} I_d, & wt \in [\beta, 30^\circ - \beta) \\ 0.5 I_d, & wt \in [30^\circ - \beta, 30^\circ + \beta) \\ \frac{1}{\sqrt{2}}\frac{2r}{2r+1} I_d, & wt \in [30^\circ + \beta, 60^\circ - \beta) \\ \frac{\sqrt{2}}{\sqrt{3}} I_d, & wt \in [60^\circ - \beta, 60^\circ + \beta) \\ \frac{2r}{2r+1} I_d, & wt \in [60^\circ + \beta, 90^\circ - \beta) \\ I_d, & wt \in [90^\circ - \beta, 90^\circ] \end{cases} \tag{4.41}$$

The second and fourth nodes of the injection circuit have DC voltages with little ripple. This injection circuit's first and third nodes both have AC voltages. This injection circuit's peak-to-peak input voltage is almost 117.4 V, while its DC output is 541.9 V. The proposed Dickson injection circuit functions as a voltage doubler, doubling the voltage while converting the AC input to DC across the load. As a result, the amount of power supplied to the load increases, as does the system's efficiency. A circulatory current I_p flows in the primary IPT due to the proposed injection circuit, which is linked to the secondary IPT and is responsible for the system's input current shaping.

4.6 Simulation Result and Discussion

4.6.1 Specification of the Converter

The specification of the converter and specification of the ZIGZAG and IPT are given in Tables 4.1 and 4.2, respectively.

4.6.2 Discussion on Various Waveforms of the Proposed Converter

The MATLAB/Simulink-based simulations are performed and the result of the proposed converter is shown in Figure 4.6. The input current I_a waveform approaches near sine wave due to the proposed injection circuit, as shown in Figure 4.6a. The input line current has 24-pulse and makes

TABLE 4.1
Specification of the Converter

	Specifications	
S. No.	Details	Value
1	Proposed system input voltage and frequency	400 V (phase to phase rms), 50 Hz
2	System average load voltage	541.9 V
3	Load power	5.55 kW
4.	Injection circuit capacitor capacitance	470 μF
5	Injection circuit diode resistance	0.001 Ω

TABLE 4.2
Specification of the ZIGZAG and IPT

	Specifications		
S. No.	Details	Zigzag Transformer	IPT
1	Winding resistance	0.002 pu	0.002 pu
2	Winding leakage inductance	0.02 pu	0.02 pu
3	Core magnetising resistance and inductance	500 pu, 500 pu	500 pu, 500 pu
4	Primary winding turns	–	25, 25
5	Secondary winding turns	–	154
6	Zigzag autotransformer turns	134,62 for K_1 and K_2 ratios	–
7	Working frequency	50 Hz	300 Hz

current as near as sinusoidal wave. The input phase-to-phase voltage V_{ab} is shown in Figure 4.6b. The input currents and voltages of DBR1 are shown in Figure 4.6c and d, respectively. The output current and voltage of DBR1 is shown in Figure 4.6e and f, respectively. The primary winding current of IPT, I_p, superimposed on DBR output current, I_{d1}, makes it unipolar triangular, as shown in Figure 4.6e. The injection current and voltage of the Dickson injection circuit are approximately triangular in nature, as shown in Figure 4.6g and 4.6h, respectively. The resultant average load current is around 10.53 A, as shown in Figure 4.6i. The load voltage is nearly 541.9 V, as shown in Figure 4.6j. For the specific turns in primary and secondary of IPT, the minimum THD obtained is 4.42% and the peak value of the fundamental component is 12%, as shown in Figure 4.6k.

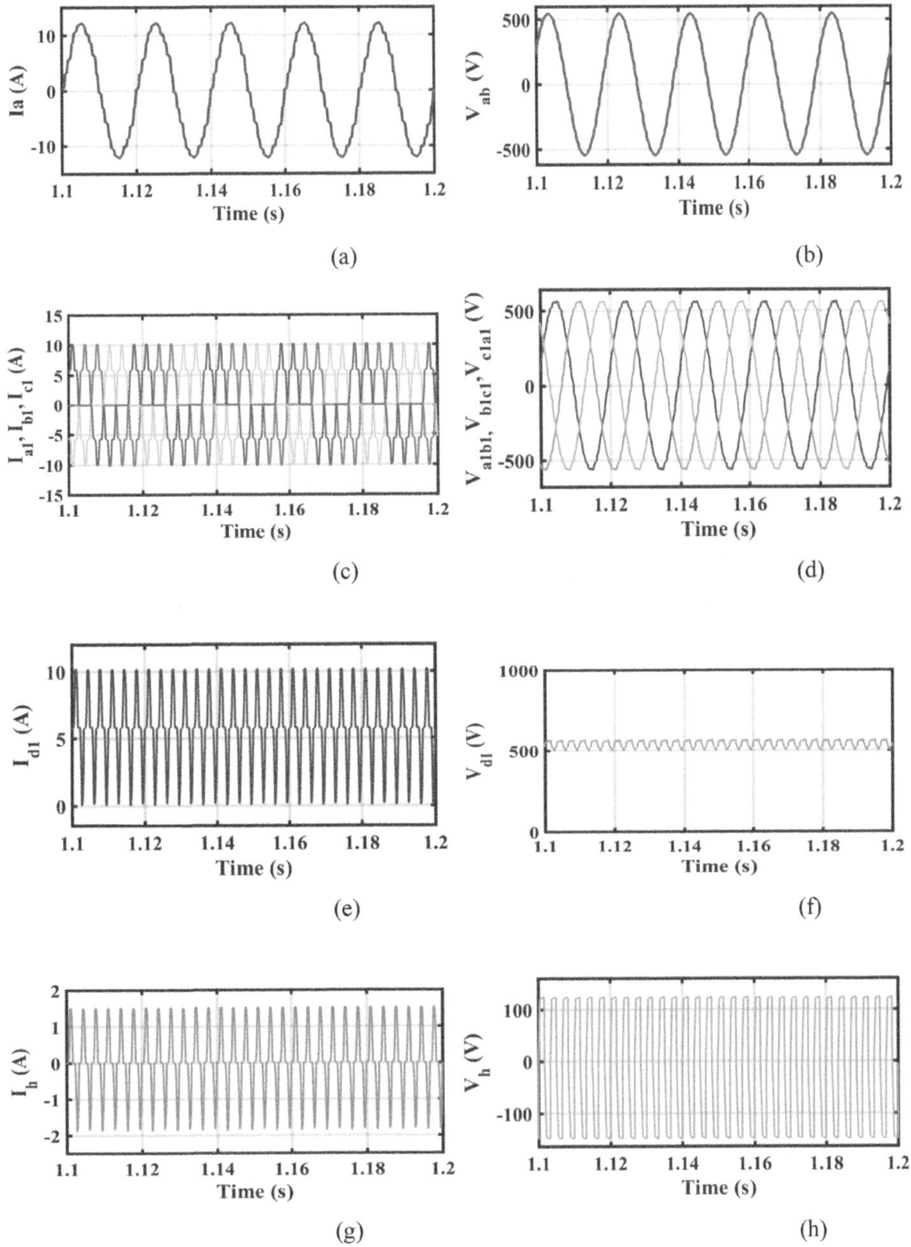

FIGURE 4.6
(a) Input current, (b) input voltage, (c) output current, and (d) output voltage of zigzag autoconnected transformer. (e) Output current and (f) output voltage of 12-pulse DBR. (g) Input current and (h) input voltage of Dickson injection circuit. (i) Output current and (j) output voltage of proposed converter. (k) FFT window of input current.

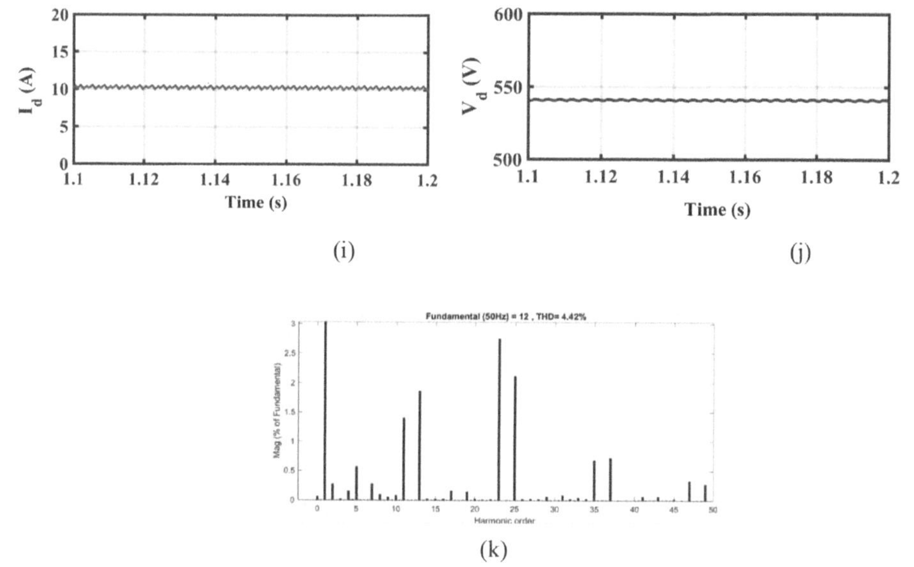

FIGURE 4.6 (***Continued***)

4.7 Hardware Result and Discussion

4.7.1 Description of Hardware Prototype

To validate the proposed design, a hardware prototype of the zigzag transformer-based 12-pulse rectifier with IPT and Dickson injection circuit was developed. The prototype was built for a 400-V, 50-Hz three-phase supply with a load rating of approximately 5.55 kW.

The hardware prototype of the proposed system was developed with a zigzag transformer configured using calculated turns ratios (K_1 = 134, K_2 = 62) to achieve a ±15° phase shift between the secondary outputs, thereby feeding two six-pulse diode bridge rectifiers (DBRs). The transformer's kVA rating was limited to about 30% of the load power, which emphasises its compactness and suitability for aircraft applications. Two three-phase DBRs were constructed and interconnected through a centre-tapped interphase transformer (IPT). The IPT was designed with a winding ratio of 25:154 turns to optimise ripple injection while keeping the kVA requirement minimal. On the DC side, a Dickson voltage multiplier injection circuit was implemented using 470 μF capacitors and low-resistance diodes ($R_{on} \approx 0.001\ \Omega$, $V_f \approx 1.35$ V). This circuit generated an auxiliary DC injection voltage of approximately 541.9 V with minimal ripple, effectively shaping the system's input current.

The experimental evaluation of the prototype demonstrated excellent performance. The input current waveform closely approximated a sinusoidal profile with 24 pulses, as verified by FFT analysis. The total harmonic distortion (THD) of the input current was measured at around 4.42%, which is well within IEEE-519 standard limits. Furthermore, the input voltage and current were observed to be in phase, confirming nearly unity power factor operation. The load voltage was ripple-free at about 541.9 V DC, while the average load current reached 10.53 A. The overall system efficiency exceeded 97%, with the IPT and injection circuit consuming less than 2% of the load power, validating the proposed configuration's effectiveness and efficiency.

4.7.2 Discussion on Various Waveforms of the Proposed Converter

Figure 4.7 shows the hardware results of the proposed system. Figure 4.7a shows the three-phase sinusoidal input voltage, and Figure 4.7b shows the three-phase 24-pulse input current. Figure 4.7c and 4.7d shows the three-phase DBR input voltage and current, respectively. Figure 4.7e shows the IPT primary voltage and triangular injection circuit input current with bipolar input voltage. Figure 4.7f shows one-phase input currents of the zigzag transformers whose summation gives sinusoidal input current of the system. Figure 4.7g shows the input voltage of the system with input voltage of DBRs. The input voltages of both the DBRs are 30° phase-shifted with each other.

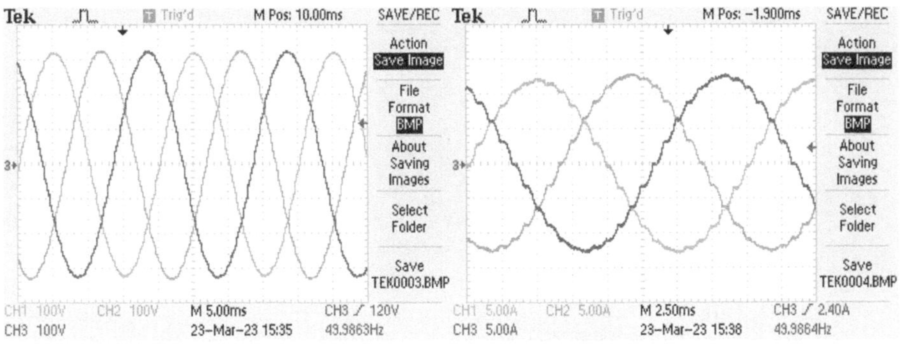

(a) (b)

FIGURE 4.7
Hardware results. (a) Input voltage of the system. (b) Input current of the system. (c) Three-phase input voltage of the DBR1. (d) Three-phase input current of the DBR1. (e) Primary voltage of IPT with its secondary current and voltage. (f) Zigzag transformer windings current for one phase with input current of the system. (g) Input voltage of the system with the phase-shifted input voltage of DBRs. (h) Input voltage and current with output voltage of the system.

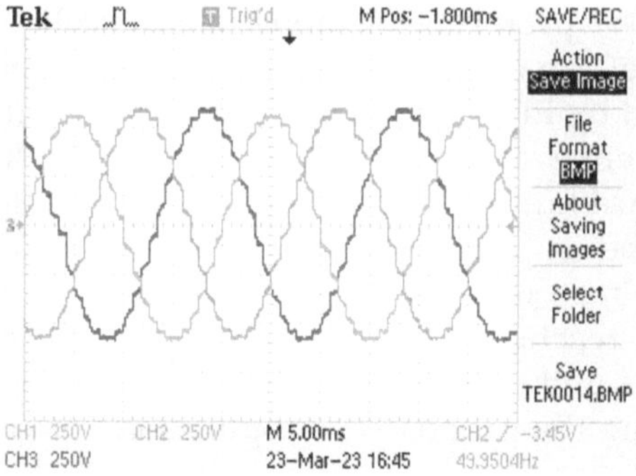

(c)

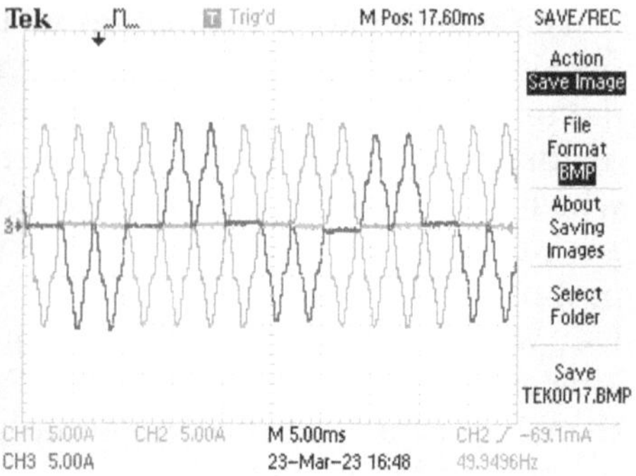

(d)

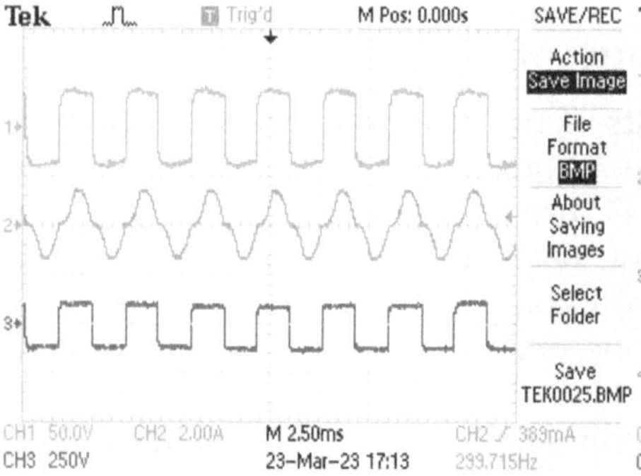

(e)

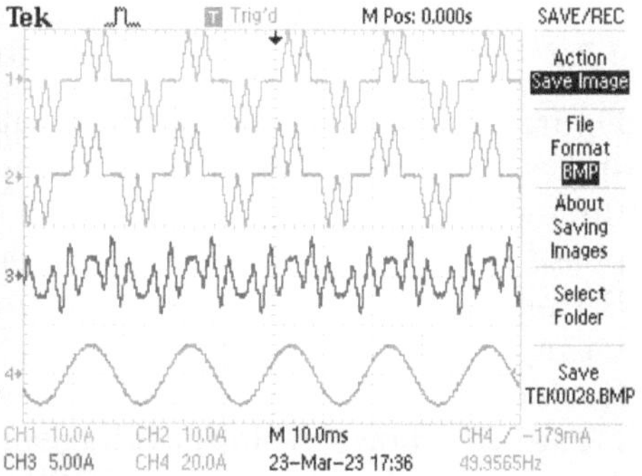

(f)

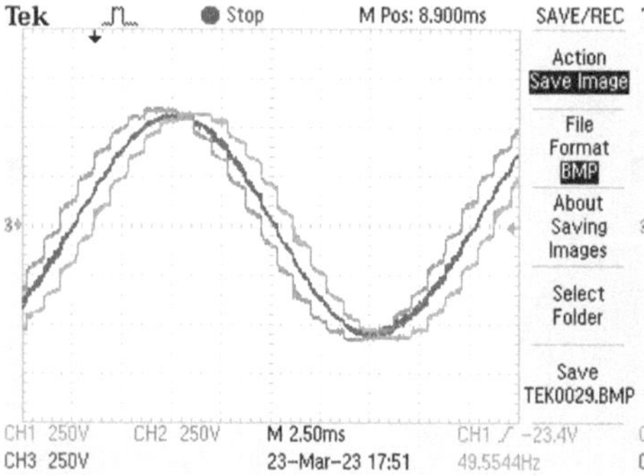

(g)

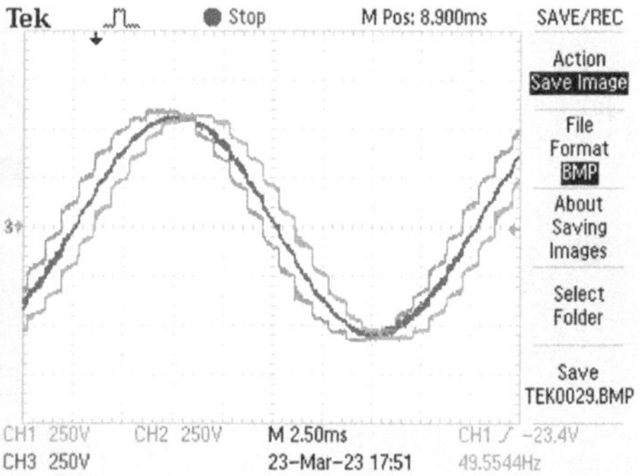

(h)

FIGURE 4.7 ***(Continued)***

The DBR1 and DBR2 input voltages are +15° and –15° phase-shifted with zigzag input voltage, respectively, in each corresponding phase. The input voltage and current of the system are in-phase with each other, and the ripple free load voltage is seen in Figure 4.7h.

4.8 Comparison of the Proposed MPR with Literature Papers

4.8.1 Comparison in Terms of kVA Rating, Transformer Type, Number of Windings, Type of IPR, Rating of IPR, Number Diodes, Number of Pulses, and Type and Complexity of Harmonic Reduction Method

In Table 4.3, the proposed converter is compared with different reference papers. It can be seen that the proposed converter has lesser THD than most of the reference papers. Only Refs. [23] and [26] have lesser THD, but the number of windings of the transformer and diodes is higher in both reference papers. In Ref. [22], the kVA rating of the transformer is higher than the proposed MPR. The kVA rating of the IPT of the proposed converter is less than that of all the reference papers, and also the kVA rating of the transformer and the number of diodes of this converter are less than or equal to that of all the reference papers. Even though the kVA rating and THD are less in Ref. [18], the presence of ZSBT makes the system bulky, and hence the proposed system is better in terms of space occupancy. Similarly, the asymmetric transformer configuration in Ref. [23] results in performance deterioration as in the case of phase outage. Further, the presence of ZSBT and the number of IPRs in Refs. [24] and [25] make the system much bulkier than the proposed system. Furthermore, ZSBT in Ref. [26] makes lesser usage in the application where space is a primary constraint. Overall, it can be concluded that the proposed system gives better power quality with a lesser component count, lesser volume, and lesser cost, and it is a better candidate for MEA applications.

4.8.2 Discussion on Harmonic Contents with Respect to Percentage of the Fundamental

From Table 4.4, it can be seen that the lower-order harmonics are of minimal magnitude as in the case of input current and voltage. Other higher-order harmonics in the input current, namely, 11th and 13th and 23rd and 25th, are less than 2% and 3%, respectively. Further, in input line voltage, 11th and 13th harmonics are lesser than 0.6% and 23rd and 25th are lesser than 1.5%, which are well within the standard limits.

4.8.3 Variation in Input Supply Voltage with Change in Different Parameters

The variation in the input voltage is applied, and the critical parameters are observed and given in Table 4.5.

TABLE 4.3

Comparison of the Proposed MPR with Different Reference Papers

Parameters		[21]	[22]	[18]	[23]	[24]	[25]	[26]	Proposed
THD		7.57	4.02	4.01	3.16	4.8	4.44	1.78	4.42
Type of harmonic reduction		Passive	Passive	Passive	Passive	Passive	Passive	Passive	Passive
No. of pulse		24	12	24	20	36	24	24	24
Phase shifting transformer	No. of windings	18	9	9	23	9	15	21	12
	Type, complexity	Isolated, Double star, Little complex	Fork Autotransformer, Simple	Star Connected, Simple	Asymmetric, Complex	Star Autotransformer, simple	Polygon Autotransformer, Simple	Differential star Autotransformer, Complex	Zigzag Autotransformer, simple
	Rating	134%	125%	30%	30.12%	–	28.65%	–	30.6%
IPT/IPR type		CTC, ITC	IPR	1-Tapped IPT	–	2-Tapped IPT	2-Tapped IPR	–	1-Tapped IPT
Rating of IPT/IPR		9.78%	10.08%	2.705%	–	2.68%	1.72%	–	1.097%
Complexity of injection circuit and IPT		Complex	Simple	Simple	–	Little complex	Little complex	–	Simple
ZSBT		–	–	Present (7.133%)	–	Present	Present (8.15%)	Present (4%)	–
No. of diodes		16	6	16	20	18	14	24	16

TABLE 4.4

Different Harmonic Contents with Respect to Percentage of the Fundamental

Harmonic Order	1	3	5	7	9	11	13	15	17	19	23	25	THD
I_a (Input current)	100	0.02	0.56	0.28	0.05	1.40	1.86	0.02	0.16	0.15	2.75	2.11	4.42
V_a (Input line voltage)	100	0.01	0.05	0.03	0.01	0.34	0.53	0	0.05	0.06	1.27	1.06	2.01

TABLE 4.5

Change in Different Parameters with Change in Input Supply Voltage

Input Line Voltage (V_{rms})	% THD of I_a	%THD of V_a	I_d (Load Current)	V_d (Load Voltage)	Power Factor	Efficiency
400	4.42	2.01	10.53	541.9	0.9996	97.77
300	4.56	1.97	7.897	406.4	0.9996	97.61
200	4.58	1.96	5.265	270.9	0.9997	97.28
100	4.66	1.93	2.632	135.5	0.9998	96.29

4.9 Summary

This chapter proposed a 12-pulse converter that uses a Dickson voltage multiplier as an injection circuit to produce a 24-pulse output. The ZSBT is skipped because the zigzag autoconnected transformer has the inherent ability to block zero-sequence components. Significantly, the THD is reduced, and the proposed 12-pulse converter has been operating as a 24-pulse converter and using its advantages. The MATLAB simulation results and hardware results have been used to validate the proposed design and analysis. Thus, the proposed multi-pulse converter is best suited for MEA application as it satisfies the prerequisites of a low magnetic rating, small size, lightweight, smaller volume, low cost, and fewer components.

References

[1] Yitao Liu, Junxiang Deng, Chao Liu, and Sen Li, "Energy optimisation analysis of the more electric aircraft," IOP Conf. Series, ICAESEE, 2018.

[2] N. Thapa, and J. Mehta, "All electric aircraft: A reality on its way," *Research Gate MATPR*, 2020.

[3] L. Setlak, Lucjan, and E. Ruda, "Review, analysis and simulation of advanced technology solutions of selected components in power electronics systems (pes) of more electric aircraft," *International Journal of Computational Systems Engineering*, vol. 9, no. 10, pp. 1194–1206, 2015.

[4] M. Wang and F. Zhang, "12-pulse auto-transformer rectifier with harmonic current injection for non-grid-connected wind power applications," 2009 World Non-Grid-Connect. Wind Power and Energy Conf., Nanjing, China, 2009, pp. 1–5.
[5] Q. Li, F. Meng, and H. Zhang, "A Series-connected 18-pulse Rectifier Using Isolated Transformer," 2019 22nd Int. Conf. Electr. Mach. Syst. (ICEMS), 2019, pp. 1–5.
[6] B. Singh, V. Garg, and G. Bhuvaneswari, "Design and development of auto-transformer based 24-pulse AC-DC converter fed induction motor drive," 2007 7th Int. Conf. Power Electron. Drive Syst., Bangkok, Thailand, 2007, pp. 73–80.
[7] F. Meng, L. Gao, S. Yang, and W. Yang, "Effect of single-phasing on multipulse rectifier with active interphase reactor," *IEEE Transactions on Power Electronics*, vol. 30, no. 5, pp. 2549–2555, May 2015.
[8] V. Sheelvant, R. Kalpana, B. Singh, and P. P. Saravana, "Improvement in harmonic reduction of a zigzag autoconnected transformer based 12-pulse diode bridge rectifier by current injection at DC side," *IEEE Transactions on Industry Applications*, vol. 53, no. 6, pp. 5634–5644, Nov.–Dec. 2017.
[9] S. Nakhaee and A. Jalilian, "Simulation of 24-pulse Autotransformer based AC – DC converter employing pulse multiplication with direct torque controlled induction motor drive load," 2011 2nd Power Electron., Drive Syst. and Technologies Conf., Tehran, Iran, 2011, pp. 593–598.
[10] B. Singh, G. Bhuvaneswari, and V. Garg, "Scott-connected autotransformer based multipulse AC-DC converters for power quality improvement in vector controlled induction motor drives," 2005 Int. Conf. Power Electron. Drives Syst., Kuala Lumpur, Malaysia, 2005, pp. 1491–1496.
[11] Q. Du, L. Gao, Q. Li, S. Wang, F. Li, and F. Meng, "Research on the DC-side passive harmonic reduction methods of the ultramultiphase parallel-connected multipulse rectifier," *IEEE Transactions on Power Electronics*, vol. 37, no. 12, pp. 14809–14819, Dec. 2022.
[12] Q. Du, L. Gao, Q. Li, T. Li, and F. Meng, "Harmonic reduction methods at DC side of parallel-connected multipulse rectifiers: A review," *IEEE Transactions on Power Electronics*, vol. 36, no. 3, pp. 2768–2782, March 2021.
[13] Q. Li, Q. Du, X. Yin, L. Gao, and F. Meng, "General injection transformer turn ratio design method and hybrid harmonic reduction methods for series-connected multipulse rectifiers," *IEEE Transactions on Power Electronics*, vol. 37, no. 12, pp. 15457–15468, Dec. 2022.
[14] A. V. Sugyana, and B. Brindha Sakthi, "Modified Dickson charge pump voltage multiplier using high voltage gain converter," *International Journal of Scientific Research IJSR SJIF*, vol. 8, no. 10, pp. 1706–1709, 2019.
[15] M. M. Al-Azawy and F. Sari, "Analysis of Dickson voltage multiplier for RF energy harvesting," 2019 1st Global Power, Energy and Commu. Conf. (GPECOM), Nevsehir, Turkey, 2019, pp. 10–14.
[16] S. P. P and A. Kumari Sinduriya, "Harmonic reduction in multi-pulse rectifier using DC injection circuit for MEA applications," 2022 IEEE International Conference on Power Electronics, Drives and Energy Systems (PEDES), Jaipur, India, 2022, pp. 1–6.
[17] B. Singh, S. Gairola, A. Chandra, and K. Al-Haddad, "Zigzag connected auto-transformer based controlled AC-DC converter for pulse multiplication," 2007 IEEE Int. Symp. Ind. Electron., Vigo, Spain, 2007, pp. 889–894.

[18] F. Meng, L. Gao, S. Yang, and W. Yang, "Effect of phase-shift angle on a delta-connected autotransformer applied to a 12-pulse rectifier," *IEEE Transactions on Industrial Electronics*, vol. 62, no. 8, pp. 4678–4690, Aug. 2015.
[19] S. Park, J. Yang, and J. Rivas-Davila, "A hybrid cockcroft–Walton/Dickson multiplier for high voltage generation," *IEEE Transactions on Power Electronics*, vol. 35, no. 3, pp. 2714–2723, March 2020.
[20] F. Meng, X. Xu, and L. Gao, "A simple harmonic reduction method in multi-pulse rectifier using passive devices," *IEEE Transactions on Industrial Informatics*, vol. 13, no. 5, pp. 2680–2692, Oct. 2017.
[21] J. Wang, T. Yu, C. Zhao, T. Liu, X. Yao, and Q. Chen, "A simple 24-pulse star rectifier combining two different tap converters," in IEEE Trans. on Transp. Electrific., Feb. 2023.
[22] F. Meng, X. Xu, L. Gao, and C. Cai, "Active harmonic reduction using DC-side current injection applied in a novel large current rectifier based on fork-connected autotransformer," *IEEE Transactions on Industrial Electronics*, vol. 64, no. 7, pp. 5250–5264, July 2017.
[23] P. Saravana Prakash, R. Kalpana, and B. Singh, "Inclusive design and development of front-end multiphase rectifier with reduced magnetic rating and improved efficiency," *IEEE Journal of Emerging and Selected Topics in Power Electronics*, vol. 8, no. 3, pp. 2989–3000, Sept. 2020.
[24] L. Gao, X. Xu, Z. Man, and J. Lee, "A 36-pulse diode-bridge rectifier using dual passive harmonic reduction methods at DC link," *IEEE Transactions on Power Electronics*, vol. 34, no. 2, pp. 1216–1226, Feb. 2019.
[25] S. Nakhaee and A. Jalilian, "Simulation of 24-pulse autotransformer based AC – DC converter employing pulse multiplication with direct torque controlled induction motor drive load," 2011 2nd Power Electronics, Drive Systems and Technologies Conference, Tehran, Iran, 2011, pp. 593–598.
[26] S. Khan *et al.*, "A novel 24-pulse rectification system," *IEEE Access*, vol. 6, pp. 59350–59361, 2018.

5

DC–DC Converters in MEA Systems

Xiaodong Li

This chapter presents a systematic classification and evaluation of DC–DC converter topologies tailored for MEA power architectures. In the domain of non-isolated topologies, multiphase interleaved Buck and three-level converters are examined, which highlights the advantages in power density and efficiency while addressing challenges related to EMI and circulating current control. Regarding isolated topologies, the analysis is focused on dual-active bridge (DAB) converter and its derivatives – such as active bridge active clamp (ABAC), phase-shifted full-bridge (PSFB), and triple-active bridge (TAB) – alongside resonant topologies, including LLC and CLLC converters. Given its widespread adoption in energy storage interfaces and bus interconnections, DAB converter is subjected to a critical analysis, with particular emphasis on inherent limitations such as backflow power and the loss of soft-switching capabilities under light-load conditions.

Furthermore, the evolution of control strategies, with a specific focus on phase-shift modulation techniques, is investigated. A comprehensive comparative analysis is conducted, which spans from the fundamental single-phase shift (SPS) modulation to advanced multi-degree-of-freedom schemes, including extended phase shift (EPS), dual phase shift (DPS), and triple phase shift (TPS). The critical trade-offs among control flexibility, soft-switching operating ranges, and implementation complexity are elucidated in the comparison. To further enhance system performance, the application of intelligent optimisation algorithms – namely genetic algorithm (GA), particle swarm optimisation (PSO), and Q-learning-based reinforcement learning – is discussed, which demonstrates significant efficacy in solving multi-objective optimisation problems. Consequently, the efficiency and adaptability to varying operating conditions are improved.

Finally, the critical design considerations essential for MEA applications are consolidated. Key aspects include the realisation of zero-voltage switching (ZVS) and zero-current switching (ZCS) to minimise switching losses, the minimisation of peak, and RMS current stresses for improved thermal management, and the mitigation of backflow power and transformer DC

 DOI: 10.1201/9781003747994-5

bias current are considered, which are presented as fundamental criteria for ensuring high efficiency and reliable system operation under complex and varying operating conditions.

5.1 The Requirements of DC–DC Converters in MEA Systems

In the paradigm of more-electric aircraft (MEA), where conventional hydraulic and pneumatic systems are progressively replaced by electrical alternatives, power electronic converters play a pivotal role. Among these, DC–DC converters serve as essential building blocks for the efficient and reliable distribution of electrical power across various on-board subsystems. Their deployment within MEA architectures must meet a comprehensive set of functional, performance, and reliability requirements, as summarised in Table 5.1.

TABLE 5.1

DC–DC Converter Requirements in the Context of MEA Systems

Requirement	Description
Bidirectional power flow	Enables energy exchange between on-board storage systems (e.g., batteries, supercapacitors) and generators or loads; supports regenerative braking, load levelling, and emergency power functionality
High power density	Essential for minimising the size and weight of the converter; facilitated by advanced topologies, effective thermal management, and wide bandgap devices (e.g., SiC, GaN)
High switching frequency	Reduces the size of passive components (inductors, capacitors), enhancing power density and transient response; requires careful design to balance switching losses and EMI
EMI reduction	Critical in the sensitive aircraft environment; achieved via filtering, soft-switching techniques, and electromagnetic shielding to comply with EMI standards
Redundancy and fault tolerance	Improves system reliability through modular redundancy, hot-swapping capabilities, and fault detection/isolation schemes
Voltage adaptation for avionics	Steps down high-voltage DC buses (e.g., 270 V or 540 V) to lower voltages (e.g., 28 V) for avionics and control electronics
Power flow regulation among DC buses	Maintains voltage stability and coordinates energy flow between distributed DC buses with different voltage levels
Energy storage interface	Manages charging/discharging of on-board storage; ensures protection and supports aircraft-wide energy management strategies

5.2 DC–DC Converter Topologies in MEA Systems

In the power architecture of MEA, DC–DC converters fulfil a range of roles that can be broadly categorised into two major functional domains: interconnection between DC buses and energy storage interfacing (such as battery charging and discharging). To meet the diverse voltage, isolation, and power flow requirements across these domains, a variety of converter topologies have been developed and applied.

5.2.1 Non-Isolated Converters

Non-isolated converters, such as multiphase interleaved Buck converters, three-level Buck converters, and three-level flying-capacitor converters, are typically used in applications where isolation is not required, such as on-board avionics power conditioning or tightly regulated intra-bus voltage conversion. These topologies offer advantages in terms of compactness, high efficiency, and simplified control, making them suitable for weight- and space-constrained subsystems.

The three-phase interleaved buck converter, as shown in Figure 5.1, is a promising non-isolated DC–DC topology. It offers reduced output current ripple, smaller filter size, and modular scalability for high-power applications. Load-sharing among phases enhances current-handling capability, although at the cost of increased hardware complexity [1].

The three-level buck converter, as shown in Figure 5.2, reduces switch voltage stress by half, enabling the use of lower-voltage devices with improved switching and conduction performance. This leads to higher efficiency and smaller magnetic components compared to the conventional two-level buck converter [2]. However, it suffers from severe EMI issues, particularly

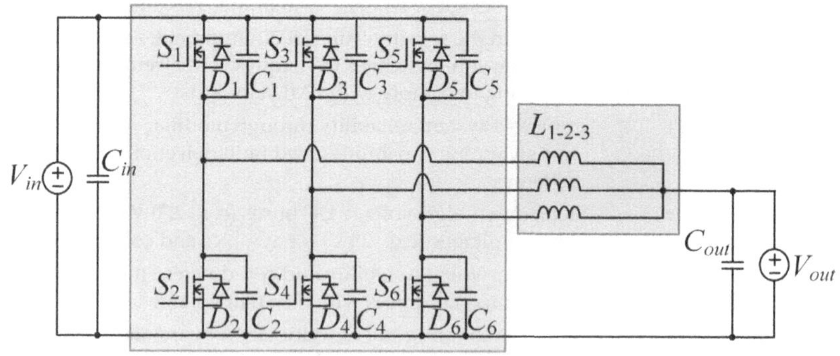

FIGURE 5.1
Three-phase interleaved buck converter.

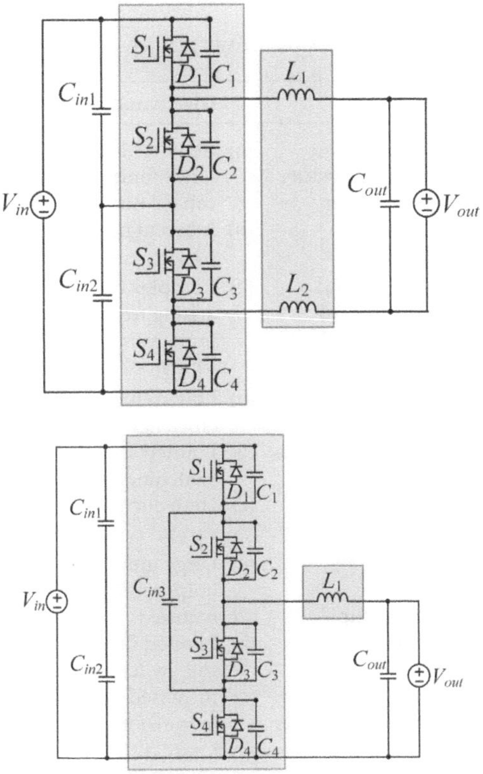

FIGURE 5.2
Three-level converter. (a) Buck converter. (b) Flying capacitor converter.

common-mode noise, which can affect battery systems and avionics [3]. Additionally, its parallelisation is challenging due to potential circulating currents without interphase inductors.

An enhanced variant, the three-level flying capacitor converter (Figure 5.2b), further reduces output inductance by minimising volt-second stress. It also exhibits lower EMI and improved scalability for interleaved operation. Nevertheless, it faces drawbacks such as complex switching paths and difficulty in implementing short-circuit protection due to the presence of the flying capacitor [4].

5.2.2 Isolated Converters

Isolated converters are commonly employed in battery charging units, energy storage interfaces, and inter-bus power transfer, especially where safety isolation or large voltage conversion ratios are required. Commonly isolated DC–DC converter topologies in MEA systems is illustrated in Table 5.2.

TABLE 5.2

Isolated DC–DC Converter Topologies in MEA Systems

Topology	Advantages	Disadvantages	Bidirectional	Resonant
Dual active bridge (DAB)	(a) Low switching loss (b) Symmetrical structure (c) High transformer utilisation factor	(a) Requires large input/output capacitance (b) Inherent backflow power	Yes	No
Active bridge active clamp (ABAC)	(a) Lower weight and volume than DAB (b) Reduced ripple current	(a) Complex control arising from numerous switches	Yes	No
Phase-shifted full-bridge (PSFB)	(a) Simple phase-shift control (b) Wide output voltage range (c) Low peak voltage stress	(a) Low conversion efficiency (b) Limited soft switching range (c) High circulating current	No	No
Triple active bridge (TAB)	(a) Fault-tolerant and highly reliable (b) Supports multiple input ports and voltage levels (c) Decoupled power flow between the ports	(a) Large input and output filters are required (b) Limited range of soft switching (c) Increased iron loss in transformers	Yes	No
LLC converter	(a) High power density (b) High efficiency	(a) Limited modularity (b) Complex control (c) High peak voltage stress	No	Yes
Full-bridge CLLC converter	(a) High efficiency (b) Moderate control (c) Low peak voltage stress	(a) Limited output voltage (b) Design difficulty due to parameter sensitivity	Yes	Yes

The dual-active bridge (DAB) converter [5], as shown in Figure 5.3, is a widely used isolated DC–DC topology. It consists of two full bridges linked by a high-frequency transformer, providing galvanic isolation to prevent LV-side faults from affecting the HV bus. The DAB enables bidirectional power flow and efficient control. Its basic modulation employs symmetrical square waves on both transformer sides, with power regulated by their phase shift. Despite its merits, the DAB converter exhibits several drawbacks, including significant backflow power under light-load conditions, inherent bidirectional current, and the requirement for bulky input/output filters due to its square-wave modulation.

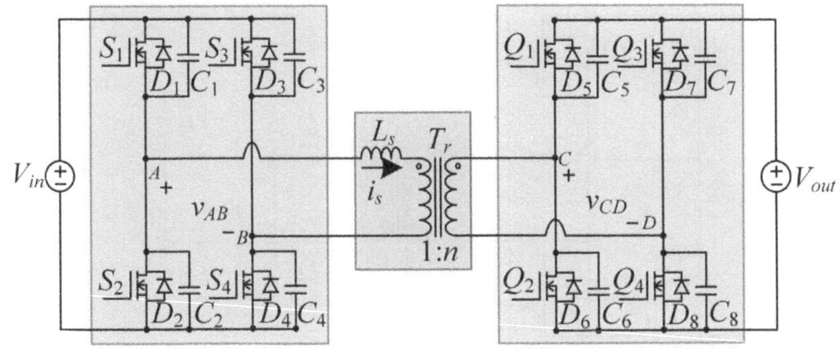

FIGURE 5.3
The topology of DAB converter.

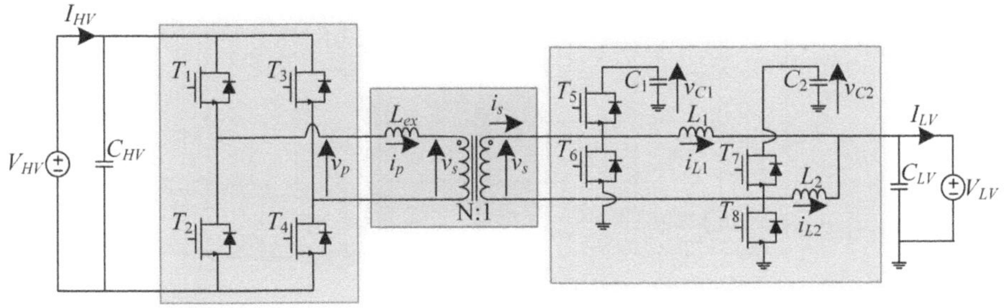

FIGURE 5.4
The topology of ABAC converter.

To address these limitations, Tarisciotti et al. [6] proposed the active bridge active clamp (ABAC) converter, as illustrated in Figure 5.4. While maintaining operational similarities with the DAB, the ABAC introduces a current-fed low-voltage (LV) stage enabled by external output inductors. Power is initially transferred through the inductor L_{ex} to the clamping circuits, and subsequently delivered to the output via L_1 and L_2, producing a smoother and low-ripple DC output. The interleaved operation of the output inductors enhances current sharing and output quality. The ABAC converter supports bidirectional power flow and achieves reduced weight and volume compared to the DAB. However, the increased number of active switches adds control complexity and elevates implementation cost.

The phase-shifted full-bridge (PSFB) converter, as shown in Figure 5.5, is widely used in fast-charging applications due to its simple control scheme, broad load-handling capability, and reduced voltage/current stress on power devices. Its modular design supports parallel operation for higher power output [7]. By employing phase-shifted pulse-width modulation (PWM),

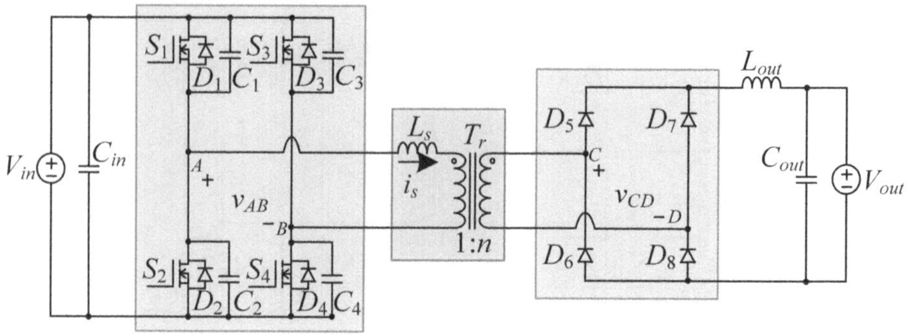

FIGURE 5.5
The topology of PSFB converter.

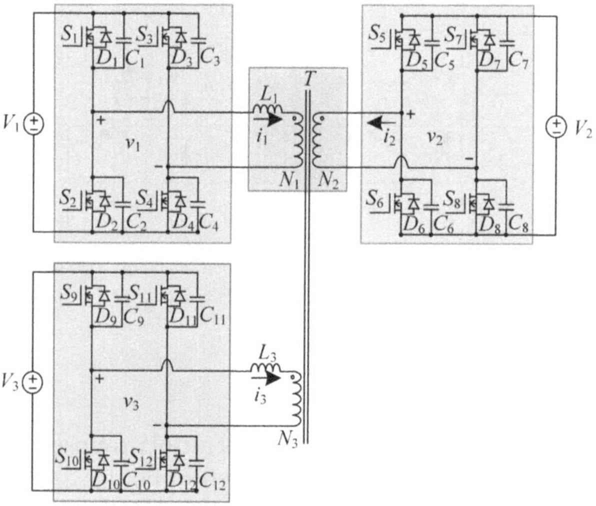

FIGURE 5.6
The topology of TAB converter.

zero voltage switching (ZVS) turn-on is realised for primary-side switches through the interaction of transformer leakage inductance and device parasitic capacitance [8]. Nevertheless, the PSFB converter suffers from high turn-off losses in the primary switches and considerable losses in the output rectifiers, thereby compromising its overall efficiency [9].

To further improve performance, Liu et al. [10] proposed a decoupled triple-active bridge (TAB) converter, depicted in Figure 5.6, as an evolution of the DAB topology. By reducing the dependency of port power transfer on phase-shift angles, the converter enables independent power regulation

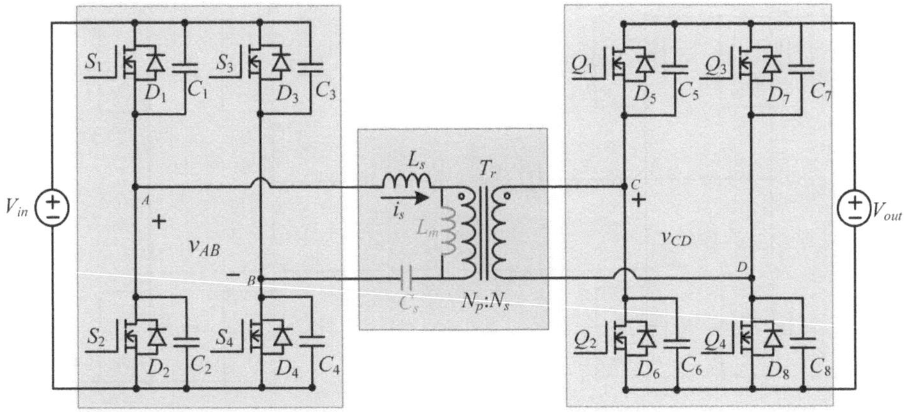

FIGURE 5.7
The topology of LLC converter.

among ports. Its multiport configuration provides high power density, superior efficiency, and robust fault tolerance, making it well-suited for high-power MEA applications. However, employing two separate two-port transformers instead of a single three-port transformer increases core losses and complicates the magnetic design.

Another widely adopted topology is the LLC resonant converter, as shown in Figure 5.7, known for its high efficiency and power density. It offers reduced current stress and improved magnetic core utilisation [11]. However, its main limitation lies in the narrow operating range, as optimal performance is achieved only near the resonant frequency. Furthermore, output voltage regulation depends on varying the switching frequency, which increases control complexity.

To enhance regulation flexibility and extend the soft-switching range, the CLLC resonant converter has been proposed (Figure 5.8), which introduces an additional resonant tank on the secondary side, forming a symmetrical topology [12]. This dual-tank structure not only enables bidirectional power flow but also improves ZVS and ZCS conditions over a wider load range. As a result, the CLLC converter is particularly attractive for isolated bidirectional applications, such as battery charging/discharging in MEA systems.

Despite offering high efficiency, reduced peak voltage stress, and relatively manageable control requirements, the CLLC converter also presents several challenges. Its performance is highly sensitive to circuit parameters, which complicates the design process and reduces tolerance to component variations. Moreover, the presence of resonant tanks on both the primary and secondary sides increases the overall component count and introduces additional complexity in magnetic design and system integration.

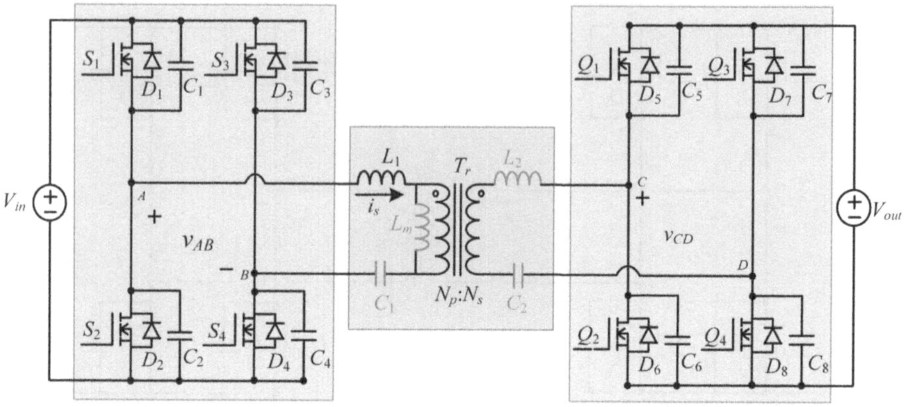

FIGURE 5.8
The topology of CLLC converter.

5.2.3 Conclusion

Various DC–DC topologies have been successfully applied in MEA systems to meet diverse operational requirements such as isolation, bidirectional power flow, high power density, and control flexibility. Isolated converters like DAB, ABAC, PSFB, and TAB are well-suited for energy routing between different voltage domains, while resonant converters such as LLC and CLLC offer high efficiency under soft-switching conditions. For non-isolated applications, interleaved and multilevel converters provide excellent scalability and reduced EMI. These topologies complement each other and can be tailored to specific MEA subsystems based on functionality, efficiency, and integration needs.

5.3 Control Methods Utilised in DC–DC Converters of MEA

5.3.1 Modulation Methods for DC–DC Converters

The development of modulation methods has progressed from single-control-variable modulations to multi-control-variable modulations, which has improved the operation characteristics in power electronics converters. There have been several modulation methods used in DC–DC converters. Based on the theory of voltage angle difference, the phase shift modulation is the mostly used method to control the transmission power through any two ports of DC–DC converters. Figure 5.3 has already shown the topology of DAB. The phase shift modulation will be discussed based on DAB converter.

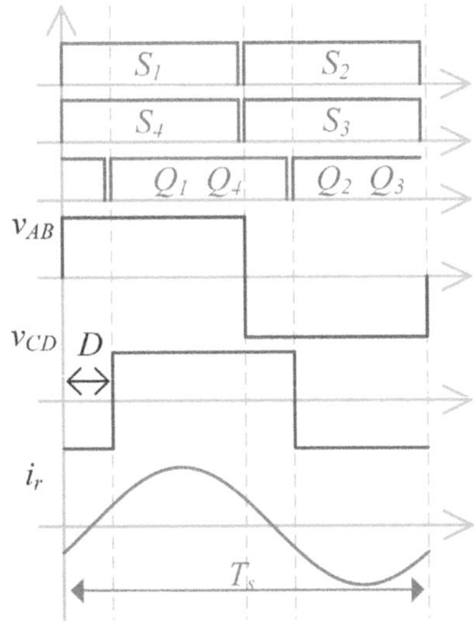

FIGURE 5.9
Single-phase shift modulation for DBSRC.

Among various modulation methods, four typical phase shift modulation methods are single-phase shift (SPS), extended-phase shift (EPS), dual-phase shift (DPS), and triple-phase shift (TPS), respectively.

The operation waveforms of SPS modulation method are depicted in Figure 5.9. The duty cycle for each switch on one bridge arm is 0.5, and the switches on the identical bridge arm are conducted complementarily during the switching cycle T_s. This means that when one switch is turned on, the other switch is turned off. For SPS modulation, there is only one control variable D, which is defined as the phase shift between the primary side and secondary side. Under SPS modulation, the leakage inductor is charged by the input power, then the energy stored in the leakage inductor is transferred to the output side when the secondary side switches are turned on.

As the simplest modulation method, the implementation of SPS is easier than other modulation methods due to only one control variable, which reduces the computational burden of the controller. When the voltage gain ratio is equal to unity, the switches can achieve ZVS easily within the dead time, which contributes fewer power loss, leading to lower voltage stress on the switching stages. Consequently, the reliability and stability of the system is improved [13, 14]. However, when the voltage gain ratio under SPS modulation is away from unity, the soft-switching operation region will become narrow, and the current stress will increase.

EPS modulation method involves two phase shift control variables, D_1 and D_2, where D_1 is the phase shift inside the primary side bridge, and D_2 is the outer phase shift between two bridges, which is shown in Figure 5.10. By adding additional phase shift control variables, the control flexibility is

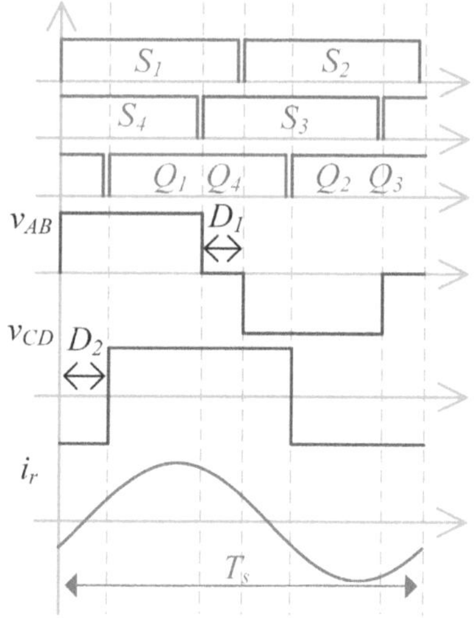

FIGURE 5.10
Extended-phase shift modulation for DBSRC.

improved. Besides, the power transfer capability is enhanced and the circulating energy is reduced under EPS modulation.

Compared to SPS modulation, the switch stress is reduced by EPS modulation under the identical power transfer capacity [15, 16]. Although EPS modulation offers improved performance, the overlap-caused power loss during switching transitions is increased. Besides, a large amount of reactive power will be generated in the high-frequency transformer when the outer phase shift angle is large.

As shown in Figure 5.11, DPS modulation is similar to EPS modulation in terms of utilising two phase shift angles, D_1 and D_2. The difference is that D_1 is the inside phase shift on both sides, which is responsible for controlling the active bridges on the input and output sides. DPS modulation inherits both the advantages and drawbacks associated with EPS modulation. Additionally, the switch stress of EPS and DPS modulation methods is optimised compared with SPS modulation [17, 18].

SPS, EPS, and DPS modulation methods can be seen as a special condition of TPS modulation, which is shown in Figure 5.12. Three control variables, D_1, D_2, and D_3, are adopted in TPS modulation, where D_1 still represents the phase shift inside the primary side bridge, D_2 denotes the outer phase shift between two bridges, and D_3 represents the phase shift inside the secondary side bridge. ZVS performance becomes better for DBSRC under TPS modulation, which can reduce switching loss and improve efficiency under light load conditions. Moreover, the flexibility and capability of controlling power transfer are enhanced, which surpasses SPS, EPS, and DPS modulation

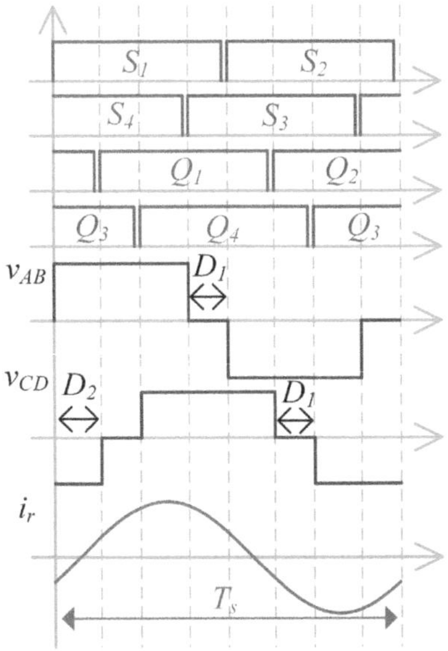

FIGURE 5.11
Dual-phase shift modulation for DBSRC.

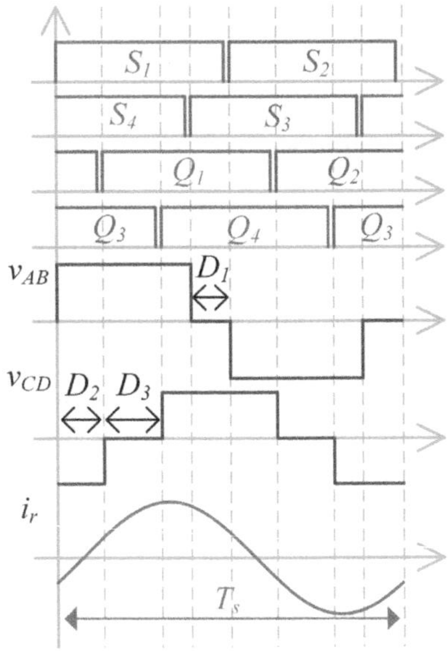

FIGURE 5.12
Triple-phase shift modulation for DBSRC.

TABLE 5.3

Comparison between SPS, EPS, DPS, and TPS Modulation Methods

Modulation Methods	Control Variables	Advantages	Disadvantages
SPS	1	• Simple implementation • Low computation burden	• Limited soft-switching range • Poor performance at non-unity voltage gain • Low flexibility
EPS	2	• Wide ZVS range • Improved power regulation capability	• High control complexity • Low total loss optimisation
DPS	2	• High control freedom • Enable ZVS under light load • Increased efficiency	• Increased control complexity • Narrow voltage gain range
TPS	3	• More flexible control • Wider ZVS range • High efficiency	• Higher control complexity • Difficult implementation

methods [19–21]. However, the increased control variables of TPS introduces noticeable complexity, making the implementation more challenging, which aggravates the computational burden of the controller.

The comparison between these modulation methods is listed in Table 5.3. On the one hand, the simplicity of SPS modulation results in less computational burden and increases the response time from the controller. On the other hand, the complexity of EPS, DPS, and TPS modulation make power electronics converters that can be adjusted for better features. Hence, adopting a suitable modulation method is beneficial to the performance of subsystems in MEA.

5.3.2 Optimisation Algorithm for DC–DC Converters

The high-efficiency operation of power electronics converters is a key challenge for MEA application. Various objectives, like inductor current stress, conduction loss, circulating energy, and soft-switching range, are commonly selected to optimise for improved transmission efficiency and better performance. In recent years, numerous intelligent algorithms, including genetic algorithm (GA), particle swarm optimisation (PSO) algorithm, and Q-learning reinforcement learning algorithm, have been adopted to solve optimisation problems under different modulation methods.

Genetic algorithm is a type of evolutionary algorithm which searches for solutions to optimisation problems by inspired principles of natural selection. To solve the optimisation problems with GA, the formula of a set of objective functions and constraints must be established first. The objective functions modelling depends on the requirements of the system, such as minimising reactive power or maximising the efficiency [22]. When it is formulated, GA can be used to find the optimal phase shift angles which meet the requirements.

The multi-objective optimisation problem of power electronics converters can be uniformly described as follows:

$$\begin{cases} \text{Obj}_1 : J_1(V_{in}, V_{out}, D_1, D_2, D_3) \\ \text{Obj}_2 : J_2(V_{in}, V_{out}, D_1, D_2, D_3) \\ \dots \\ \text{Obj}_n : J_n(V_{in}, V_{out}, D_1, D_2, D_3) \\ \text{Equcon}_1 : P(V_{in}, V_{out}, D_1, D_2, D_3) = P_{out} \\ \text{Inequcon}_1 : \text{Operation mode}(D_1, D_2, D_3) \\ \text{Inequcon}_2 : \text{ZVS}(V_{in}, V_{out}, D_1, D_2, D_3) \end{cases} \tag{5.1}$$

where the objective functions, Obj_n, can represent the converter characteristics like root-mean-square (rms) current, circulating power, and current stress. The equality constraint, Equcon_1, commonly denotes the actual transmission power, which is expected to match the desired power. The equality constraints, Inequcon_1 and Inequcon_2, represent the constraints under different operation modes and ZVS constraints for each switch, respectively. The optimisation equation (5.1) is crucial in constructing the fitness function, which is utilised to evaluate and select the generated offspring, making it a key step during the optimisation process [23].

Like GA, the particle swarm optimisation algorithm is regarded as swarm intelligence search algorithm. For the optimisation problems of power electronics converters, the core of PSO algorithm, the fitness function, is constructed based on the optimisation equation (5.1) as well. In PSO algorithm, each solution to be optimised can be seen as a particle. Its motion state can be denoted by position and velocity, where the position represents the parameter values, which is expressed as $x(D_1, D_2, D_3)$. The velocity, expressed as v, denotes the search direction and step size in the solution space. The motion state of a particle is affected by the historical best position of itself and the swarm [24, 25].

During the search procedure, the positions and velocities of the particles are continuously updated for more advantageous solutions. The formulas for position and velocity update are expressed as follows:

$$\begin{cases} v_{i,d}^{t+1} = \omega v_{i,d}^{t} + c_1 r_{1,d}\left(p_{best,i,d} - x_{i,d}^{t}\right) + c_2 r_{2,d}\left(g_{best,i,d} - x_{i,d}^{t}\right) \\ x_{i,d}^{t+1} = x_{i,d}^{t} + v_{i,d}^{t+1} \end{cases} \tag{5.2}$$

where ω is the inertia weight, $v_{i,d}^{t}$ represents the velocity of the ith particle in the dth dimension, c_1 and c_2 denote the individual and social learning factors, $r_{1,d}$ and $r_{2,d}$ are random numbers in the range of 0 to 1, $p_{best,i,d}$ represents the historical best position of the swarm, $x_{i,d}^{t}$ is the position of the ith particle

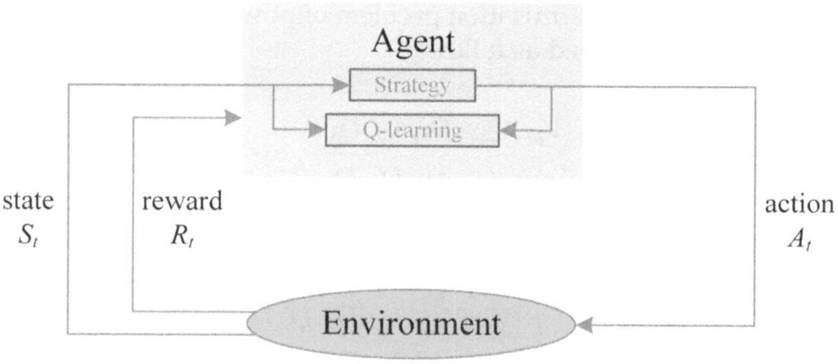

FIGURE 5.13
The schematic diagram of Q-learning reinforcement learning algorithm.

in the dth dimension, and $g_{best,i,d}$ represents the historical best position of the particle. With the help of Equation (5.2), PSO can search for optimal solutions by continuously updating the positions and velocities.

The convergence speed of PSO algorithm is faster and the parameter dependence is lower when compared with GA. However, the optimisation results when utilising PSO are easy to trap in local optima. Besides, PSO requires abundant computation resources, particularly in multi-control-variable and complex modulation methods and high-dimensional optimisation problems. This requires longer computation time and higher computing capacity of the controller, which is commonly addressed by offline calculation and look-up table method in existing works.

Q-learning reinforcement learning algorithm is a classic method that uses temporal differences to solve reinforcement learning control problems. The basic framework of Q-learning reinforcement learning algorithm is illustrated in Figure 5.13.

As the figure shows, Q-learning reinforcement learning algorithm is a trial-and-error method, which pursues in improving the decision-making capabilities of the agent gradually. Through each training process, the agent will select an action A_t depending on the current state S_t. Then it will receive a corresponding reward and turns to the next state S_{t+1}. By conducting the procedures, the agent can discover the optimal policy in each discrete state, which contributes to the objective, maximising the expected discounted cumulative reward [26]. During continuous interactions between the agent and the environment, the optimal solution with the maximum cumulative reward can be provided.

Compared with other reinforcement-learning-based algorithms, Q-learning is simpler, which features a Q-function that is used in the online interaction between the agent and the environment. For optimisation problems in power electronics converters, Q-learning algorithm

TABLE 5.4
Comparison between GA, PSO, and Q-Learning Algorithms

Algorithm	Algorithm Application Complexity	Optimisation Speed	Environmental Adaptability	Multiobjective Optimisation Capability	Online Continuous Optimisation
GA	Medium	Medium	Medium	Medium	×
PSO	High	Slow	High	Medium	×
Q-learning	High	Fast	High	High	×

aims to optimise various performance metrics, which are primarily corresponded to the phase-shift angles D_1, D_2, and D_3. Based on the indices of Q-value table, which is utilised to record the experience, Q-learning algorithm can retrieve the optimal phase-shift solution [27]. It can learn modulation methods through trial-and-error and automatic optimisation, which can achieve optimal performance for power electronics converters.

The performance comparison between the aforementioned advanced algorithms are listed in Table 5.4. From Table 5.4, it can be observed that GA algorithm demonstrates moderate performance under various operation conditions. Hence it is applicable to simple modulation methods such as EPS and DPS. PSO algorithm exhibits significant environmental adaptability. Compared with GA, PSO can adjust to the changes in circuit parameters. However, the tendency to converge to local optima makes PSO need to be combined with other methods for utilisation, which increases the complexity. In addition, Q-learning reinforcement learning algorithm demonstrates excellent characteristics, which may be the most competitive among these optimisation algorithms.

5.3.3 Conclusion

This section presents a review of two aspects concerning the research on power electronics converters for MEA, including the modulation methods and optimisation algorithms. Four mostly used modulation methods, SPS, EPS, DPS, and TPS modulation, are discussed in detail. It is found that when the voltage gain ratio is at unity, SPS modulation exhibits superiority. However, when the voltage gain ratio is away from unity, EPS, DPS, and TPS modulation perform better than SPS. Additionally, three typical optimisation algorithms, GA, PSO, and Q-learning reinforcement learning algorithms, are introduced to optimise the galvanic features of power electronics converters. According to the actual situation, appropriate modulation methods and optimisation algorithms can be chosen to control the power electronics converters in MEA for better performance.

5.4 Design Considerations

For improving the performance in the DAB converter used in MEA, some design considerations need to be considered, such as ZVS, current stress, and so on, which will cause conduction loss and reduce the power density. So, it is important to optimise performance, minimise losses, and ensure reliable operation across varying load conditions. This section will list common considerations and principles for DAB design.

5.4.1 ZVS

Zero-voltage switching (ZVS) is a soft-switching technique that enables power semiconductor devices to turn on/off when the voltage across them is zero, effectively eliminating switching losses and improving converter efficiency [28–30]. In DAB converters, ZVS operation is particularly valuable for high-frequency, high-power applications where reduced switching losses and minimised electromagnetic interference (EMI) are critical.

Figure 5.14 shows the state of voltage and current across the switch before the switch is turned on. It takes some time for voltage to return to zero and for current to rise. During this period of time, the voltage and current both are not equal to zero, which will generate useless power. However, the useless power will influence the performance of DAB converter, which is considered as switching loss.

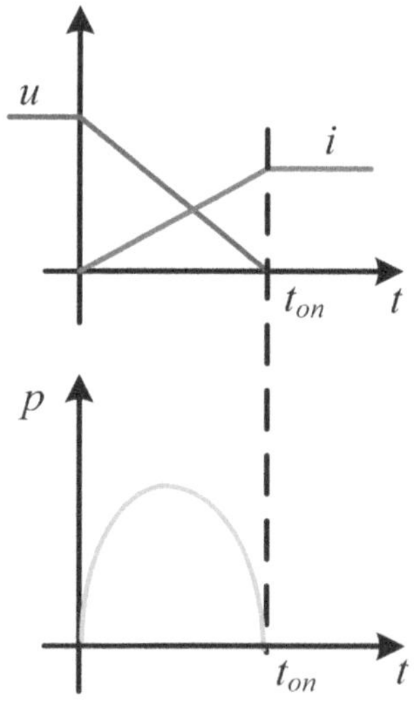

FIGURE 5.14
The status of switch before turning on.

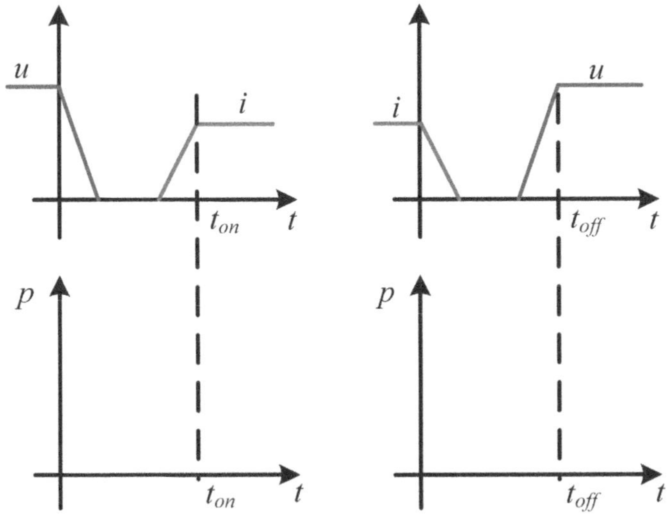

FIGURE 5.15
Soft-switching technology. (a) ZVS. (b) ZCS.

Figure 5.15 illustrates the principles of ZVS and ZCS in DAB converter. ZVS is achieved by leveraging the transformer's leakage inductance and the converter's phase-shift control. Before a switch turns on, the current through the leakage inductance discharges the switch's parasitic capacitance, reducing its voltage to zero, which ensures that the switch is turned on at zero voltage, minimising switching losses and improving efficiency. ZVS typically occurs within a specific load and phase-shift range, where the inductor current remains sufficient to charge or discharge the capacitances during dead-time intervals. Proper design of leakage inductance, dead-time, and switching frequency is critical to maintain ZVS across varying operating conditions.

Zero-current switching (ZCS) is achieved by ensuring that the current through a power switch naturally reaches zero before the switch is turned off, which is facilitated by the phase-shift control between the two active bridges and the energy transfer dynamics of the leakage inductance. When the primary and secondary bridge voltages are aligned, which can reduce the voltage difference across the inductor, the inductor current decays to zero. If the switch is turned off at this instant, ZCS eliminates turn-off losses, reducing switching stress and improving efficiency.

Table 5.5 shows the comparison between ZVS and ZCS, both of which can reduce the switching loss. However, ZCS in DAB converters is often load-dependent and may only occur under specific conditions, such as light loads and certain phase-shift angles, unlike ZVS, which is more commonly prioritised for wide-range soft switching.

TABLE 5.5
Comparison between ZVS and ZCS

Feature	ZVS	ZCS
Switching condition	Turn-on at zero voltage	Turn-off at zero current
Primary loss reduction	Eliminates turn-on losses	Eliminates turn-off losses
Load range	Effective at medium-high loads	Better at light loads
Practical challenges	ZVS loss at light loads	Difficult to maintain ZCS for all switches

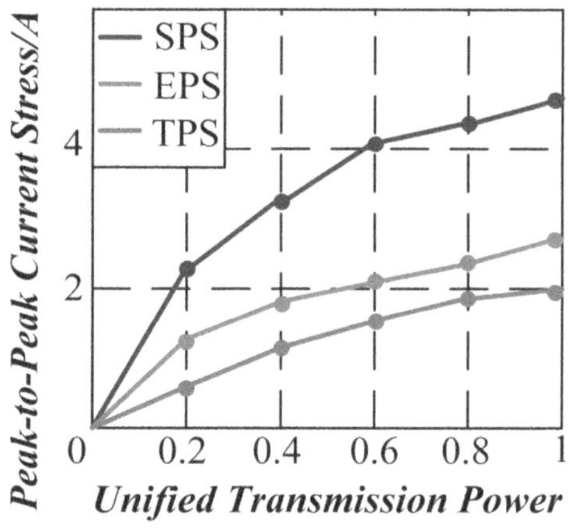

FIGURE 5.16
Peak current stress among different modulation schemes.

5.4.2 Current Stress

In DAB converters, current stress refers to the peak current flowing through semiconductor devices and passive components. The DAB converter relies on phase-shift modulation for power transfer, leading to hard switching and significant current stress under conditions such as large phase-shift angles, voltage mismatches, or light loads [31–33]. Figure 5.16 shows the current stress among different modulation schemes, in which TPS has lower current stress at the same power. This stress arises primarily from reactive power circulation and deviations from the ideal transformer turns ratio, increasing conduction losses and thermal demands on components.

Optimising current stress is essential to enhance efficiency, reliability, and cost-effectiveness. High current stress elevates conduction losses in switches and magnetics, reduces soft-switching capability, and exacerbates thermal management challenges. Over time, unmitigated stress degrades semiconductor lifespan. Effective optimisation strategies include advanced modulation techniques, variable frequency control, and careful magnetic design to balance power transfer and stress reduction.

Addressing current stress directly impacts DAB performance, enabling higher efficiency, improved power density, and broader operational flexibility in applications. By minimising unnecessary current stress, designers can achieve more compact, reliable, and economical high-power converters.

5.4.3 RMS Current

The RMS current stress in DAB converters represents a critical design consideration, fundamentally affecting converter performance and reliability [34–36]. This stress originates primarily from phase-shift modulation characteristics, voltage ratio mismatches, and varying load conditions, all of which contribute to elevated current circulation through power semiconductors and magnetic components. The quadratic relationship between RMS current and conduction losses makes its optimisation essential for achieving high efficiency, particularly in high-power applications where even marginal reductions in RMS current can yield significant efficiency improvements.

Optimisation of RMS current stress delivers multiple system benefits: it enhances overall efficiency by minimising conduction losses, improves thermal performance by reducing component heating, and extends device lifetime by mitigating thermal stress. Furthermore, proper RMS current management enables more compact designs through reduced cooling requirements and prevents the loss of soft-switching capabilities at light loads.

Figure 5.17 illustrates the RMS current among different modulation methods. Various techniques prove effective for optimisation, including advanced

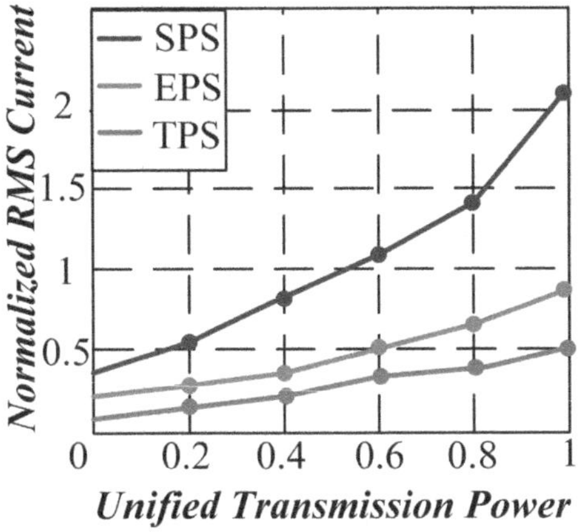

FIGURE 5.17
RMS current among different modulation schemes.

modulation strategies like DPS and TPS control, careful transformer design to achieve voltage matching, and leakage inductance optimisation. These methods collectively work to reduce reactive power circulation while maintaining effective power transfer.

The consequences of unoptimised RMS current stress are substantial, potentially leading to decreased system efficiency below 95% in practical implementations, increased thermal management challenges, and higher overall system costs due to the need for overdesigned components.

In contrast, effective RMS current optimisation contributes to achieving the full potential of DAB converters in demanding applications, where both high efficiency and power density are paramount requirements. The comprehensive management of RMS current stress thus represents a crucial aspect of DAB converter design, directly impacting the converter's performance metrics, reliability, and economic viability.

5.4.4 Backflow Power

Backflow power represents a fundamental challenge in DAB converters, characterised by the circulation of reactive power between converter bridges without contributing to useful energy transfer [37–39]. This phenomenon primarily stems from phase-shift modulation characteristics and voltage mismatches between input and output stages, manifesting as increased current stress and additional losses throughout the system. The presence of significant backflow power directly impacts converter performance by elevating RMS currents, which in turn generates substantial conduction losses proportional to the square of the current magnitude.

The optimisation of backflow power proves essential for multiple aspects of converter design and operation. From an efficiency perspective, even moderate levels of backflow power can degrade overall system efficiency by 3–5% in practical implementations.

Thermally, the resulting circulating currents produce excessive heating in both semiconductor devices and magnetic components, potentially compromising long-term reliability and necessitating oversized cooling solutions. Furthermore, backflow power influences the converter's soft-switching capabilities, where excessive circulation can disrupt zero-voltage switching conditions, particularly during light-load operation, leading to increased switching losses and reduced performance.

Figure 5.18 shows the backflow power among different modulation schemes. TPS can significantly reduce backflow power. Effective mitigation of backflow power involves a combination of strategic approaches. Advanced modulation schemes help decouple power transfer mechanisms from reactive power circulation. Careful attention to voltage matching through transformer design or additional regulation stages minimises the fundamental causes of backflow.

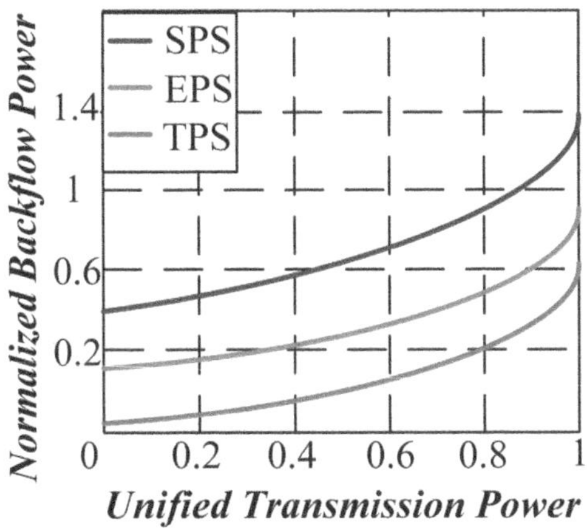

FIGURE 5.18
Backflow power among different modulation schemes.

Magnetic component optimisation, particularly in leakage inductance design, plays a crucial role in balancing energy transfer requirements against circulating current reduction. These optimisation efforts yield significant benefits across multiple performance metrics, including improved efficiency, enhanced thermal performance, and increased power density.

The comprehensive management of backflow power thus represents a vital consideration in DAB converter design, requiring careful analysis of operating conditions and implementation of appropriate mitigation strategies to achieve optimal system performance.

5.4.5 DC Bias Current

Ideally, there is no DC bias current in the HF transformer of DAB converter in the steady states, because the inductor current is continuous and the average inductor voltage is zero in each switching period. The transient state may happen when the phase-shift angles of the gating signals need to be adjusted to respond to the command of load change. During transient state, it can be shown that a DC bias current may result from the improper change of phase shift [40–42].

The presence of DC bias current creates multiple detrimental effects that necessitate careful optimisation. Most critically, it risks transformer core saturation which severely limits power transfer capability and may lead to catastrophic failure. Additionally, DC bias increases core losses through elevated hysteresis and eddy current effects, reduces system efficiency, and causes unbalanced current stress distribution across power switches. The phenomenon also generates undesirable audible noise through degraded current waveform symmetry and potentially interfering with control system

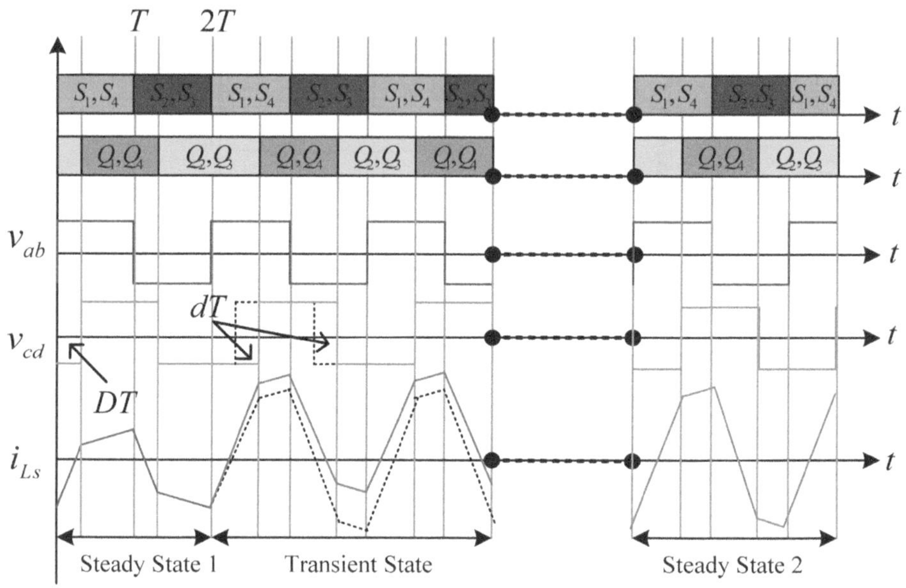

FIGURE 5.19
The DC bias current in DAB converter under SPS control.

operation. These effects become particularly pronounced in high-power applications where even small DC offsets can produce substantial impacts due to large operating currents.

Figure 5.19 shows the DC bias current in DAB converter under SPS control. The original phase shift is *DT*, where *2T* is the switching period. It is assumed that the phase shift is requested to increase by *dT* to transfer more power from the primary side to the secondary side. The gating signals of the primary bridge is assumed fixed and the gating signals of the secondary bridge is delayed by *dT*. The dashed lines represent the original waveforms. The transient duration is defined as the time period between two minimum values of the inductor current, in which the gating signals are modified. It can be derived that the inductor current has a non-zero average value (or a DC bias current) right after the phase-shift angle is changed.

Phase-shift modulation is an effective and widely adopted method to eliminate DC bias current in DAB converters. By ensuring symmetrical switching patterns and incorporating closed-loop compensation, it prevents transformer saturation and improves efficiency without additional hardware.

5.4.6 Conclusion of Design Considerations

The design of a DAB converter involves several critical considerations to ensure optimal performance, efficiency, and reliability. ZVS is essential for minimising switching losses and improving efficiency, particularly under varying load conditions. Current stress must be carefully managed to prevent excessive component ratings and ensure long-term durability. Additionally, RMS current plays a significant role in determining conduction losses and thermal performance, directly impacting the converter's efficiency.

Another crucial aspect is the backflow power, which, if unmitigated, can lead to increased circulating energy and reduced system efficiency. Proper modulation techniques and phase-shift control are necessary to minimise this effect. Furthermore, the DC bias current must be considered, as it influences transformer saturation and overall converter stability. Addressing these factors collectively ensures a balanced trade-off between efficiency, power density, and operational reliability in DAB converter design.

References

[1] P. Chatterjee and M. Hermwille, "Tackling the challenges of electric vehicle fast charging," *ATZelectronics worldwide*, vol. 15, no. 10, pp. 18–22, Oct. 2020, https://doi.org/10.1007/s38314-020-0263-6.

[2] S. Dusmez, A. Hasanzadeh and A. Khaligh, "Comparative analysis of bidirectional three-level DC-DC converter for automotive applications," *IEEE Transactions on Industrial Electronics*, vol. 62, no. 5, pp. 3305–3315, May 2015, https://doi.org/10.1109/TIE.2014.2336605.

[3] Z. Zhang et al., "High-efficiency Silicon Carbide (SiC) converter using paralleled discrete devices in energy storage systems," *2019 IEEE Energy Conversion Congress and Exposition (ECCE)*, 2019, pp. 2471–2477, https://doi.org/10.1109/ECCE.2019.8912733.

[4] L. Tan, N. Zhu, and B. Wu, "An integrated inductor for eliminating circulating current of parallel three-level DC-DC converter-based EV fast charger," *IEEE Transactions on Industrial Electronics*, vol. 63, no. 3, pp. 1362–1371, March 2016, https://doi.org/10.1109/TIE.2015.2496904.

[5] B. Zhao, Q. Song, W. Liu and Y. Sun, "Overview of dual-active-bridge isolated bidirectional DC-DC converter for high-frequency-link power-conversion system," *IEEE Transactions on Power Electronics*, vol. 29, no. 8, pp. 4091–4106, Aug. 2014, https://doi.org/10.1109/TPEL.2013.2289913.

[6] L. Tarisciotti, A. Costabeber, L. Chen, A. Walker, and M. Galea, "Current-fed isolated DC/DC converter for future aerospace microgrids," *IEEE Transactions on Industry Applications*, vol. 55, no. 3, pp. 2823–2832, May–June 2019, https://doi.org/10.1109/TIA.2018.2889845.

[7] H. Ramakrishnan and J. Rangaraju, "Power topology considerations for electric vehicle charging stations," Texas Instrum., Dallas, TX, USA, Appl. Rep. SLLA497, 2020.

[8] J. A. Sabate, V. Vlatkovic, R. B. Ridley, and F. C. Lee, "High-voltage, high-power, ZVS, full-bridge PWM converter employing an active snubber," *Proc. APEC 6th Annu. Appl. Power Electron. Conf. Exhib.*, 1991, pp. 158–163, https://doi.org/10.1109/APEC.1991.146157.

[9] H. Tu, H. Feng, S. Srdic, and S. Lukic, "Extreme fast charging of electric vehicles: A technology overview," *IEEE Transactions on Transportation Electrification*, vol. 5, no. 4, pp. 861–878, Dec. 2019, https://doi.org/10.1109/TTE.2019.2958709

[10] R. Liu, L. Xu, Y. Kang, Y. Hui, and Y. Li, "Decoupled TAB converter with energy storage system for HVDC power system of more electric aircraft," *The Journal of Engineering*, vol. 2018, no. 15, pp. 593–602, 2018, https://doi.org/10.1049/joe.2018.0033.

[11] Q. Zhao, J. Zhang, Y. Gao, D. Wang, and Q. Yang, "Hybrid variable frequency LLC resonant converter with wide output voltage range," *IEEE Transactions on Power Electronics*, vol. 38, no. 9, pp. 11038–11049, Sept. 2023, https://doi.org/10.1109/TPEL.2023.3283548.

[12] X. Zhang et al., "Design and modeling of CLLC converter for bidirectional on-board charger," *IEEE Transactions on Industry Applications*, vol. 59, no. 5, pp. 6095–6102, Sept.–Oct. 2023, https://doi.org/10.1109/TIA.2023.3281300.

[13] X. Chen, G. Xu, H. Han, D. Liu, Y. Sun, and M. Su, "Light-load efficiency enhancement of high-frequency dual-active-bridge converter under SPS control," *IEEE Transactions on Industrial Electronics*, vol. 68, no. 12, pp. 12941–12946, Dec. 2021, https://doi.org/10.1109/TIE.2020.3044803.

[14] C. Sun, X. Jiang, J. Liu, L. Cao, Y. Yang, and K. H. Loo, "A unified design approach of optimal transient single-phase-shift modulation for nonresonant dual-active-bridge converter with complete transient DC-offset elimination," *IEEE Transactions on Power Electronics*, vol. 37, no. 11, pp. 13217–13237, Nov. 2022, https://doi.org/10.1109/TPEL.2022.3182966.

[15] J. Tian, F. Wang, F. Zhuo, Y. Wang, H. Wang, and Y. Li, "A zero-backflow-power EPS control scheme with multiobjective coupled-relationship optimization in DAB-based converter," *IEEE Journal of Emerging and Selected Topics in Power Electronics*, vol. 10, no. 4, pp. 4128–4145, Aug. 2022, https://doi.org/10.1109/JESTPE.2021.3126435.

[16] G. Xu, L. Li, X. Chen, Y. Liu, Y. Sun, and M. Su, "Optimized EPS control to achieve full load range ZVS with seamless transition for dual active bridge converters," *IEEE Transactions on Industrial Electronics*, vol. 68, no. 9, pp. 8379–8390, Sept. 2021, https://doi.org/10.1109/TIE.2020.3014562.

[17] N. Fu, J. Deng, Z. Wang, and D. Chen, "Dual-phase-shift control strategy with switch-controlled capacitor for overall efficiency optimization in wireless power transfer system," *IEEE Transactions on Vehicular Technology*, vol. 72, no. 6, pp. 7304–7317, June 2023, https://doi.org/10.1109/TVT.2023.3241695.

[18] Z. Guo and M. Li, "An optimized DPS control strategy for LCL resonant dual active bridge converter for wide voltage conversion ratio," *IEEE Journal of Emerging and Selected Topics in Industrial Electronics*, vol. 2, no. 4, pp. 501–512, Oct. 2021, https://doi.org/10.1109/JESTIE.2021.3088383.

[19] T. Luo et al., "An optimal VF-TPS modulation scheme for CLLC converter with minimized RMS current," *IEEE Transactions on Power Electronics*, vol. 40, no. 3, pp. 4170–4185, March 2025, https://doi.org/10.1109/TPEL.2024.3495828.

[20] F. An, W. Song, K. Yang, S. Yang and L. Ma, "A simple power estimation with triple phase-shift control for the output parallel DAB DC–DC converters in power electronic traction transformer for railway locomotive application," *IEEE Transactions on Transportation Electrification*, vol. 5, no. 1, pp. 299–310, March 2019, https://doi.org/10.1109/TTE.2018.2876057.

[21] F. Lin, X. Zhang, X. Li, C. Sun, W. Cai and Z. Zhang, "Automatic triple phase-shift modulation for DAB converter with minimized power loss," *IEEE Transactions on Industry Applications*, vol. 58, no. 3, pp. 3840–3851, May–June 2022, https://doi.org/10.1109/TIA.2021.3136501.

[22] L. Zhou, Y. Gao, H. Ma, and P. T. Krein, "Wide-load range multiobjective efficiency optimization produces closed-form control solutions for dual active bridge converter," *IEEE Transactions on Power Electronics*, vol. 36, no. 8, pp. 8612–8616, Aug. 2021, https://doi.org/10.1109/TPEL.2021.3054970.

[23] L. Meng, T. Dragicevic, J. C. Vasquez, and J. M. Guerrero, "Tertiary and secondary control levels for efficiency optimization and system damping in droop controlled DC–DC converters," *IEEE Transactions on Smart Grid*, vol. 6, no. 6, pp. 2615–2626, Nov. 2015, https://doi.org/10.1109/TSG.2015.2435055.

[24] H. Shi, H. Wen, Y. Hu, and L. Jiang, "Reactive power minimization in bidirectional DC–DC converters using a unified-phasor-based particle swarm optimization," *IEEE Transactions on Power Electronics*, vol. 33, no. 12, pp. 10990–11006, Dec. 2018, https://doi.org/10.1109/TPEL.2018.2811711.

[25] O. M. Hebala, A. A. Aboushady, K. H. Ahmed, and I. Abdelsalam, "Generic closed-loop controller for power regulation in dual active bridge DC–DC converter with current stress minimization," *IEEE Transactions on Industrial Electronics*, vol. 66, no. 6, pp. 4468–4478, June 2019, https://doi.org/10.1109/TIE.2018.2860535.

[26] Y. Tang et al., "Reinforcement learning based efficiency optimization scheme for the DAB DC–DC converter with triple-phase-shift modulation," *IEEE Transactions on Industrial Electronics*, vol. 68, no. 8, pp. 7350–7361, Aug. 2021, https://doi.org/10.1109/TIE.2020.3007113.

[27] Y. Tang et al., "RL-ANN-based minimum-current-stress scheme for the dual-active-bridge converter with triple-phase-shift control," *IEEE Journal of Emerging and Selected Topics in Power Electronics*, vol. 10, no. 1, pp. 673–689, Feb. 2022, https://doi.org/10.1109/JESTPE.2021.3071724.

[28] M. Mahdavifard, N. Mazloum, F. Zahin, A. KhakparvarYazdi, A. Abasian, and S. A. Khajehoddin, "An asymmetrical DAB converter modulation and control systems to extend the ZVS range and improve efficiency," *IEEE Transactions on Power Electronics*, vol. 37, no. 10, pp. 12774–12792, Oct. 2022, https://doi.org/10.1109/TPEL.2022.3177401.

[29] G. Xu, D. Sha, Y. Xu, and X. Liao, "Dual-transformer-based DAB converter with wide ZVS range for wide voltage conversion gain application," *IEEE Transactions on Industrial Electronics*, vol. 65, no. 4, pp. 3306–3316, April 2018, https://doi.org/10.1109/TIE.2017.2756601.

[30] X. Yang, J. Wang, Y. Du, C. Liu, T. Zhang, and J. Zhang, "Bidirectional ZVS operation of all switches for a DAB converter over a full range of loads with optimized current stress," *IEEE Transactions on Industry Applications*, vol. 60, no. 1, pp. 1183–1195, Jan.–Feb. 2024, https://doi.org/10.1109/TIA.2023.3292322.

[31] J. Tian, F. Wang, F. Zhuo, X. Cui, and D. Yang, "An optimal primary-side duty modulation scheme with minimum peak-to-peak current stress for DAB-based EV applications," *IEEE Transactions on Industrial Electronics*, vol. 70, no. 7, pp. 6798–6808, July 2023, https://doi.org/10.1109/TIE.2022.3206698.

[32] Q. Gu, L. Yuan, J. Nie, J. Sun, and Z. Zhao, "Current stress minimization of dual-active-bridge DC–DC converter within the whole operating range," *IEEE Journal of Emerging and Selected Topics in Power Electronics*, vol. 7, no. 1, pp. 129–142, March 2019, https://doi.org/10.1109/JESTPE.2018.2886459.

[33] D. Guo, P. Wang, C. Ren, and J. M. Guerrero, "Linearized minimum current stress modulation scheme of single-phase bidirectional DAB AC–DC converters," *IEEE Transactions on Industrial Electronics*, vol. 70, no. 12, pp. 12410–12420, Dec. 2023, https://doi.org/10.1109/TIE.2023.3239863.
[34] A. Tong, L. Hang, G. Li, X. Jiang, and S. Gao, "Modeling and analysis of a dual-active-bridge-isolated bidirectional DC/DC converter to minimize RMS current with whole operating range," *IEEE Transactions on Power Electronics*, vol. 33, no. 6, pp. 5302–5316, June 2018, https://doi.org/10.1109/TPEL.2017.2692276.
[35] H. Xie, F. Cai, J. Jiang, and D. Li, "A fundamental harmonic analysis-based optimized scheme for DAB converters with lower RMS current and wider ZVS range," *IEEE Transactions on Transportation Electrification*, vol. 9, no. 3, pp. 4045–4058, Sept. 2023, https://doi.org/10.1109/TTE.2022.3231865.
[36] G. Xu, J. Tang, L. Zhang, W. Xiong, Y. Sun, and M. Su, "A hybrid extended phase shift modulation strategy for DAB converter with DC blocking capacitor to extend ZVS range and reduce RMS current," *IEEE Journal of Emerging and Selected Topics in Power Electronics*, vol. 10, no. 5, pp. 6192–6207, Oct. 2022, https://doi.org/10.1109/JESTPE.2022.3163290.
[37] J. Tian, F. Wang, F. Zhuo, Y. Wang, H. Wang, and Y. Li, "A zero-backflow-power EPS control scheme with multiobjective coupled-relationship optimization in DAB-based converter," *IEEE Journal of Emerging and Selected Topics in Power Electronics*, vol. 10, no. 4, pp. 4128–4145, Aug. 2022, https://doi.org/10.1109/JESTPE.2021.3126435.
[38] B. Zhou, X. Yang, Z. Zhang, Z. Li, T. Q. Zheng, and X. You, "Backflow power optimization of DAB with gradient descent algorithm based extended-phase-shift control in EER application," 2019 IEEE 3rd International Electrical and Energy Conference (CIEEC), Beijing, China, 2019, pp. 1737–1742, https://doi.org/10.1109/CIEEC47146.2019.
[39] F. Xie, H. Wang, and J. Shu, "Backflow power optimization of DAB based on asymmetric duty cycle and internal phase shift control," *IEEE Access*, vol. 11, pp. 25682–25688, 2023, https://doi.org/10.1109/ACCESS.2023.3257057.
[40] X. Li and Y. -F. Li, "An optimized phase-shift modulation for fast transient response in a dual-active-bridge converter," *IEEE Transactions on Power Electronics*, vol. 29, no. 6, pp. 2661–2665, June 2014, https://doi.org/10.1109/TPEL.2013.2294714.
[41] Q. Bu, H. Wen, H. Shi, and Y. Zhu, "A comparative review of high-frequency transient DC bias current mitigation strategies in dual-active-bridge DC–DC converters under phase-shift modulations," *IEEE Transactions on Industry Applications*, vol. 58, no. 2, pp. 2166–2182, March–April 2022, https://doi.org/10.1109/TIA.2021.3136498.
[42] Q. Bu, H. Wen, J. Wen, Y. Hu, and Y. Du, "Transient DC bias elimination of dual-active-bridge DC–DC converter with improved triple-phase-shift control," *IEEE Transactions on Industrial Electronics*, vol. 67, no. 10, pp. 8587–8598, Oct. 2020, https://doi.org/10.1109/TIE.2019.2947809.

6

Towards Sustainable Aviation: Power Inverters Enabling Aircraft Electrification

Sze Sing Lee, Sara Hasanpour, Kazem Varesi, Felipe B. Grigoletto, and Akshay Kumar Rathore

6.1 Introduction to Aircraft Electrical Power Systems

The electrification of aircraft power systems represents one of the most transformative developments in modern aviation. Historically, aircraft have relied on conventional fuel-based propulsion systems and extensive hydraulic and pneumatic subsystems to manage flight operations. However, the demand for higher fuel efficiency, lower emissions, reduced maintenance, and enhanced reliability has driven a shift towards more-electric aircraft (MEA) and ultimately all-electric aircraft (AEA). This evolution is primarily enabled by power electronics, which allows for efficient conversion, distribution, and management of electrical energy across propulsion and non-propulsion subsystems.

This section provides an introduction of the aircraft electric power system, detailing the transition from traditional fuel-based architectures to increasingly electrified configurations. Three key figures are used to illustrate this progression:

- Figure 6.1: Classification of the aircraft power system, showing the division into propulsion and non-propulsive subsystems.
- Figure 6.2: Evolution of propulsion technologies, highlighting the increasing role of electric drive systems.
- Figure 6.3: Variations in on-board power distribution architectures, from conventional AC-based distribution to high-voltage DC architectures supporting electrified propulsion.

DOI: 10.1201/9781003747994-6

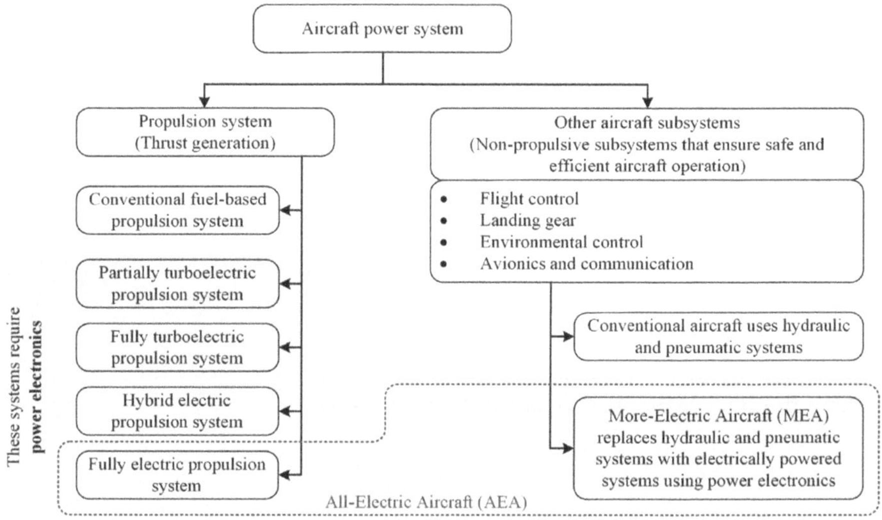

FIGURE 6.1
Overview and classification of propulsion and non-propulsion components in the aircraft power system.

6.1.1 Propulsion Systems

As depicted in Figure 6.1 and further detailed in Figure 6.2, aircraft propulsion systems can be categorised into five levels of electrification:

- *Conventional Fuel-Based Propulsion System*

  In the conventional propulsion system, energy from fuel is directly converted into mechanical thrust by combustion-driven jet engines. The fundamental principle behind these engines is the thermodynamic Brayton cycle, where air is compressed, mixed with fuel, and ignited in a combustion chamber to produce high-temperature, high-pressure gases. These gases expand through turbines, driving a fan or propeller to generate thrust. This propulsion system is entirely mechanical, relying on direct shaft-driven propulsion, where the turbine's rotational energy is mechanically transmitted to the propulsion system. Despite their reliability, conventional propulsion systems face key limitations, including high fuel consumption, energy losses, heavy reliance on hydraulic/pneumatic systems, and rigid power distribution constrained by mechanical transmission. These drawbacks drive the shift towards electrified propulsion, which offers improved efficiency, reduced emissions, and greater flexibility.

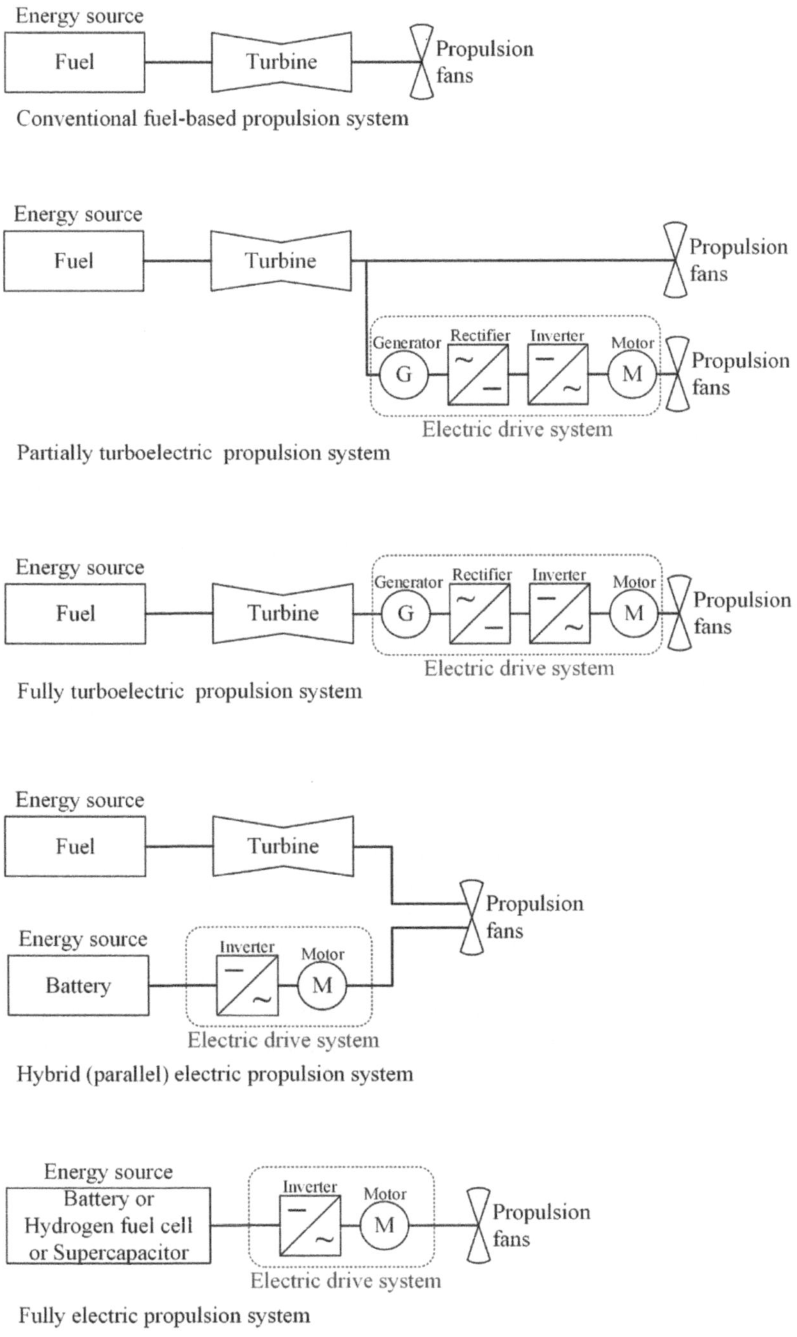

FIGURE 6.2
Types of aircraft propulsion system [1].

- *Partially Turboelectric Propulsion System*

  This system integrates an electric drive alongside the conventional turbine-driven propulsion, creating a hybrid power architecture. A generator extracts mechanical power from the turbine and converts it into electrical energy via rectifiers and inverters, which then supply an auxiliary electric motor to assist in propulsion. This propulsion system enables dynamic power-sharing between mechanical and electrical components, improving operational flexibility, fuel efficiency, and load management. By partially electrifying propulsion, this system reduces mechanical constraints, optimises energy utilisation, and serves as a stepping-stone towards fully turboelectric and hybrid-electric aircraft architectures.
- *Fully Turboelectric Propulsion System*

  In this system, the gas turbine does not directly drive the propulsion fans. Instead, it functions as a power source, where an electric machine (generator) extracts mechanical power from the turbine and converts it into electrical power. This electricity is then processed by power electronic converters (rectifiers and inverters) and supplied to electric motors that drive the propulsion fans. The decoupling of mechanical and electrical power enables distributed propulsion, where multiple electric motors can be strategically placed along the airframe to enhance aerodynamic efficiency, thrust distribution, and system redundancy.
- *Hybrid-Electric Propulsion System*

  This system combines conventional fuel-based propulsion with a battery-powered electric drive, creating a versatile dual-energy architecture. The integration of both energy sources allows for multiple operational modes, such as battery-assisted take-off for reduced fuel consumption and fuel-based cruising for extended range. This configuration enhances overall efficiency, reduces emissions, and optimises power management by enabling smart energy distribution between the turbine and the electric drive. Hybrid-electric propulsion serves as a key transitional technology towards fully electric aircraft, balancing performance, sustainability, and operational feasibility.
- *Fully Electric Propulsion System*

  This system completely eliminates combustion engines, relying exclusively on electrical energy from batteries, fuel cells, or supercapacitors to power electric motors. At the core of this architecture are high-performance power electronics, particularly inverters, which efficiently convert and regulate electrical energy to drive propulsion motors. By removing fuel dependency, fully electric propulsion significantly reduces emissions, noise, and maintenance complexity, making it a key enabler of sustainable aviation. However, advancements in energy storage, power density, and thermal management remain crucial for scaling this technology to larger commercial aircraft.

6.1.2 Non-Propulsive Subsystems

Apart from propulsion, aircraft rely on non-propulsive subsystems to support safe and efficient operations. These systems include flight control, landing gear operation, environmental control, and avionics, all of which have historically depended on hydraulic and pneumatic actuation powered by bleed air from engines or mechanical pumps. However, with the advancement of more-electric aircraft (MEA) technology, these traditional systems are being gradually replaced by electrically driven alternatives, as illustrated in Figure 6.1.

The transition to MEA architecture aims to eliminate or significantly reduce reliance on hydraulic and pneumatic actuation by implementing power electronics–based electrical solutions:

- *Electric Flight Control Actuation:* Traditional hydraulic actuators used for operating ailerons, elevators, rudders, and spoilers are replaced with electro-hydrostatic actuators (EHA) and electromechanical actuators (EMA). These systems enhance precision, reliability, and energy efficiency, while reducing maintenance and eliminating hydraulic fluid leaks.
- *Electric Landing Gear Systems:* MEA replaces hydraulic landing gear retraction and braking systems with electrically actuated systems, improving controllability and reducing system weight.
- *Electric Environmental Control Systems (ECS):* Cabin pressurisation, air conditioning, and anti-icing systems traditionally powered by engine bleed air are now electrically operated using high-efficiency electric compressors and thermal management systems. This shift improves fuel efficiency by eliminating bleed-air losses.
- *Advanced Avionics and Power Management:* MEA employs power electronic converters to ensure precise voltage regulation and distribution, supporting highly integrated avionics, radar, communication, and navigation systems.

By transitioning to fully electric non-propulsive systems, MEA enhances efficiency, reliability, and maintainability, contributing to the long-term goal of all-electric aircraft (AEA).

6.1.3 Aircraft Power Distribution Architectures

As aircraft propulsion and non-propulsive systems become increasingly electrified, the power distribution architecture must evolve to efficiently manage electrical energy across various subsystems. Historically, aircraft have relied on AC power distribution, but advancements in power electronics have enabled the transition towards high-voltage DC (HVDC) architectures, which offer improved efficiency and compatibility with electric propulsion.

In conventional aircraft, power is primarily distributed using constant frequency AC (CF-AC) systems, as shown in the top-right diagram of Figure 6.3. Here, electrical power is generated by constant-speed generators

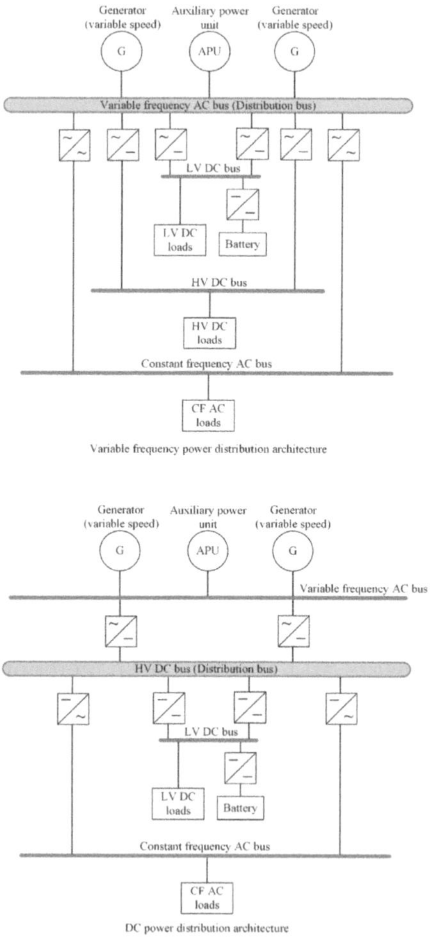

FIGURE 6.3
Types of aircraft power distribution architecture [2, 3].

and supplied through an AC distribution bus. To eliminate the need for constant-speed generators, variable frequency AC (VF-AC) power distribution has been introduced, as seen in the top-left diagram of Figure 6.3. This system allows variable-speed generators to dynamically adjust frequency, enhancing fuel efficiency and reducing mechanical complexity.

Common AC voltage buses in aircraft power systems include the following:

- *Constant-Frequency AC Bus:* 115 V/230 V AC, 400 Hz
- *Variable-Frequency AC Bus:* 115 V/230 V AC, 350–800 Hz

To support aircraft electrification, DC buses are integrated to efficiently distribute electrical energy across various on-board systems. These buses provide stable and efficient power transmission, reducing energy losses and optimising overall system performance.

Common DC voltage buses in modern aircraft power systems include the following:

- *Low-Voltage DC Bus (28 V DC):* Powers avionics, lighting, communication systems, and small auxiliary loads
- *High-Voltage DC Bus (270 V DC):* Supports high-energy subsystems, including electrified propulsion and large electrical loads

The HVDC bus plays a crucial role in hybrid-electric and fully electric propulsion systems by supplying high-voltage power to electric motors via inverters. It also facilitates efficient energy transmission, minimising resistive losses and cable weight. Additionally, HVDC is essential for high-power subsystems such as electric environmental control systems (ECS), high-power radar, and advanced actuation systems. Future aircraft may adopt ultra-high-voltage DC (≥540 V DC) to further optimise power efficiency and propulsion scalability.

The LVDC bus, on the other hand, ensures reliable power for non-propulsive subsystems. It is particularly critical for avionics, flight control actuation, cabin lighting, and auxiliary loads, where stable and noise-free power is essential. The LVDC bus also serves as a redundant power source, ensuring continued operation of safety-critical systems in case of a power failure.

By integrating HVDC and LVDC buses, modern aircraft achieve greater efficiency, weight savings, and improved power management, marking a key step towards the realisation of more-electric aircraft and all-electric aircraft (AEA).

The evolution of aircraft electric power systems marks a paradigm shift in aviation, driven by the increasing adoption of power electronics as the key enabler of electrification. This section has explored the transformation from conventional fuel-based architectures to more-electric and all-electric aircraft, emphasising the role of DC power distribution, high-voltage DC buses, and electrified propulsion systems. A critical component in this electrification journey is the power inverter, which enables the efficient conversion of power from DC buses into AC power for electric propulsion and on-board systems. Power inverters facilitate precise control of electric motors, optimise energy usage, and enhance the overall efficiency of aircraft power systems.

As aircraft electrification advances, the subsequent sections of this chapter will explore power inverters, covering conventional two-level inverters, advanced multilevel inverter topologies, and emerging power conversion technologies designed to enhance voltage gain and efficiency. A deep understanding of these power electronics innovations is crucial for enabling the next generation of sustainable aviation, accelerating the transition towards fully electric aircraft.

6.2 Conventional DC–AC Power Inverters

6.2.1 Two-Level Inverter

The two-level inverter is one of the most fundamental and widely used power electronic converters in aircraft electrification and motor drive applications. It serves as the primary interface for converting DC power from an energy source (e.g., batteries, fuel cells, or HVDC buses) into AC power, which is essential for driving electric motors in MEA and AEA. The name "two-level" originates from its ability to produce two distinct voltage levels at the output. Despite its simplicity, the two-level inverter remains a crucial technology due to its high reliability, ease of control, and compatibility with various modulation techniques.

The circuit diagram of a two-level inverter is shown in Figure 6.4. It operates by switching the power semiconductor devices to generate an AC voltage waveform from the DC supply. The inverter generates two voltage levels ($\pm 0.5V_{dc}$) by controlling the conduction states of the power switches. The basic switching states for single-phase (phase A is considered as an example) are shown in Figure 6.5, demonstrating how different switching combinations result in positive and negative voltage levels.

- State [+1]: When the upper switch (S_{a1}) is ON and the lower switch (S_{a2}) is OFF, the output voltage (v_{an}) is $+0.5V_{dc}$.
- State [–1]: When the lower switch (S_{a2}) is ON and the upper switch (S_{a1}) is OFF, the output voltage (v_{an}) is $-0.5V_{dc}$.

By alternating between these states at a high frequency using pulse width modulation (PWM), a quasi-sinusoidal AC voltage waveform is synthesised. As illustrated in Figure 6.6, the PWM technique regulates the switching of

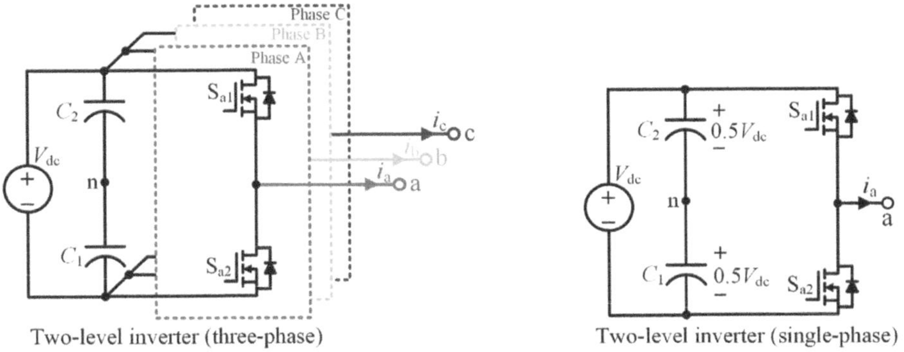

FIGURE 6.4
Two-level inverter.

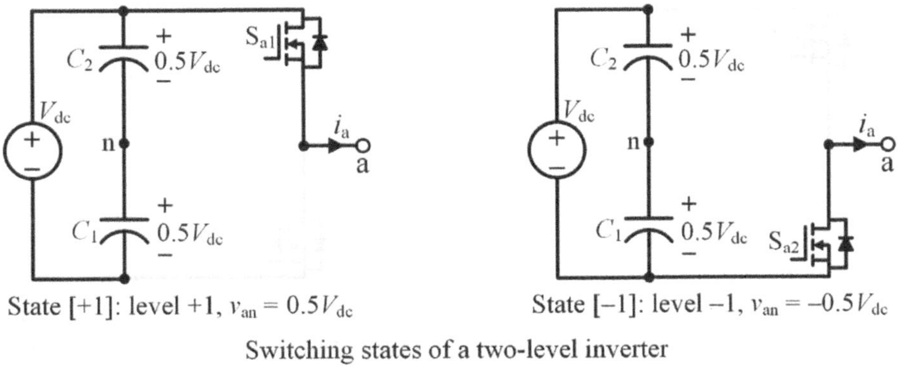

FIGURE 6.5
Switching states of a two-level inverter.

power switches to control the inverter output voltage. The figure shows the following key aspects of the PWM process:

- Sinusoidal Reference and Triangular Carrier Waveforms: The dashed waveform represents the sinusoidal reference signal, which defines the desired fundamental AC output voltage. The solid triangular waveform acts as the high-frequency carrier signal.
- Switching Signals (S_{a1} *and* S_{a2}): The switching signals of S_{a1} (upper switch) and S_{a2} (lower switch) are determined by comparing the sinusoidal reference with the triangular carrier.
- When the sinusoidal reference is higher than the carrier signal, S_{a1} is ON, and the output voltage is $+0.5V_{dc}$.

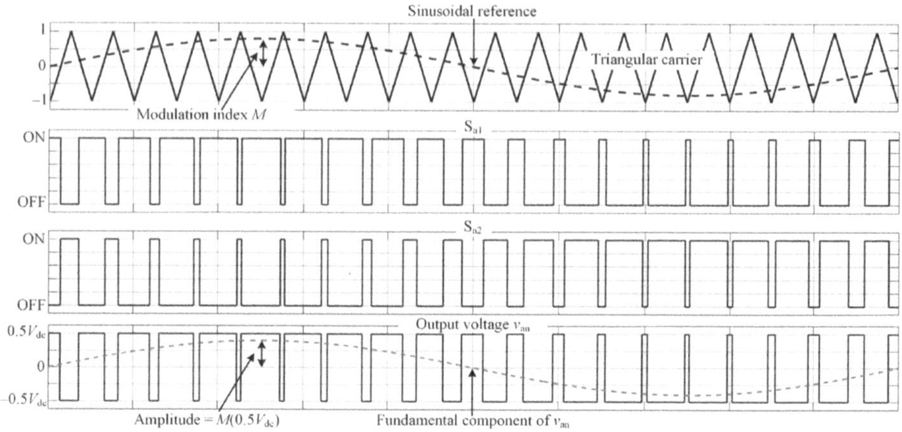

FIGURE 6.6
PWM for a two-level inverter (single-phase).

- When the sinusoidal reference is lower than the carrier signal, S_{a2} is ON, and the output voltage is $-0.5V_{dc}$.
- Modulation Index (M): Defines the amplitude relationship between the reference and carrier signals.
- Output Voltage (v_{an}): The bottom waveform in Figure 6.6 shows the resulting PWM-controlled output voltage, which is a high-frequency switched waveform that approximates a sinusoidal AC voltage.
- Fundamental Component of v_{an}: The red dashed curve represents the fundamental AC voltage component, which is extracted through filtering. The amplitude of the fundamental output voltage is given by $M \times (0.5V_{dc})$.

The two-level inverter is widely used due to its simplicity, reliability, and ease of control, making it an essential power conversion technology in aircraft electrification. Its fast-switching capability, especially with wide-bandgap semiconductors (SiC/GaN), allows for high efficiency and compact designs. However, its high-voltage stress on power switches and significant harmonic distortion present limitations. These challenges have driven the development of multilevel inverter topologies, which offer enhanced voltage handling, reduced harmonics, and improved overall efficiency.

6.2.2 Multilevel Inverters

6.2.2.1 Neutral Point Clamped Multilevel Inverter

In a two-level inverter, each power switch must block the full DC-link voltage when in the OFF state. When the system operates at voltages higher than the blocking capability of available power switches, alternative solutions must be considered. One approach is to connect multiple power switches in series to achieve the required blocking voltage. However, this method introduces control complexity, uneven voltage sharing among switches, and limitations on the maximum switching frequency of the inverter.

A more innovative and practical solution is to divide the DC voltage into multiple sections (i.e., levels) and use these lower-voltage sections to generate an output waveform with multiple steps. This principle forms the foundation of multilevel power converters, which offer significant improvements in system performance through the following:

- Reducing voltage stress on individual switches, allowing the use of lower-rated power switches.
- Improving waveform quality through multi-stepped voltage synthesis, which reduces harmonic distortion.
- Enhancing power conversion efficiency, especially in medium- and high-voltage applications.

The neutral point–clamped (NPC) multilevel inverter [4], as shown in Figure 6.7, is one of the earliest and most widely adopted multilevel inverter topologies. It builds upon this concept by introducing the following:

- An additional neutral point, which divides the DC-link voltage into smaller voltage steps, reducing voltage stress on switches.
- Clamping diodes, which provide voltage balancing and prevent excessive voltage stress across power switches.
- Additional switching states, allowing finer control over the output voltage and improving waveform quality.

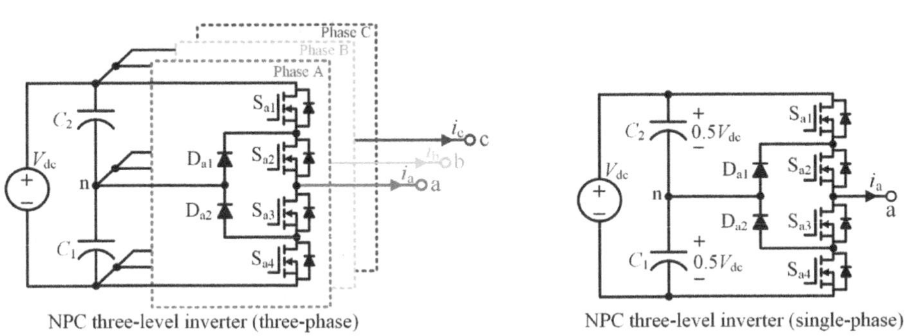

FIGURE 6.7
NPC three-level inverter.

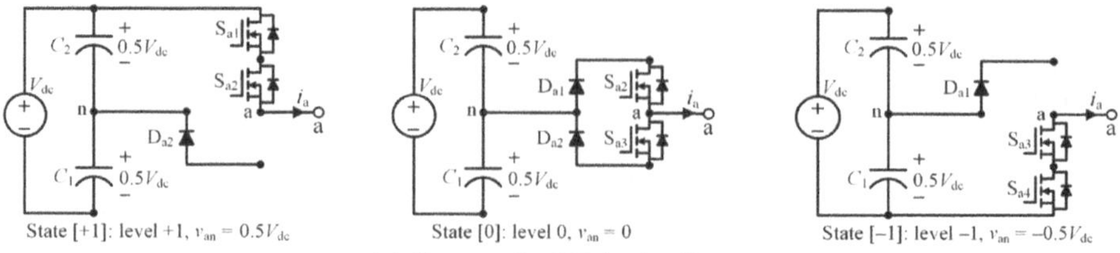

FIGURE 6.8
Switching states of an NPC three-level inverter.

The NPC three-level inverter generates three distinct voltage levels at the output terminal ($+0.5V_{dc}$, 0, $-0.5V_{dc}$) by switching its power switches appropriately. As illustrated in Figure 6.8, the three possible switching states for each phase are as follows:

- State [+1]: Output voltage $v_{an} = +0.5V_{dc}$

  Switches S_{a1} and S_{a2} are ON, connecting the output terminal a to the $+0.5V_{dc}$ level.

  Switches S_{a3} and S_{a4} are OFF, isolating the lower part of the circuit.

  Current flows through S_{a1} and S_{a2} to the output phase a, providing a positive voltage.
- State [0]: Output voltage $v_{an} = 0$

  Switches S_{a2} and S_{a3} are ON, while S_{a1} and S_{a4} are OFF.

  The output terminal a is connected to the neutral point n via clamping diodes D_{a1} and D_{a2}.

  This switching state creates an intermediate zero-voltage level, improving waveform quality and reducing harmonic distortion.
- State [−1]: Output voltage $v_{an} = -0.5V_{dc}$

  Switches S_{a3} and S_{a4} are ON, connecting the output terminal a to the $-0.5V_{dc}$ level.

  Switches S_{a1} and S_{a2} are OFF, isolating the upper part of the circuit.

  Current flows through S_{a3} and S_{a4}, delivering a negative voltage to the output.

To generate a three-level output voltage waveform, the NPC inverter requires a PWM technique that differs from conventional two-level inverter PWM. Unlike the two-level inverter, which uses a single triangular carrier waveform, the three-level inverter requires two triangular carriers to define the additional voltage level. There are two primary PWM strategies used for multilevel inverters:

- *Level-Shifted PWM (LSPWM):* Multiple carriers are vertically shifted to define voltage levels.
- *Phase-Shifted PWM (PSPWM):* Multiple carriers are phase-shifted to balance the switching among switches.

Figure 6.9 illustrates a LSPWM approach, where two triangular carrier waves are employed to generate the switching signals for the NPC three-level inverter. The key waveforms are as follows:

- Sinusoidal Reference Signal: The dashed waveform represents the fundamental AC waveform that the inverter aims to generate.
- Triangular Carrier 1 and Triangular Carrier 2: These two triangular waves are vertically shifted (level-shifted PWM) to define the three voltage levels.
- Switching Signals: The modulation process determines the ON/OFF states of the power switches:

  S_{a1} turns ON when the sinusoidal reference is higher than Carrier 1.

  S_{a4} turns ON when the sinusoidal reference is lower than Carrier 2.

  S_{a3} is complementary to S_{a1}.

  S_{a2} is complementary to S_{a4}.
- Output Voltage (v_{an}): The bottom waveform illustrates the resulting three-level output voltage, switching between $+0.5V_{dc}$, 0, and $-0.5V_{dc}$.
- The fundamental AC component is shown as the dashed red curve, demonstrating how PWM approximates a sinusoidal output. The amplitude of the fundamental output voltage is given by $M \times (0.5V_{dc})$.

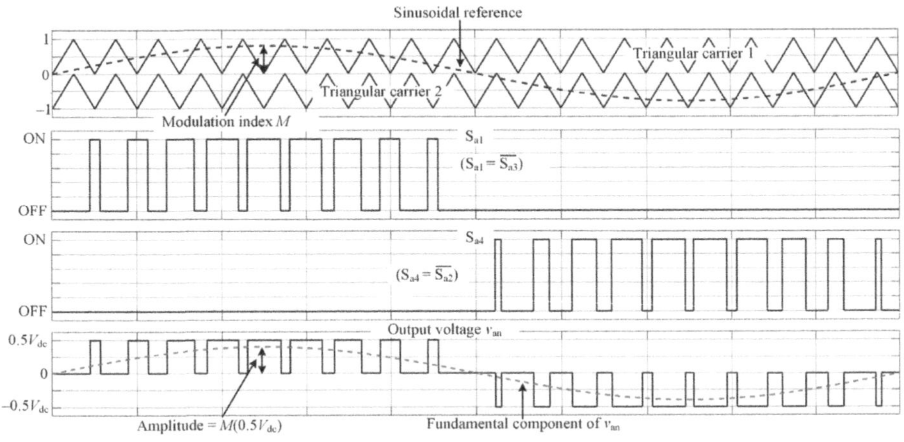

FIGURE 6.9
PWM for an NPC three-level inverter.

The diodes in the NPC inverter can be replaced with power switches, forming the active neutral point–clamped (ANPC) inverter, which allows for better control flexibility, optimised switching sequences, and reduced conduction losses. This enhancement improves efficiency and thermal distribution, making ANPC a more effective solution for high-performance aircraft power systems.

6.2.2.2 T-Type Multilevel Inverter

The T-type multilevel inverter [5] is an advanced three-level topology that was introduced as an alternative to the NPC inverter for lower voltage applications. The name "T-Type" originates from the T-shaped structure of its switches' arrangement, which differentiates it from other three-level topologies. The main distinction of the T-type inverter compared to the NPC topology is that instead of using clamping diodes, it introduces two additional switches, S_{a3} and S_{a4}, which form a bidirectional conduction path at the neutral point. These switches enable the inverter to generate the intermediate zero-voltage level while reducing conduction losses and improving overall switching performance. Compared to the NPC inverter, the number of conducting switches in states [+1] and [–1] is reduced from two to one, significantly lowering conduction losses and improving overall efficiency. However, this advantage comes with a trade-off – the voltage stress on switches S_{a1} and S_{a2} increases to V_{dc}, instead of $0.5V_{dc}$ in the NPC topology.

The T-type inverter provides three output voltage levels ($+0.5V_{dc}$, 0, $-0.5V_{dc}$) similar to the NPC inverter but with fewer power switches and reduced conduction losses. The circuit topology and switching states of the T-type three-level inverter are depicted in Figures 6.10, and 6.11, respectively. It generates three voltage levels by switching the four power switches as follows:

- State [+1]: Output voltage $v_{an} = +0.5V_{dc}$

  S_{a1} and S_{a4} are ON, while S_{a2} and S_{a3} are OFF.

  The output terminal a is connected to $+0.5V_{dc}$, supplying positive voltage to the load.
- State [0]: Output voltage $v_{an} = 0$

  S_{a3} and S_{a4} are ON, while S_{a1} and S_{a2} are OFF.

  The output terminal a is connected to the neutral point n, producing a zero-voltage level.
- State [–1]: Output voltage $v_{an} = -0.5V_{dc}$

  S_{a2} and S_{a3} are ON, while S_{a1} and S_{a4} are OFF.

  The output terminal a is connected to $-0.5V_{dc}$, generating a negative voltage.

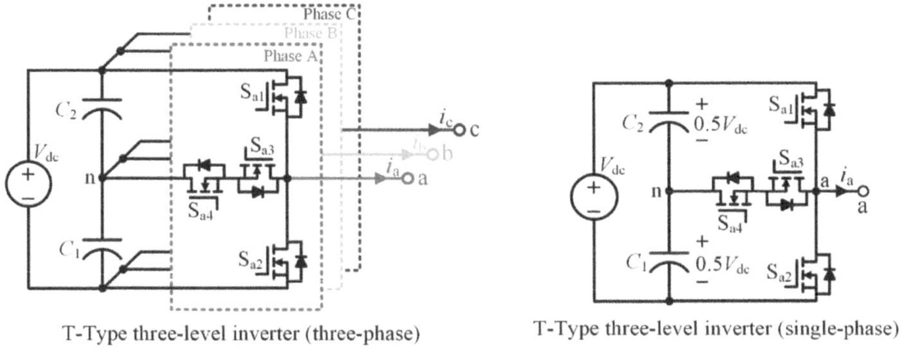

FIGURE 6.10
T-type three-level inverter.

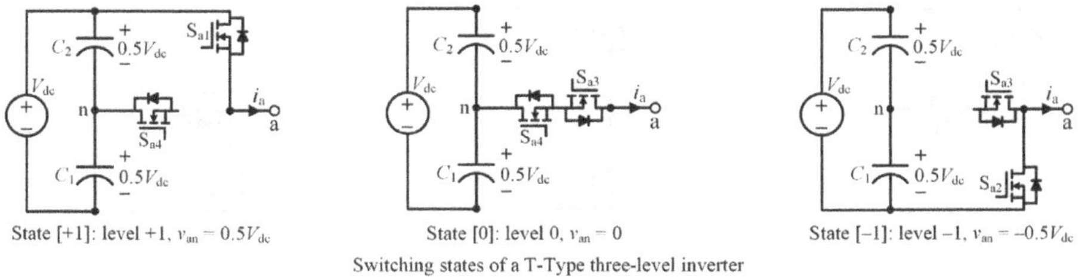

FIGURE 6.11
Switching states of a T-type three-level inverter.

6.2.2.3 Flying Capacitor Multilevel Inverter

The flying capacitor (FC) multilevel inverter utilises a series of flying (floating) capacitors to generate multiple voltage levels, reducing harmonic distortion and enabling better power quality for high-power applications, including aircraft electrification and propulsion systems [6]. Unlike NPC inverters, which rely on clamping diodes to create intermediate voltage levels, the FC inverter employs capacitors as energy storage elements to balance voltage levels dynamically.

Figure 6.12 shows a generalised circuit diagram of the FC multilevel inverter. For N flying capacitors in a FC multilevel inverter, a control algorithm actively regulates the voltage across each capacitor, ensuring balanced charge distribution with a step voltage magnitude of $V_{dc}/(N + 1)$. Therefore, the inverter generates $N + 2$ distinct voltage levels, ranging from $+V_{dc}/2$ to $-V_{dc}/2$. The available output voltage levels are $+V_{dc}/2, \ldots, +2V_{dc}/(N+1)$, $+V_{dc}/(N+1)$, 0, $-V_{dc}/(N+1)$, $-2V_{dc}/(N+1)$, $-V_{dc}/2$.

To understand the operation of a FC inverter, a three-level configuration serves as a representative example. The three-level FC inverter generates output voltages at $+0.5V_{dc}$, 0, and $-0.5V_{dc}$, utilising controlled switching

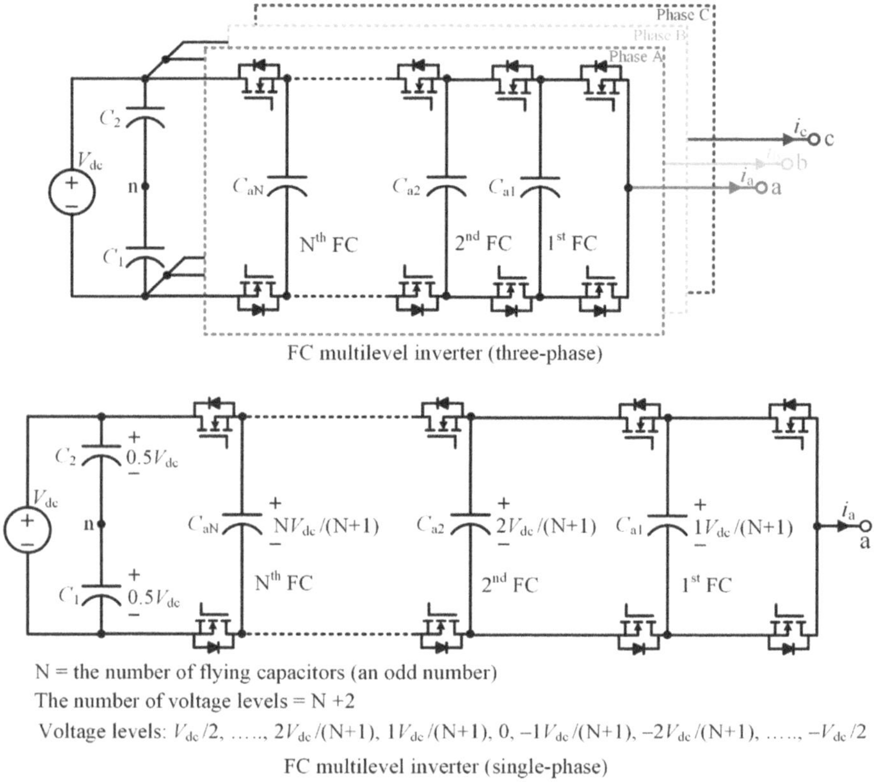

FIGURE 6.12
FC multilevel inverter.

sequences and capacitor voltage balancing techniques. Figure 6.13 illustrates the various switching states and the corresponding charge/discharge behaviour of the flying capacitor (C_{a1}). The operation of the inverter is defined by four distinct switching states:

- State [+1]: Output voltage $v_{an} = +0.5V_{dc}$
  S_{a1} and S_{a3} are ON, connecting the output terminal *a* to $+0.5V_{dc}$.
  S_{a2} and S_{a4} are OFF, preventing current flow through the lower part of the circuit.
  The flying capacitor C_{a1} maintains its charge, with no significant voltage variation.
- State [–1]: Output voltage $v_{an} = -0.5V_{dc}$
  S_{a2} and S_{a4} are ON, connecting the output terminal *a* to $-0.5V_{dc}$.
  S_{a1} and S_{a3} are OFF, isolating the upper part of the circuit.
  The flying capacitor C_{a1} maintains its charge, with no significant voltage variation.

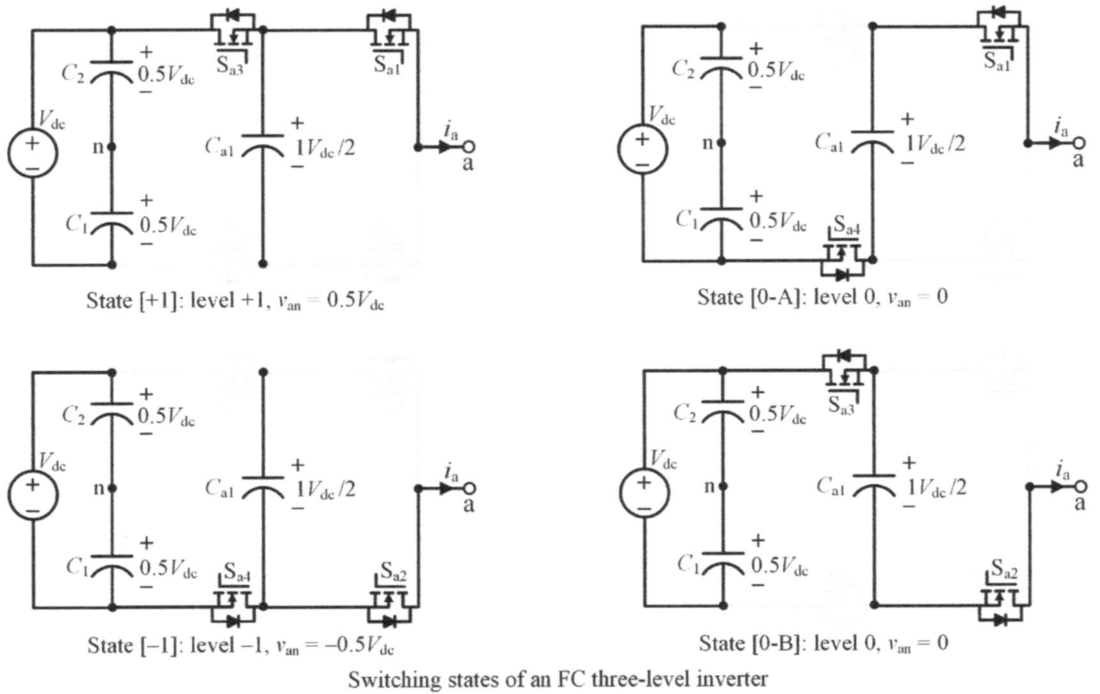

State [0-A]

i_a	Capacitor C_{a1}
positive	discharging
negative	charging

State [0-B]

i_a	Capacitor C_{a1}
positive	charging
negative	discharging

FIGURE 6.13
Switching states of an FC three-level inverter.

- State [0–*A*]: Output voltage $v_{an} = 0$

  S_{a1} and S_{a4} are ON, while S_{a2} and S_{a3} are OFF.

  The DC-link capacitor and the flying capacitor are connected in series but in opposite polarity, causing their voltages to offset each other, resulting in a net-zero output voltage.
- State [0–*B*]: Output voltage $v_{an} = 0$

  S_{a2} and S_{a3} are ON, while S_{a1} and S_{a4} are OFF.

  The DC-link capacitor and the flying capacitor are connected in series but in opposite polarity, causing their voltages to offset each other, resulting in a net-zero output voltage.

One of the critical aspects of the FC inverter is maintaining the charge balance across the flying capacitor (C_{a1}). The table in Figure 6.13 outlines how C_{a1} charges and discharges depending on the direction of the load current (i_a):

- When i_a is positive, C_{a1} discharges in state [0–*A*] and charges in state [0–*B*].
- When i_a is negative, C_{a1} charges in state [0–*A*] and discharges in state [0–*B*].

By alternating between state [0–*A*] and state [0–*B*], a control algorithm actively regulates the flying capacitor voltage, ensuring balanced charge distribution, minimising voltage drift, and maintaining high-quality AC output voltage.

6.2.3 Limitations

All conventional inverter topologies, including two-level, NPC, T-type, and FC multilevel inverters, inherently exhibit limited voltage gain. This limitation arises because their maximum output voltage level is restricted to $0.5V_{dc}$, which is only half of the DC-link voltage (V_{dc}). Consequently, these inverters are unable to generate an output voltage higher than the DC-link voltage, significantly restricting their applicability in systems that require a wide operating voltage range or higher output voltage levels without additional voltage-boosting stages. The inability of conventional inverters to achieve high voltage gain presents several drawbacks, particularly in aircraft electrification and propulsion systems:

6.2.3.1 Compromised Motor Drive Performance in Electric Propulsion

The torque produced by an electric motor is directly related to the applied voltage and current. The limited voltage gain of conventional inverters negatively affects motor performance in several ways:

- A lower inverter output voltage necessitates a higher current to maintain the same torque output, leading to increased conduction losses (copper losses) and core losses, ultimately reducing overall efficiency.
- Excessive current draw at start-up, potentially causing overheating and increasing stress on power components.
- Restricted speed range, as inadequate voltage levels limit torque production at high speeds, especially during flux weakening control.

By extending voltage gain, the inverter can provide the same torque at a lower current, improving overall system efficiency and thermal performance. This reduction in current minimises copper and core losses, enhances power density, and improves the reliability of the motor drive system.

6.2.3.2 Reduced Performance across a Wide Speed Range

Aircraft electric propulsion systems operate across a wide speed range, from take-off to cruise. The inverter's voltage gain directly affects the motor's speed-torque characteristics, particularly in flux weakening control at high speeds. A higher inverter voltage gain provides the following:

- Higher starting torque without excessive current, improving take-off performance.
- Consistent torque production across a broad speed range, ensuring stable operation from low-speed acceleration to high-speed cruising.
- Better flux weakening capability, enabling the motor to sustain high torque at high speeds while minimising losses caused by excessive current. Significant research has been conducted on extending voltage gain of conventional inverters through overmodulation techniques, allowing for an extended torque-speed characteristic during flux weakening, thereby improving motor performance at high speeds [7].

6.2.3.3 Incompatibility with Low-Voltage Energy Storage Systems

Modern aircraft propulsion and power distribution systems increasingly rely on batteries, fuel cells, and supercapacitors, many of which operate at low-voltage levels. Conventional inverters cannot efficiently interface with these sources because of the following:

- Inability to step up voltage to match motor drive or power system requirements
- Need for bulky and complex DC-link voltage boosting circuits, adding weight and reducing reliability

6.3 Active Neutral Point–Clamped Multilevel Inverters with Extended Voltage Gain

In recent years, significant research efforts have been dedicated to developing novel multilevel inverter topologies that offer enhanced voltage gain. This section introduces the derivation of an example ANPC five-level inverter with extended voltage gain, highlighting its structural evolution and operational principles. The derivation of this topology originates from the NPC inverter, as shown in Figure 6.14. An active NPC (ANPC) three-level topology replaces the clamping diodes with active power switches.

To extend both the number of voltage levels and the voltage gain, two additional diodes and a switched capacitor are integrated into the ANPC inverter. The capacitor, denoted as C_{a1}, plays a crucial role in extending voltage gain. It is strategically charged to V_{dc} during the switching states [+1] and [−1], where the output voltage levels correspond to $0.5V_{dc}$ and $-0.5V_{dc}$, respectively. Subsequently, by discharging the capacitor, higher voltage levels of V_{dc} and $-V_{dc}$ can be achieved. This operation effectively extends the output voltage range from a conventional ANPC three-level structure to a five-level inverter, achieving voltage levels at V_{dc}, $0.5V_{dc}$, 0, $-0.5V_{dc}$, and $-V_{dc}$. The integration of the switched-capacitor technique not only increases the number of output voltage levels but also enhances the maximum voltage gain. Unlike conventional ANPC inverters, which are constrained to a peak voltage of $0.5V_{dc}$, this topology depicted in Figure 6.14 effectively doubles the voltage gain, reaching V_{dc}.

Figure 6.15 shows the switching states of the ANPC five-level inverter with extended gain. The states are summarised as follows:

- State [+2]: Output voltage $v_{an} = +V_{dc}$

  S_{a2} and S_{a6} are ON, connecting the output terminal a to $+V_{dc}$ which is the voltage across the capacitor C_{a1}.
- State [+1]: Output voltage $v_{an} = +0.5V_{dc}$

  S_{a1} and S_{a2} are ON, allowing the output terminal a to be connected to $+0.5V_{dc}$.

  S_{a4} is ON, allowing the capacitor C_{a1} to be charged by the DC source V_{dc}.
- State [0–*A*]: Output voltage $v_{an} = 0$

  S_{a2} and S_{a5} are ON, connecting the output terminal a to the neutral point.

  The switched capacitor C_{a1} maintains its charge, with no significant voltage variation.
- State [0–*B*]: Output voltage $v_{an} = 0$

  S_{a3} and S_{a6} are ON, connecting the output terminal a to the neutral point.

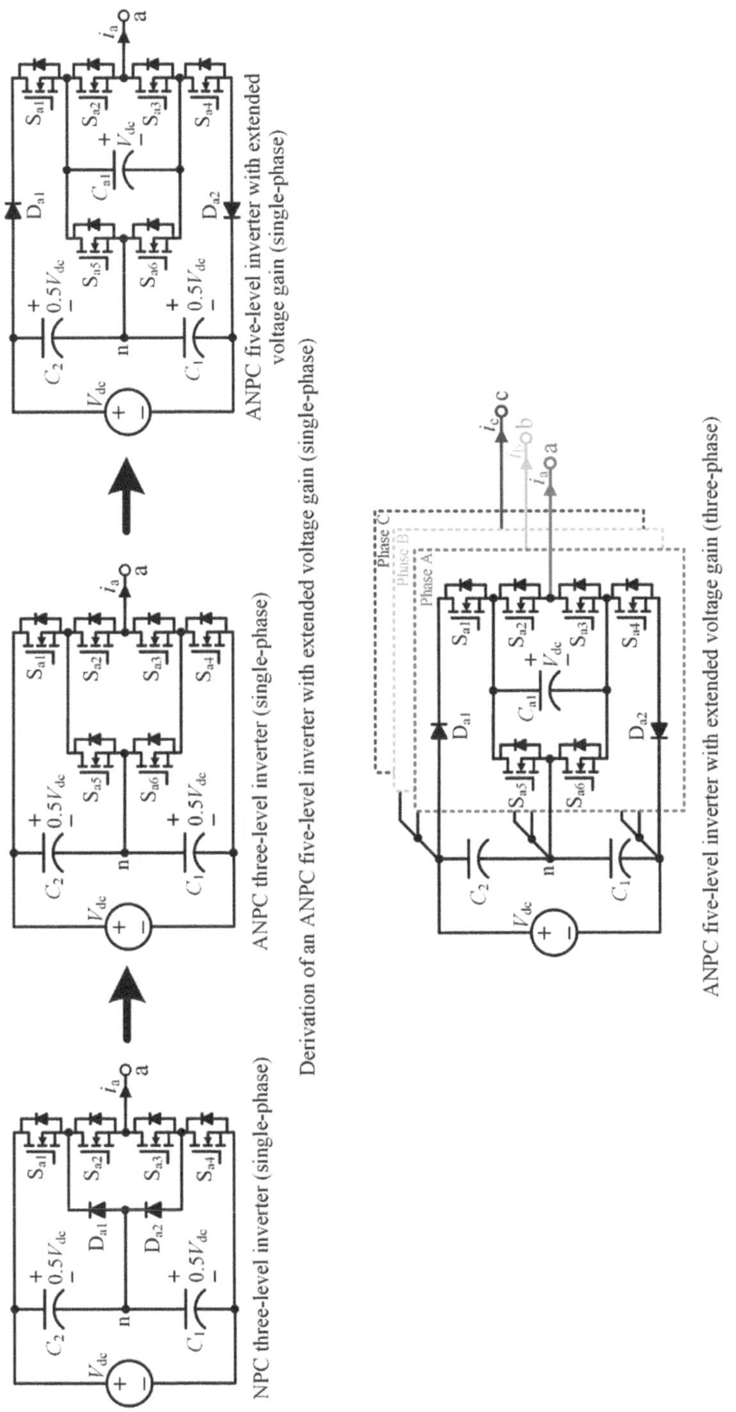

FIGURE 6.14
ANPC five-level inverter with extended voltage gain [8, 9].

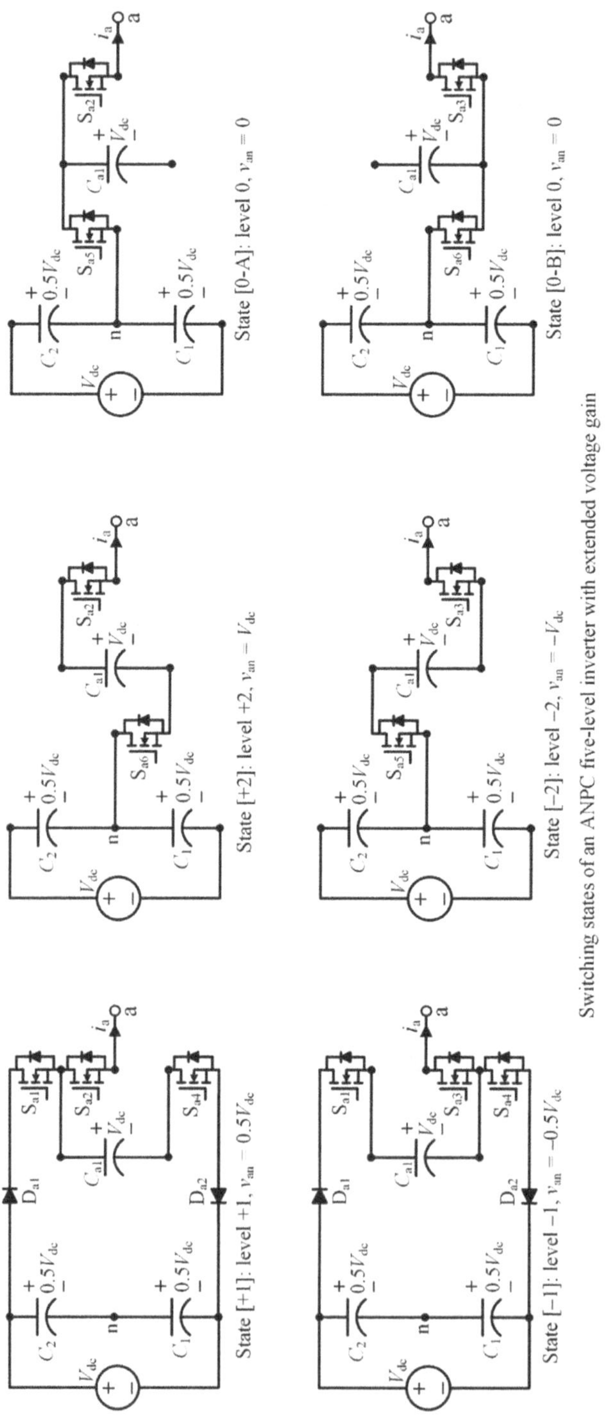

FIGURE 6.15
Switching states of an ANPC five-level inverter with extended voltage gain.

The switched capacitor C_{a1} maintains its charge, with no significant voltage variation.

- State [–1]: Output voltage $v_{an} = -0.5V_{dc}$

  S_{a3} and S_{a4} are ON, allowing the output terminal a to be connected to $-0.5V_{dc}$.

  S_{a1} is ON, allowing the capacitor C_{a1} to be charged by the DC source V_{dc}.
- State [–2]: Output voltage $v_{an} = -V_{dc}$

  S_{a3} and S_{a5} are ON, connecting the output terminal a to $-V_{dc}$ which is the voltage across the capacitor C_{a1} in negative polarity.

6.4 Switched-Midpoint Boost Inverters (SMBI)

6.4.1 Concept

The switched-midpoint boost inverter (SMBI) [10] is an innovative power conversion topology that integrates a voltage-boosting mechanism directly into the inverter structure, eliminating the need for a separate DC–DC boost stage. As illustrated in Figure 6.16, the SMBI concept evolves from the conventional two-stage power conversion system, where a boost converter is used to step up the voltage before feeding a multilevel inverter. In contrast, the SMBI combines voltage boosting and multilevel AC voltage generation into a single-stage operation, improving efficiency and reducing system complexity.

The core principle behind the SMBI is the utilisation of a switched-midpoint circuit that actively regulates the DC-link capacitors, achieving both voltage boosting and balancing while simultaneously generating the required AC output voltage. The topology consists of the following:

- *Boosting Element (Inductor or Switched Capacitor Cell):* Provides the required voltage step-up capability.
- *Switch-Midpoint Circuit:* Dynamically controls the voltage magnitude and ensures voltage balancing across the DC-link capacitors.
- *Multilevel Inverter:* Operates in synchronisation with the switched-midpoint circuit, ensuring efficient power conversion and generating a high-quality AC output waveform.

Compared to the conventional two-stage power conversion structure, the SMBI achieves several key advantages:

- *Natural Voltage Balancing:* The SMBI inherently balances the DC-link capacitor voltages without requiring additional controllers or sensors, simplifying system design.

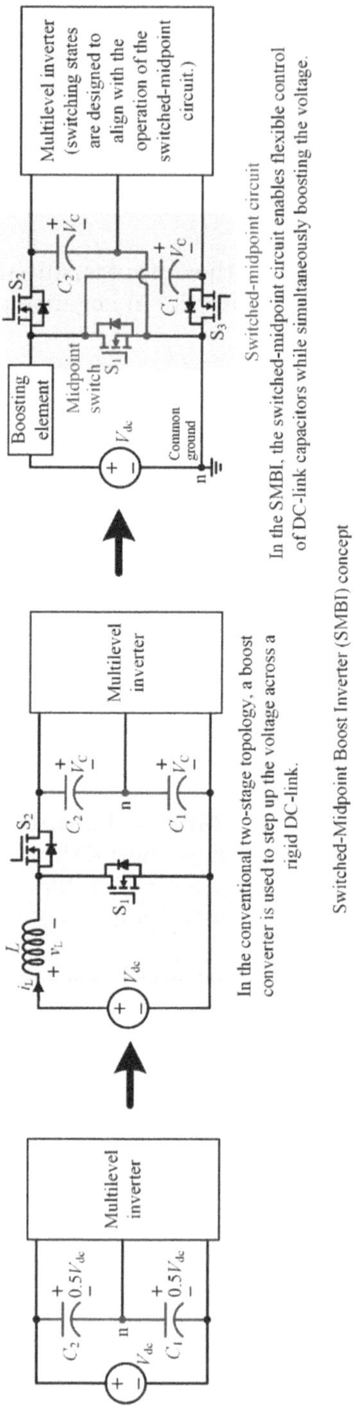

FIGURE 6.16
SMBI concept.

- *Enhanced Voltage Gain:* The integrated boosting mechanism through switched-midpoint circuit achieves higher voltage gain.
- *Reduced Current Stress:* The power switches in the switched-midpoint circuit experience lower current stress compared to those in a traditional boost converter.
- *Reduced Voltage Stress:* The power switches in the switched-midpoint circuit also experience lower voltage stress compared to those in a boost converter.

6.4.2 SMBI with Inductive Boosting

An example ANPC-based three-level SMBI with inductive boosting is depicted in Figure 6.17. It utilises an inductive boosting element to enhance voltage gain while maintaining balanced DC-link capacitor voltages and generating three-level AC voltage. The operation of this topology involves switching states illustrated in Figure 6.18, summarised as follows:

- State [+0|C]: Output voltage $v_{\mathrm{an}} = 0$

  S_1 is ON, allowing the DC source to charge the inductor L.

  S_{a3} and S_{a6} are ON, connecting the output terminal a to the neutral point.
- State [+0|*D*]: Output voltage $v_{\mathrm{an}} = 0$

  S_1 is OFF, while S_2 and S_3 are ON, allowing the inductor to release its stored energy, which charges C_1 and C_2.

  The voltage across the capacitors is naturally balanced and controlled by the inductor charging duty-cycle D.

  S_{a3} and S_{a6} are ON, connecting the output terminal a to the neutral point.
- State [−0|*C*]: Output voltage $v_{\mathrm{an}} = 0$

  Similar to state [+0|C] but occurring during the negative half-cycle.

  S_1 is ON, allowing the DC source to charge the inductor L.

  S_{a2} and S_{a5} are ON, connecting the output terminal a to the neutral point.
- State [−0|*D*]: Output voltage $v_{\mathrm{an}} = 0$

  Similar to state [+0|*D*] but occurring during the negative half-cycle.

  S_1 is OFF, while S_2 and S_3 are ON, allowing the inductor to release its stored energy, which charges C_1 and C_2.

  The voltage across the capacitors is naturally balanced and controlled by the inductor charging duty-cycle D.

  S_{a2} and S_{a5} are ON, connecting the output terminal a to the neutral point.

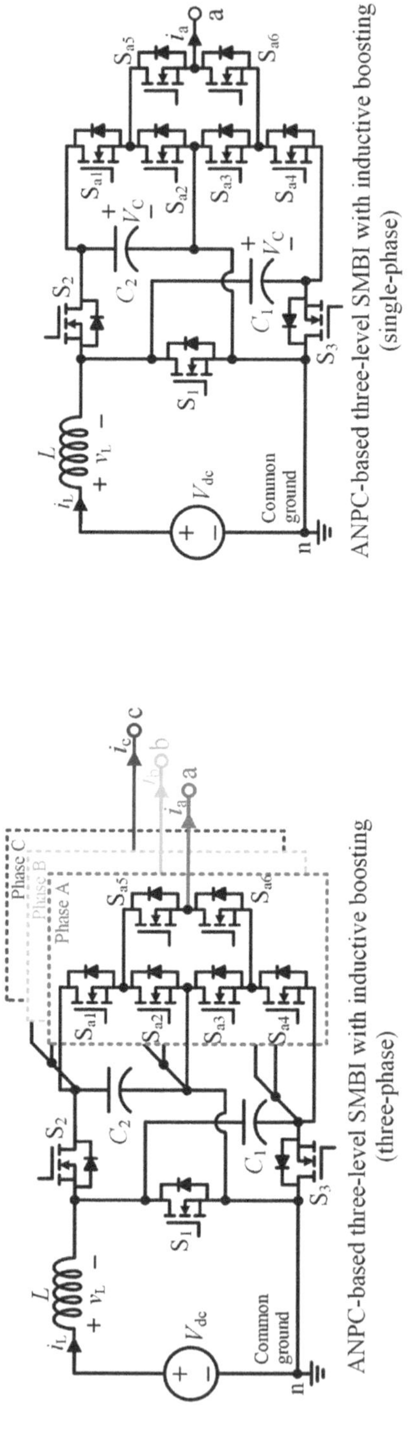

FIGURE 6.17
ANPC-based three-level SMBI with inductive boosting.

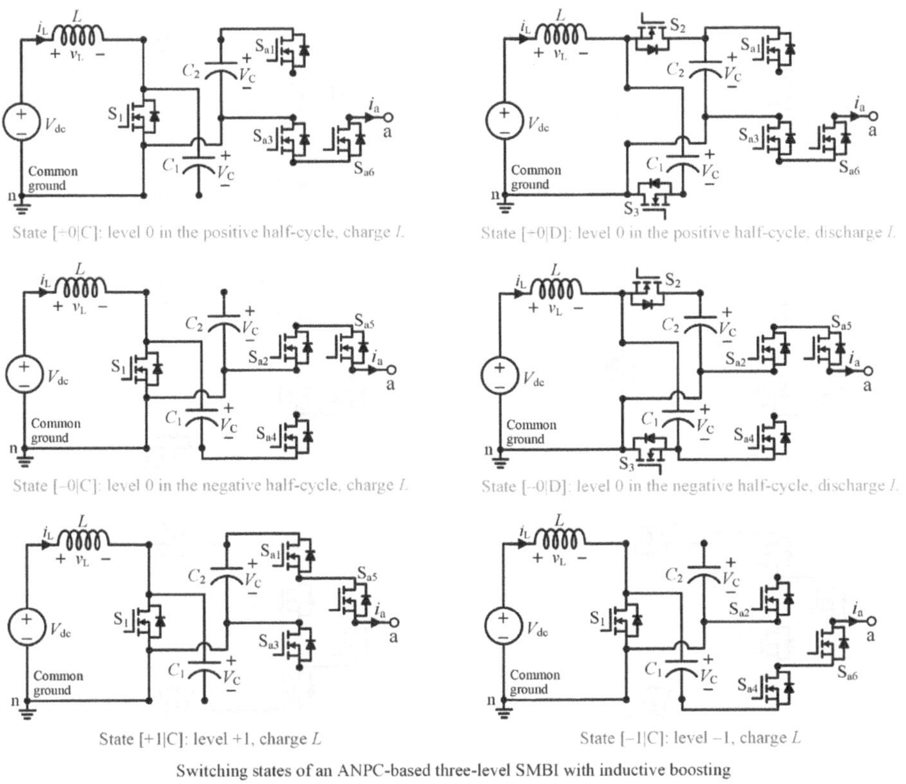

FIGURE 6.18
Switching states of an ANPC-based three-level SMBI with inductive boosting.

- State [+1|*C*]: Output voltage $v_{an} = V_C$

  S_1 is ON, allowing the DC source to charge the inductor *L*.

  S_{a1} and S_{a5} are ON, connecting the output terminal *a* to $+V_C$ which is the voltage across the capacitor C_2.
- State [–1|*C*]: Output voltage $v_{an} = -V_C$

  Similar to state [+1|C], but for the negative half-cycle.

  S_1 is ON, allowing the DC source to charge the inductor *L*.

  S_{a4} and S_{a6} are ON, connecting the output terminal *a* to $-V_C$ which is the voltage across the capacitor C_1 in negative polarity.

These switching states are coordinated with the modulation strategy shown in Figure 6.19, where the duty-cycle *D* of S_1 is controlled to optimise inductor charging and discharging, ensuring effective voltage boosting and efficient power conversion. The voltage across the DC-link capacitors in the

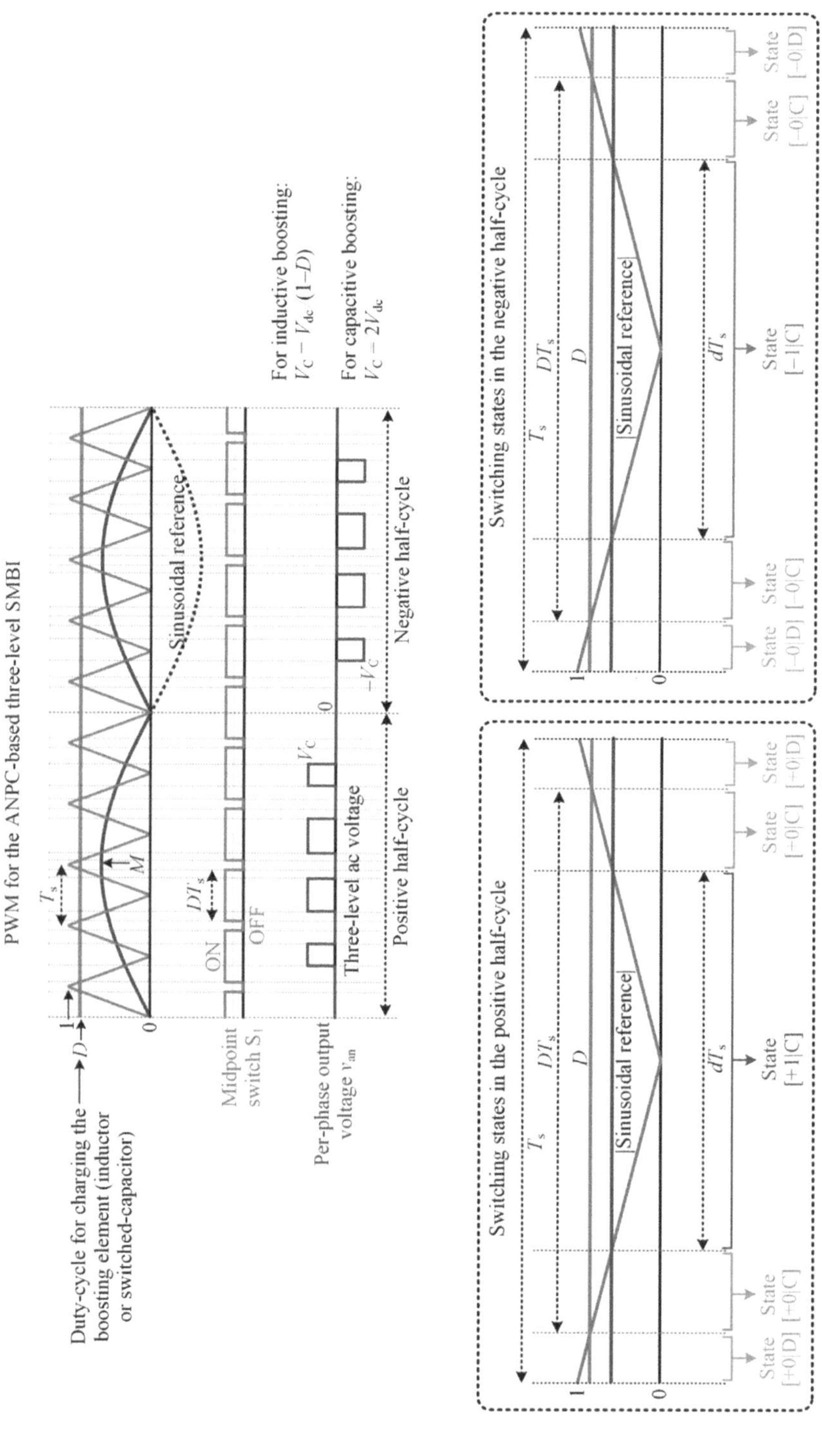

FIGURE 6.19
PWM method for ANPC-based three-level SMBIs.

ANPC-based three-level SMBI with inductive boosting is naturally balanced due to the controlled operation of the switched-midpoint circuit. During operation, the boosting mechanism regulates the voltage across C_1 and C_2 to $V_C = V_{dc}/(1–D)$, where D is the duty-cycle of the midpoint switch. As a result, the AC output voltage generated by the inverter follows a three-level pattern: $+V_C$, 0, and $–V_C$.

6.4.3 SMBI with Capacitive Boosting

The boosting inductor in Figure 6.17 can be replaced by a switched-capacitor cell, consisting of two power switches, two diodes, and a capacitor to implement the capacitive boosting method. This configuration enables voltage step-up using capacitor charging and discharging cycles, eliminating the need for an inductor. As depicted in Figure 6.20, the ANPC-based three-level SMBI with capacitive boosting retains the multilevel inverter stage while incorporating a switched-capacitor circuit to simplify voltage boosting operation.

The switching states of the ANPC-based three-level SMBI with capacitive boosting, as shown in Figure 6.21, define its AC voltage generation and boosting functionality. The operation is described as follows:

- State [+0|*C*]: Output voltage $v_{an} = 0$

  S_1 is ON, allowing the DC source to charge the capacitor C_{in} to V_{dc}.

  S_{a3} and S_{a6} are ON, connecting the output terminal a to the neutral point.
- State [+0|*D*]: Output voltage $v_{an} = 0$

  S_1 is OFF, while S_2, S_3, S_4, and S_5 are ON, allowing the capacitor C_{in} to release its stored energy, which charges C_1 and C_2 to $V_C = 2V_{dc}$.

  The voltages across the capacitors C_1 and C_2 are naturally balanced.

  S_{a3} and S_{a6} are ON, connecting the output terminal *a* to the neutral point.
- State [–0|*C*]: Output voltage $v_{an} = 0$

  Similar to state [+0|*C*], but occurring during the negative half-cycle.

  S_1 is ON, allowing the DC source to charge the capacitor C_{in} to V_{dc}.

  S_{a2} and S_{a5} are ON, connecting the output terminal *a* to the neutral point.
- State [–0|*D*]: Output voltage $v_{an} = 0$

  Similar to state [+0|*D*], but occurring during the negative half-cycle.

  S_1 is OFF, while S_2, S_3, S_4, and S_5 are ON, allowing the capacitor C_{in} to release its stored energy, which charges C_1 and C_2 to $V_C = 2V_{dc}$.

  The voltages across the capacitors C_1 and C_2 are naturally balanced.

  S_{a2} and S_{a5} are ON, connecting the output terminal a to the neutral point.

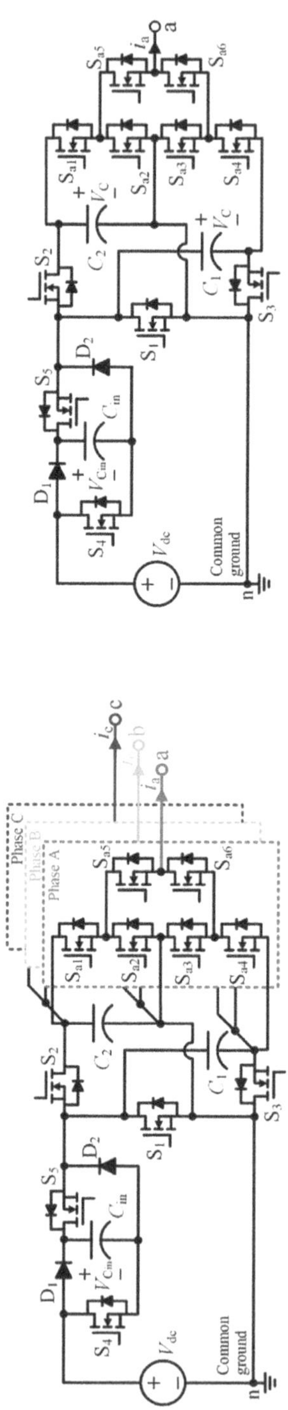

ANPC-based three-level SMBI with capacitive boosting (three-phase)

ANPC-based three-level SMBI with capacitive boosting (single-phase)

FIGURE 6.20
ANPC-based three-level SMBI with capacitive boosting.

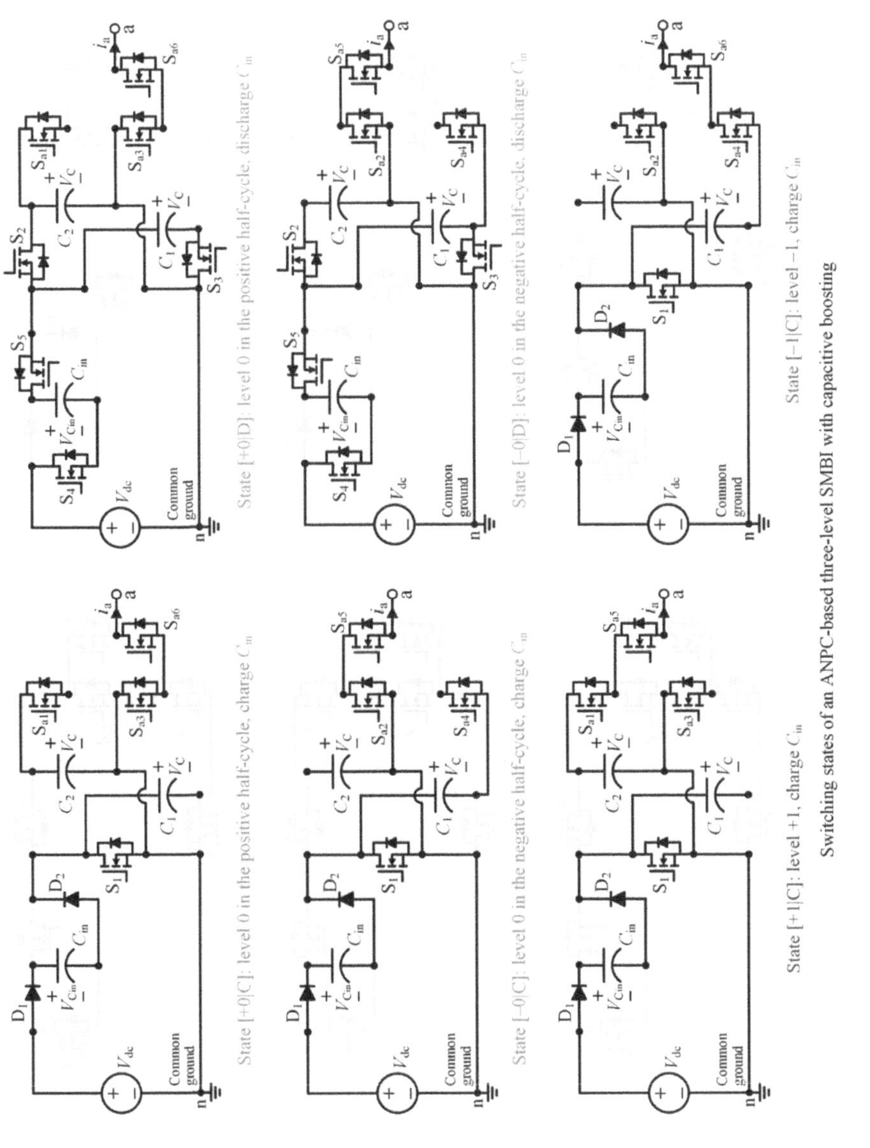

FIGURE 6.21
Switching states of an ANPC-based three-level SMBI with capacitive boosting.

- State [+1|C]: Output voltage $v_{an} = +V_C = +2V_{dc}$
  S_1 is ON, allowing the DC source to charge the capacitor C_{in} to V_{dc}.
  S_{a1} and S_{a5} are ON, connecting the output terminal a to $+V_C$, which is the voltage across C_2.
- State [−1|C]: Output voltage $v_{an} = -V_C = -2V_{dc}$
  Similar to state [+1|C], but for the negative half-cycle.
  S_1 is ON, allowing the DC source to charge the capacitor C_{in} to V_{dc}.
  S_{a4} and S_{a6} are ON, connecting the output terminal a to $-V_C$, which is the voltage across C_1 in negative polarity.

In the ANPC-based three-level SMBI with capacitive boosting, the switched capacitor C_{in} is balanced at V_{dc}, while the DC-link capacitors C_1 and C_2 are naturally balanced at $2V_{dc}$. Consequently, the inverter generates three output voltage levels: $+2V_{dc}$, 0, and $-2V_{dc}$, as illustrated in Figure 6.22, effectively

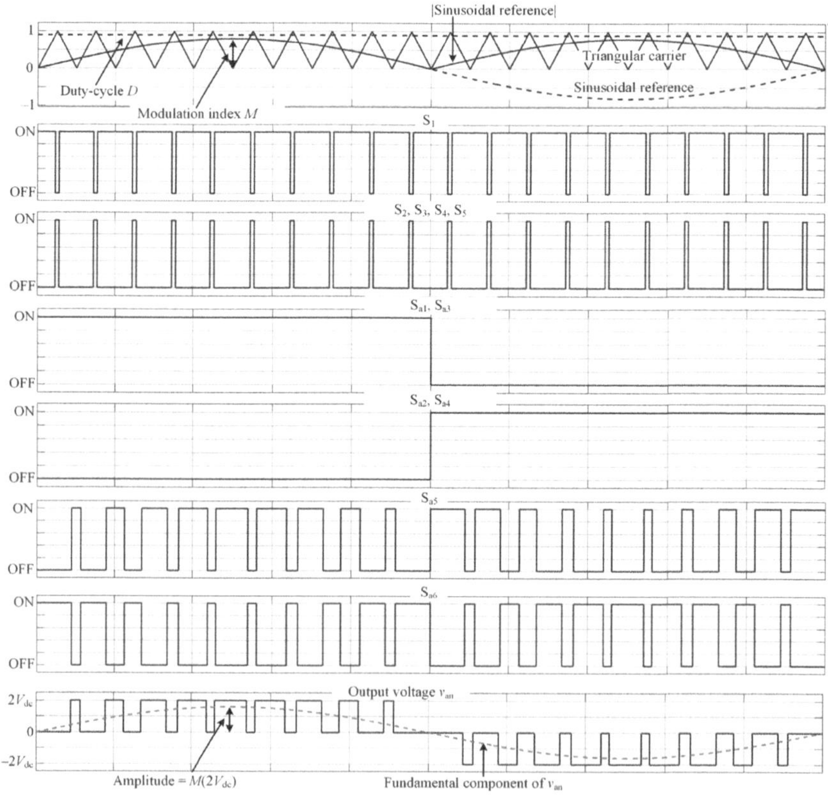

FIGURE 6.22
PWM for an ANPC-based three-level SMBI with capacitive boosting.

extending the voltage gain of the conventional ANPC inverter by a factor of 4, increasing it from $0.5V_{dc}$ to $2V_{dc}$.

6.5 Summary

This chapter has provided an overview of power inverters for aircraft electrification, detailing the transition from conventional inverter topologies to advanced high-gain inverter solutions. Conventional two-level and multilevel inverters were examined, highlighting their voltage gain limitations and the necessity for enhanced power conversion strategies. The introduction of ANPC-based inverters with extended voltage gain demonstrated an effective solution for improving efficiency and power conversion performance. Furthermore, the switched-midpoint boost inverter (SMBI) was explored, offering a single-stage voltage boosting and inversion solution that eliminates the need for an external boost converter. Both inductive and capacitive boosting SMBI topologies were analysed, showcasing their ability to achieve higher voltage levels, reduced voltage and current stresses, and simplified control with natural capacitor voltage balancing. The innovations presented in this chapter pave the way for the next generation of power inverters, contributing to the realisation of sustainable, all-electric aviation by optimising power electronics for electric propulsion and on-board power distribution.

References

[1] R. Dyson, R. Jansen, and N. Madavan, "Benefits of electrified propulsion for large aircraft," in: K. Haran, N. Madavan, and T. C. O'Connell, eds. *Electrified Aircraft Propulsion: Powering the Future of Air Transportation*. Cambridge University Press, 2022, pp. 1–28.

[2] B. Rahrovi and M. Ehsani, "A review of the more electric aircraft power electronics," 2019 IEEE Texas Power and Energy Conference (TPEC), College Station, TX, USA, 2019, pp. 1–6, https://doi.org/10.1109/TPEC.2019.8662158.

[3] D. Wang et al., "Multilevel inverters for electric aircraft applications: Current status and future trends," *IEEE Transactions on Transportation Electrification*, vol. 10, no. 2, pp. 3258–3282, June 2024, https://doi.org/10.1109/TTE.2023.3296284.

[4] H. Akagi, "Multilevel converters: Fundamental circuits and systems," *Proceedings of the IEEE*, vol. 105, no. 11, pp. 2048–2065, Nov. 2017, https://doi.org/10.1109/JPROC.2017.2682105.

[5] M. Schweizer and J. W. Kolar, "Design and implementation of a highly efficient three-level T-type converter for low-voltage applications," *IEEE Transactions on Power Electronics*, vol. 28, no. 2, pp. 899–907, Feb. 2013, https://doi.org/10.1109/TPEL.2012.2203151.

[6] T. Modeer, N. Pallo, T. Foulkes, C. B. Barth, and R. C. N. Pilawa-Podgurski, "Design of a GaN-based interleaved nine-level flying capacitor multilevel inverter for electric aircraft applications," *IEEE Transactions on Power Electronics*, vol. 35, no. 11, pp. 12153–12165, Nov. 2020, https://doi.org/10.1109/TPEL.2020.2989329.

[7] F. Guo et al., "An overmodulation algorithm with neutral-point voltage balancing for three-level converters in high-speed aerospace drives," *IEEE Transactions on Power Electronics*, vol. 37, no. 2, pp. 2021–2032, Feb. 2022, https://doi.org/10.1109/TPEL.2021.3105752.

[8] Y. P. Siwakoti, A. Palanisamy, A. Mahajan, S. Liese, T. Long, and F. Blaabjerg, "Analysis and design of a novel six-switch five-level active boost neutral point clamped inverter," *IEEE Transactions on Industrial Electronics*, vol. 67, no. 12, pp. 10485–10496, Dec. 2020, https://doi.org/10.1109/TIE.2019.2957712.

[9] S. S. Lee and K.-B. Lee, "Dual-T-type seven-level boost active-neutral-point-clamped inverter," *IEEE Transactions on Power Electronics*, vol. 34, no. 7, pp. 6031–6035, July 2019, https://doi.org/10.1109/TPEL.2019.2891248.

[10] S. S. Lee, "Switched-Midpoint Boost Inverter (SMBI)," *IEEE Transactions on Industry Applications*, vol. 61, no. 1, pp. 463–471, Jan.–Feb. 2025, https://doi.org/10.1109/TIA.2024.3476449.

7

Magnetic Gears: Principles, Topologies, and Applications in Rotary and Linear Machines

Sashidhar Sampathirao and Neel Shrivastava

7.1 Introduction

Gearboxes are widely used in engineering to bridge the mismatch between the high-speed, low-torque characteristics of prime movers (e.g., electric motors) and the low-speed, high-torque demands of loads. Although mechanical gears offer high torque density, they are constrained by several drawbacks, including frictional losses, acoustic noise, vibration, lubrication requirements, and mechanical wear and fatigue [1, 2].

Magnetic gears (MGs) provide a contactless alternative by transmitting torque through magnetic field interaction. This fundamental distinction confers several advantages: inherent overload protection, physical isolation between input and output shafts, minimal maintenance, quiet operation, and enhanced reliability [1, 2, 3]. The concept of magnetic gearing is not new – patents were filed as early as the beginning of the twentieth century [4, 5]. However, the limited performance of permanent magnet (PM) materials of that era, especially ferrites, resulted in low torque densities and restricted practical deployment. The development of high-energy rare-earth magnets, such as neodymium–iron–boron (NdFeB), in the 1980s rekindled research interest. A major breakthrough occurred in 2001, when Atallah and Howe introduced the coaxial magnetic gear based on the flux-modulation principle [6]. This topology enabled all permanent magnets to actively participate in torque transmission, achieving torque densities comparable to mechanical gears and marking a departure from earlier, less efficient "converted" designs [1, 7].

Magnetic gears can be broadly categorised into two groups: non-modulated MGs (e.g., spur, worm, planetary, cycloid, and harmonic gears) and modulated MGs (e.g., coaxial, axial, linear, and hybrid-flux topologies). While some non-modulated variants demonstrated competitive performance, the

DOI: 10.1201/9781003747994-7

flux-modulated topologies offer the highest torque densities and efficiency, making them the focus of current research [1]. Over the past two decades, novel variants such as Halbach-array MGs, ferrite-based flux-focusing gears, axial-flux MGs, bearingless MGs, and linear or tubular configurations (including high-temperature superconducting designs) have significantly advanced the state of the art.

This chapter provides a comprehensive overview of magnetic gear technology. It first introduces the fundamental operating principles and classifies the major MG topologies. The discussion then focuses on the design and performance of both rotary and linear MGs, with comparisons to their mechanical counterparts. Subsequent sections examine the integration of MGs with electrical machines, identify key design challenges, and highlight emerging applications across renewable energy, transportation, and industrial sectors.

7.1.1 Magnetically Geared Permanent Magnet Machines (MGPMs)

Magnetically geared (MG) technology has emerged as a promising innovation in modern electric machine design, enabling torque transmission and speed conversion without physical contact. A key application of this technology is its integration with permanent magnet (PM) brushless machines to create a compact, single device called a magnetically geared permanent magnet (MGPM) machine. This integration combines the high-speed performance and power density of PM machines with the torque amplification of magnetic gearing, resulting in compact machines that can deliver high torque at low speeds for demanding applications.

The primary advantage of MGPM machines lies in their ability to decouple rotor speed from load speed. This feature allows the rotor to operate at higher speeds – enhancing power density – while directly supplying low-speed, high-torque output to the load. In doing so, MGPMs overcome one of the major challenges in conventional direct-drive systems: the trade-off between torque density and machine size [1].

7.1.2 Topologies of MGPM Machines

MGPM machines are typically classified according to the number of airgaps in their design, as illustrated in Figure 7.1. The number of airgaps strongly affects both electromagnetic performance and mechanical complexity.

Magnetically geared permanent magnet (MGPM) machines can be realised in three main topologies – three-airgap, two-airgap, and one-airgap designs – each offering distinct advantages and trade-offs. The three-airgap topology integrates an outer-rotor PM machine with a concentric magnetic gear (CMG), delivering high torque density and a good power factor, which makes it suitable for compact, high-torque applications. However, the presence of two rotating elements and three airgaps increases mechanical complexity, leading to challenges in rotor balancing, assembly, and

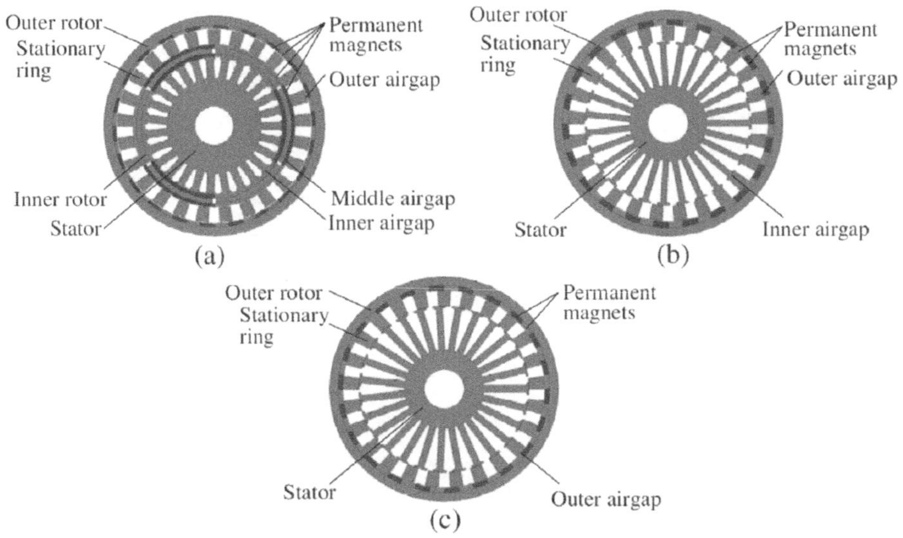

FIGURE 7.1
MGPM machines. (a) Three-airgap topology. (b) Two-airgap topology. (c) One-airgap topology [1].

manufacturing. The two-airgap topology addresses this by simplifying the structure through torque transmission via harmonic magnetic field modulation using a stationary ring. While this reduces complexity compared to the three-airgap design, it typically requires additional copper windings to compensate for the lower flux density, offering a balanced trade-off between performance and manufacturability.

The simplest among these is the one-airgap topology, where the stator and stationary modulator are merged into a single structure. This configuration achieves high electromagnetic efficiency and maximum torque per unit of PM material, with an operating principle similar to that of a PM Vernier machine. Although it tends to suffer from higher torque ripple due to magnetic saturation, the one-airgap MGPM is often regarded as the most practical choice for direct-drive systems, thanks to its combination of high torque density, reduced PM usage, and straightforward construction.

7.1.3 Performance Comparison and Design Considerations

7.1.3.1 Comparison with Mechanical Gears

Compared to traditional mechanical gear systems, MGs offer several distinct advantages:

- *Torque Density:* Modern MGs achieve torque densities of 100–150 kN-m/m^3, comparable to spur gears.

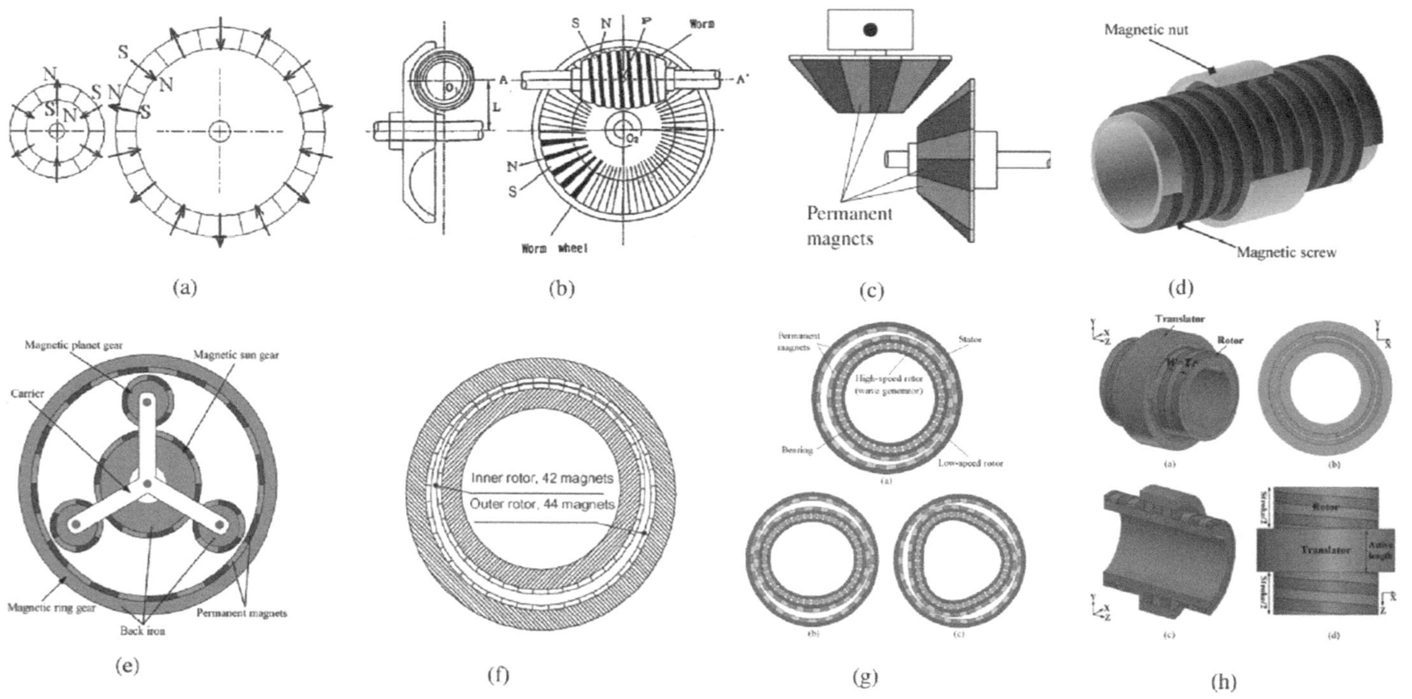

FIGURE 7.2
Schemes of various conventional MGs. (a) Spur MG. (b) Worm MG. (c) Perpendicular MG. (d) Magnetic screw. (e) Planetary MG. (f) Cycloid MG. (g) Harmonic MG. (h) Trans-rotary MG [2].

- *Efficiency:* With efficiencies exceeding 95–97%, MGs avoid the frictional losses of mechanical contact.
- *Service Life:* Their contactless nature ensures operational lifetimes exceeding ten years without wear-related degradation.
- *Noise and Vibration:* MGs operate quietly with minimal vibration, making them ideal for noise-sensitive environments.
- *Cost:* The main drawback is the high initial cost due to reliance on rare-earth PMs, although reduced maintenance requirements can offset this over the system lifetime [1, 2].

7.1.3.2 Key Design Challenges

Despite these benefits, several challenges limit the widespread adoption of MGs:

- *Mechanical Integrity:* Supporting modulator pole-pieces and securing PMs on high-speed rotors requires robust design solutions such as non-magnetic sleeves or structural bridges [2, 3].
- *Losses and Thermal Management:* High-speed operation can cause eddy-current losses in PMs, often mitigated by magnet segmentation. Iron losses in yokes and modulators must be minimised through laminated steels and effective cooling strategies.
- *Cogging Torque:* Torque ripple can be reduced by designing aperiodic MGs with fractional gear ratios and ensuring a small greatest common divisor (GCD) between pole and modulator counts [2].
- *Cost of PMs:* The dependence on expensive, rare-earth materials remains a barrier. Alternatives under study include ferrite-based MGs and variable-flux designs using low-coercivity magnets such as Alnico [2, 3].

7.1.3.3 Applications

The unique capabilities of MG technology make it suitable for diverse applications:

- *Wind Turbine Generators (WTGs):* MG-based pseudo-direct-drive systems eliminate mechanical gearboxes, offering high torque density with improved reliability [1, 3].
- *Electric and Hybrid Vehicles (EVs/HEVs):* MGs can be integrated into traction motors or in-wheel drives, providing high torque, overload protection, and quiet operation. Companies like *Magnomatics* have developed MG-based traction systems [1, 3].

- *Marine Propulsion:* MGs provide inherent physical isolation, protecting motors from underwater shocks and ensuring overload protection in thrusters and propulsion drives [3].
- *Aerospace:* Their lightweight, lubrication-free, and reliable operation makes them attractive for UAVs and next-generation electric aircraft [2, 3].
- *Industrial Pumps and Robotics:* MGs are well-suited for hermetically sealed or clean-room environments, offering contactless, maintenance-free operation [2].

7.2 Classification and Working Principle

Magnetic gears can be broadly classified into two families based on their operating principle.

7.2.1 Conventional Non-Modulated Magnetic Gears

The earliest magnetic gear (MG) designs were direct adaptations of mechanical gears, where the teeth were replaced by permanent magnet (PM) poles arranged on the input and output rotors. Common examples include spur, worm, perpendicular-axis, and planetary magnetic gears [8, 11].

Conventional non-modulated magnetic gears mimic the mechanical counterparts by relying directly on the interaction of permanent magnets arranged in different geometries. The simplest of non-modulated MGs is the spur MG (Figure 7.2), where parallel magnetised poles interact in a manner analogous to spur gears. The worm MG introduces a helical magnetic arrangement to emulate worm gearing, while the perpendicular MG couples shafts at right angles using radial flux interaction. The magnetic screw adopts a helical PM arrangement to provide high force density in linear or rotary actuators. More complex are planetary arrangements, such as the magnetic planetary gear, which employs multiple magnetic planets to distribute torque, and the cycloid MG, where a cycloidal motion modulates the airgap field to achieve large gear ratios. The harmonic MG, analogous to a mechanical harmonic drive, uses elliptical modulation of the airgap to enable very high gear ratios with smooth torque transmission. Finally, the trans-rotary MG converts linear motion into rotary motion via magnetically coupled components, extending the versatility of MGs beyond purely rotary drives

In these topologies, torque transfer relies on the direct interaction of magnetic fields: as the magnets on one rotor rotate past those on the other, alternating regions of attraction and repulsion generate a tangential force that drives the second rotor. The transmission ratio is analogous to that of mechanical gears and is determined by the ratio of the number of magnetic poles on the two rotors, namely, P_1, P_2.

Although conceptually simple, this operating principle inherently limited performance. Only the magnets directly facing each other contributed to torque transmission, while the remainder of the magnetic material remained underutilised. As a result, the torque density achieved by these designs was relatively low, typically around 15 kN-m/m^3 [1, 2]. This poor utilisation of magnetic material, combined with bulky structures, restricted their suitability for practical applications and largely confined them to experimental and demonstrative use.

7.2.2 Modulated Magnetic Gears

Modulated magnetic gears (MMGs) represent the state-of-the-art MG technology and are distinguished by their ability to achieve high torque densities. Their key innovation lies in the introduction of a stationary flux-modulation ring (or set of ferromagnetic pole pieces) positioned between two permanent-magnet (PM) rotors.

With reference to Figures 7.2 and 7.3, the coaxial magnetic gear has attracted the most attention due to its high torque density and structural simplicity. It consists of inner and outer PM rotors separated by iron pole

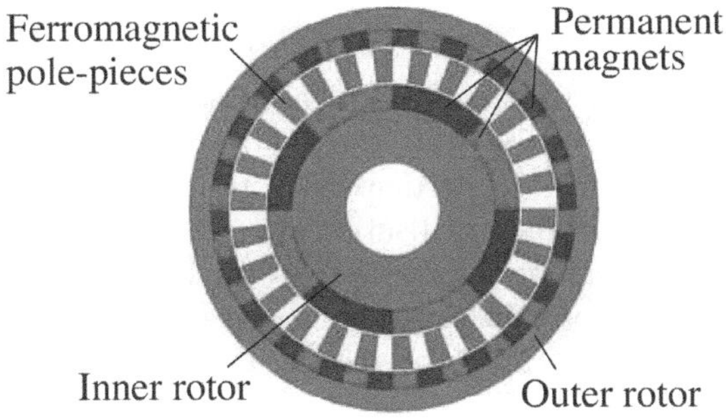

FIGURE 7.3
Coaxial magnetic gear [1].

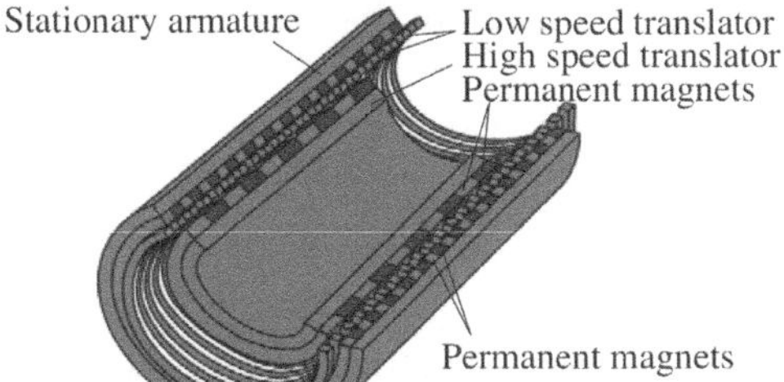

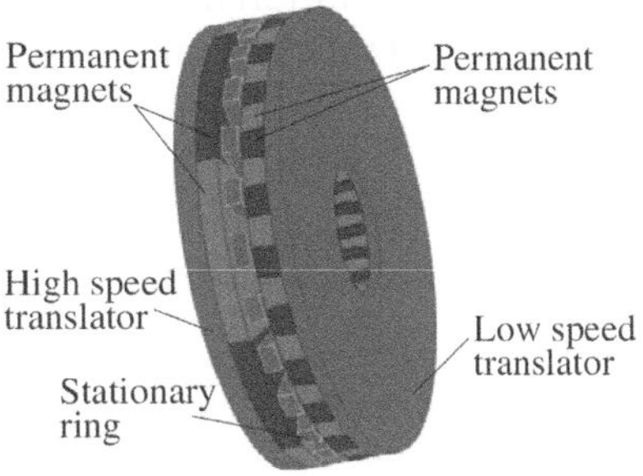

FIGURE 7.4
Linear- and axial-type coaxial magnetic gear topologies [1].

pieces (flux modulators), enabling all magnets to participate in torque transmission. Variants such as spoke-type, interior PM, and Halbach-array configurations have been developed to further enhance torque density and efficiency. The axial magnetic gear (AMG) (Figure 7.4) employs an axial flux path, making it well-suited for applications with diameter constraints or where hermetic isolation is required, such as pumps and aerospace systems. Advanced designs, including L-shaped and T-shaped flux modulators, have achieved torque densities exceeding 200 kN-m/m^3 in simulations. The linear magnetic gear (LMG) (Figure 7.4) extends the flux-modulation concept to translational motion, providing direct thrust transmission for wave energy converters and linear actuators. Tubular configurations, including those using high-temperature superconductors, have demonstrated thrust force densities above 1.7 MN/m^3, highlighting their potential for high-force, low-maintenance linear drives.

In this configuration, the ferromagnetic pole pieces act as modulators of the magnetic field produced by one rotor. As the rotor's magnetic field interacts with the pole pieces, a spectrum of spatial harmonics is generated in the airgap. By carefully choosing the number of pole pairs on the inner rotor (P_i), outer rotor (P_o), and the number of modulation segments (n_s),

$$n_s = p_i + p_o$$

a synchronous space harmonic is produced that couples effectively with the permanent magnet field of the other rotor. This harmonic interaction enables all magnets on both rotors to participate in torque transmission simultaneously, in contrast to the limited utilisation seen in early, non-modulated topologies. The result is a substantial increase in torque density and efficiency [6, 7].

The kinematic relationship between the three components – the inner rotor (ω_i), outer rotor (ω_o), and the modulation ring (ω_s) – is expressed as follows:

$$p_i \omega_i + p_o \omega_o = n_s \omega_s$$

In the most common operating mode, the modulation ring is held stationary ($\omega_s = 0$), and the gear ratio (G_r) is derived directly from the ratio of pole pairs on the rotors.

$$G_r = \frac{\omega_{\text{high speed}}}{\omega_{\text{low speed}}} = \frac{P_o}{P_i}$$

Depending on which element is fixed, different gear ratios can be achieved, giving modulated magnetic gears flexibility comparable to their mechanical counterparts. The high-speed and low-speed rotors rotate in opposite directions [1, 2, 7].

7.3 Topologies of Rotary Magnetic Gears

7.3.1 Radial-Flux Concentric Magnetic Gears (RF-CMGs)

Among the various magnetic gear configurations proposed to date, the radial-flux concentric magnetic gear (RF-CMG) has emerged as the most widely studied and practically implemented. Its architecture consists of two concentric cylindrical rotors – an inner rotor and an outer rotor – separated by a stationary flux-modulation ring. The flux-modulation ring, made of ferromagnetic pole pieces, plays a crucial role in modulating the spatial harmonics of the magnetic field, thereby enabling synchronous torque transfer between the two PM rotors.

This class of gears is attractive because of its compact geometry, scalable design, and comparatively high torque density, making it suitable for diverse applications ranging from renewable energy systems to transportation and robotics. Over the years, several RF-CMG variants have been developed to optimise torque density, mechanical integrity, and material utilisation. The most prominent configurations are described in the following sections.

7.3.1.1 Surface-Mounted PM (SPM) CMG

The surface-mounted CMG is the baseline and simplest topology, where radially magnetised permanent magnets are mounted directly on the surfaces of both rotors (Figure 7.5). The airgap flux density is shaped primarily by the number of pole pairs and the arrangement of the flux-modulation ring.

- *Advantages:* Simple to construct and cost-effective; suitable for prototyping.
- *Limitations:* At high rotational speeds, surface-mounted magnets experience significant centrifugal forces, leading to mechanical integrity concerns. Torque density is moderate (~70–110 kN-m/m^3).
- *Applications:* Low- to medium-speed applications prioritising manufacturability and cost.

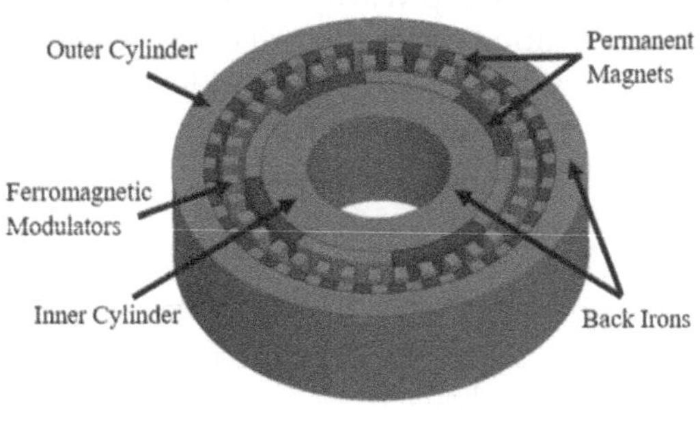

FIGURE 7.5
Surface-mounted permanent magnet coaxial magnetic gear (CMG) [3].

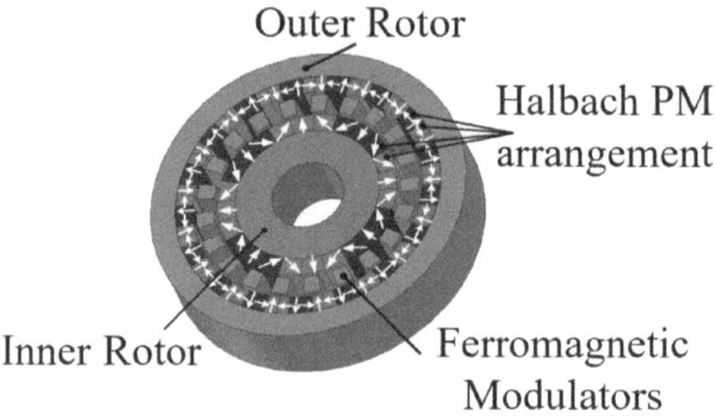

FIGURE 7.6
Halbach magnetised CMG [3].

7.3.1.2 Halbach PM (HPM) CMG

This configuration employs Halbach magnet arrays, where the magnetisation vector is spatially varied around the rotor circumference (Figure 7.6). This produces a nearly sinusoidal airgap flux density [12].

- *Advantages:* High torque density (85–280 kN-m/m^3), reduced cogging torque, and lower iron losses.
- *Limitations:* Complex and expensive magnetisation process; requires precise assembly to avoid demagnetisation.
- *Applications:* High-performance systems such as aerospace propulsion, robotics, and renewable energy.

7.3.1.3 Flux-Focusing (FF) or Spoke-Type CMG

The fundamental principle of the Spoke-type design is right in its name: flux-focusing (Figure 7.7). Instead of the magnets being placed on the surface of the rotor (surface-mounted PMSM) or buried just underneath it (interior PMSM), they are embedded deep within the rotor in a radial, spoke-like pattern, with their magnetisation direction oriented radially. This arrangement allows the rotor structure itself to act as a guide, "focusing" or "concentrating" the magnetic flux from the magnets as it travels through the rotor iron and into the airgap, where it does useful work.

- *Advantages:* High torque density (~150–240 kN-m/m^3), robust mechanical design, especially for high-speed operation.
- *Limitations:* Increased flux leakage; requires careful optimisation using finite element analysis.
- *Applications:* Electric vehicles, traction drives, and aerospace systems requiring robustness and efficiency.

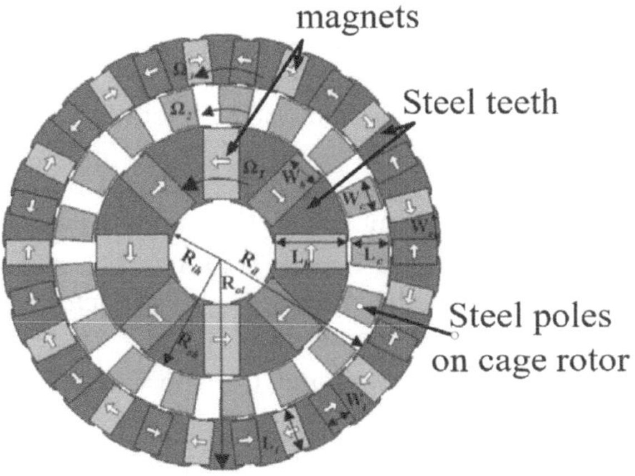

FIGURE 7.7
Spoke-type CMG [3].

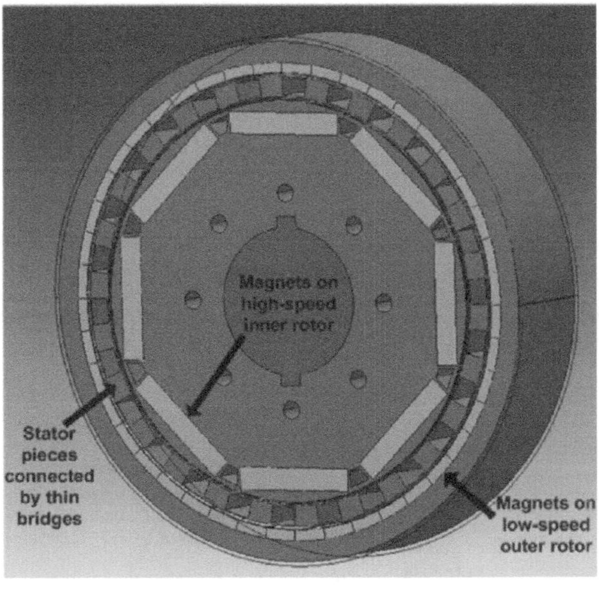

FIGURE 7.8
Interior permanent magnet CMG [3].

7.3.1.4 Interior PM (IPM) CMG

Interior permanent magnet CMGs feature magnets embedded within the rotor structure, rather than on its surface (Figure 7.8). This configuration allows for the creation of strategic flux barriers, which guide the magnetic field efficiently to optimise torque production. A key advantage of designs like the V-shaped IPM is their ability to achieve high torque density with a lower mass of permanent magnets compared to spoke-type designs. Furthermore, innovative geometries, such as using non-rectangular magnets and slotted yokes, contribute to reducing overall iron usage and weight, enhancing the power-to-weight ratio of the CMG.

- *Advantages:* Strong mechanical integrity and efficient PM utilisation.
- *Limitations:* Lower torque density (40–160 kN m/m^3 compared to flux-focusing or Halbach designs; manufacturing complexity increases with advanced geometries.
- *Applications:* Industrial drives and transport systems where structural strength is critical.

7.3.1.5 Unipolar (UP) or Consequent-Pole CMG

The unipolar permanent magnet CMG employs interior-mounted permanent magnets of identical polarity along the rotor circumference, simplifying construction and reducing magnet usage (Figure 7.9). This homopolar configuration lowers material cost and eases manufacturing while still enabling flux modulation through iron pole pieces. However, the achievable torque density is typically lower than that of surface-mounted or Halbach-based CMGs. This design uses PMs of the same polarity, with iron pole pieces completing the magnetic circuit.

- *Advantages:* Significantly reduced PM cost; simplified assembly.
- *Limitations:* Low torque density (~50–55 kN-m/m^3).
- *Applications:* Low-cost products and auxiliary drives where efficiency is less critical.

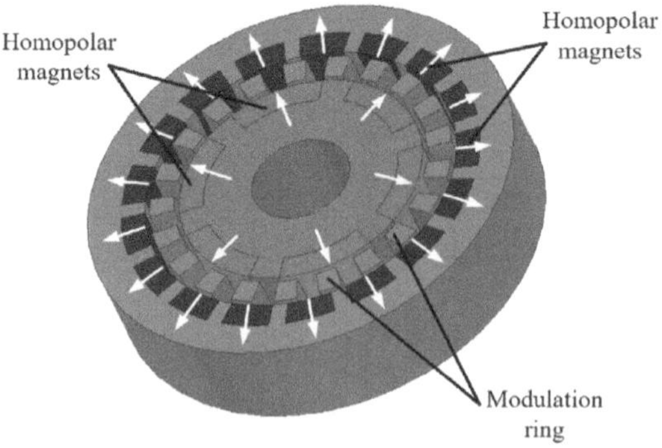

FIGURE 7.9
Unipolar permanent magnet CMG [3].

TABLE 7.1

Comparison of Radial-Flux CMG Topologies [3]

Topology	Key Feature	Torque Density (kN-m/m³)	Advantages	Disadvantages
SPM-CMG	Radially mounted PMs	~70–110	Simple to manufacture	Reduced mechanical strength at high speed
HPM-CMG	Halbach magnetisation	~85–280	High torque density, low cogging torque, reduced iron loss	Complex and costly magnetisation process
FF-CMG	Spoke-type flux concentration	~150–240	High torque density, robust at high speeds	Higher flux leakage, complex design
IPM-CMG	Buried magnets (V-shaped or straight)	~40–160	Strong mechanical integrity, efficient PM use	Moderate torque density, harder to manufacture
UP-CMG	Same-polarity PMs with iron pole pieces	~50–55	Lower PM cost, simplified assembly	Low torque density

7.3.1.6 Comparative Insights

The RF-CMG topologies highlight important trade-offs:

- SPM designs are simple but limited in torque density.
- Halbach and flux-focusing designs achieve high performance but at a higher cost and complexity.
- IPM topologies balance robustness and efficiency.
- Unipolar gears prioritise cost-savings over performance.

7.4 Alternative Rotary Topologies

In addition to radial-flux concentric magnetic gears, several alternative rotary topologies have been proposed to address specific design requirements, such as compact geometry, higher torque density, or ease of integration with specialised applications. The most notable configurations are outlined in the following sections.

7.4.1 Axial-Flux CMGs (AF-CMGs)

Axial flux CMGs (AF-CMGs) implement the magnetic gear principle in the axial direction, where the magnetic flux flows parallel to the axis of rotation

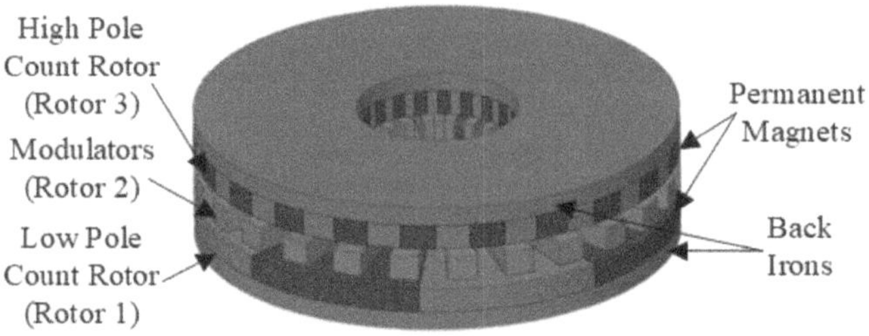

FIGURE 7.10
Axial flux permanent magnet CMG [3].

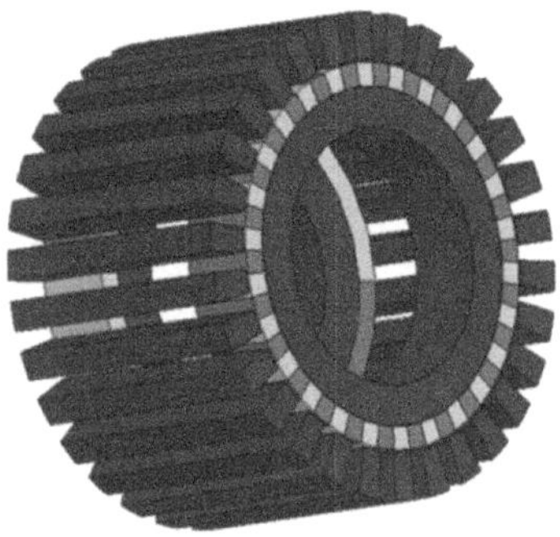

FIGURE 7.11
Transverse-flux CMG [3].

(Figure 7.10). This configuration is particularly advantageous for applications requiring hermetic isolation between the input and output shafts, as it allows for a sealed barrier. In their structure, magnets are mounted on facing rotor disks, and a stationary ring of ferromagnetic pole pieces is axially sandwiched between them to modulate the magnetic field.

A key benefit of this layout is the significant torque density it can achieve, with advanced flux-focusing (FF) designs reaching over 250 kN-m/m^3. Furthermore, these topologies can exhibit very low torque ripple and acceptable levels of axial force, leading to smooth and reliable operation.

- *Performance:* Reported torque densities exceed 70 kN-m/m^3, with advanced axial designs achieving values above 250 kN-m/m^3 [1, 2, 3].
- *Applications:* Pumps, compressors, and compact rotary drives.

7.4.2 Transverse-Flux MGs (TF-MGs)

The transverse-flux CMG is a unique design where the magnetic flux path changes direction, flowing radially in some components and axially in others (Figure 7.11). This key feature allows the high-speed and low-speed rotors to be physically separated, significantly simplifying manufacturing compared to conventional radial or axial designs. However, early TF prototypes suffered from very low torque density. This limitation was overcome by integrating advanced topologies like the flux-focusing (FF) arrangement and innovative pole pieces (L-shaped and T-shaped), which drastically improved torque density by mitigating flux leakage and iron saturation. The trade-off for this high performance in modern TF designs is increased structural complexity, as they now require very short airgaps in both radial and axial directions, posing a significant manufacturing challenge.

- *Limitations of Early Designs:* Initial prototypes suffered from relatively low torque density due to poor flux utilisation.
- *Advancements:* Novel designs employing L-shaped *or* T-shaped modulators have demonstrated significantly improved torque densities and efficiency [2, 3].
- *Applications:* Suitable for cases where relaxed tolerances and manufacturing simplicity are desired.

7.4.3 Emerging Topologies

Ongoing research continues to expand the diversity of MG configurations to meet demanding performance targets. Notable directions include the following:

- *Dual-Flux-Modulator MGs:* These employ two modulator rings to create a more complex flux path, significantly boosting torque transmission capability and density beyond single-modulator designs.
- *Triple-Speed MGs:* Featuring three rotating members, these MGs provide multiple, fixed gear ratios from a single device. This makes them ideal for applications like wind turbines, where power can be managed across different speed regimes without a complex mechanical gearbox.
- *Variable Gear-Ratio MGs (m-CVTs):* Functioning as magnetic continuously variable transmissions, these systems allow for seamless, real-time adjustment of the gear ratio. This enables optimal efficiency in highly dynamic environments, such as electric vehicles, by continuously matching the drive conditions.

7.5 Conclusion

Since the introduction of the flux-modulation principle, magnetic gear technology has undergone remarkable evolution. From basic converted topologies, the field now encompasses high-performance concentric, axial, and linear magnetic gears, achieving torque densities on par with mechanical systems. Integrating magnetic gears with PM machines further enhances their potential for direct-drive applications, offering compact, high-efficiency, and reliable solutions for renewable energy systems, electric transportation, and industrial machinery.

While challenges related to cost, mechanical design, and high-speed losses remain, ongoing research continues to address these limitations. Overall, MGs provide a compelling alternative to conventional mechanical gears, combining high efficiency, reliability, and contactless torque transmission, making them a promising technology for the modern industrial landscape.

References

[1] X. Li, K.-T. Chau, M. Cheng, and W. Hua, "Comparison of magnetic- geared permanent-magnet machines," *Progress in Electromagnetics Research*, vol. 133, pp. 177–198, 2013.

[2] Y. Wang, M. Filippini, N. Bianchi, and P. Alotto, "A review on magnetic gears: Topologies, computational models, and design aspects," *IEEE Transactions on Industry Applications*, vol. 55, no. 5, pp. 4557–4566, 2019.

[3] A. Shoaei and Q. Wang, "A comprehensive review of concentric magnetic gears," *IEEE Transactions on Transportation Electrification*, vol. 10, no. 3, pp. 5581–5598, 2024.

[4] A. H. Neuland, "Apparatus for transmitting power," U.S. Patent 1 171 351, 1916.

[5] H. T. Faus, "Magnet gearing," U.S. Patent 2 243 555, 1941.

[6] K. Atallah and D. Howe, "A novel high-performance magnetic gear," *IEEE Transactions on Magnetics*, vol. 37, no. 4, pp. 2844–2846, 2001.

[7] K. Atallah, S. D. Calverley, and D. Howe, "Design, analysis and realisation of a high-performance magnetic gear," *IEE Proceedings – Electric Power Applications*, vol. 151, no. 2, pp. 135–143, 2004.

[8] K. Ikuta, S. Makita, and S. Arimoto, "Non-contact magnetic gear for micro transmission mechanism," *Proc. IEEE Micro Electro Mechanical Systems*, 1991, pp. 125–130.

[9] S. Kikuchi and K. Tsurumoto, "Design and characteristics of a new magnetic worm gear using permanent magnet," *IEEE Transactions on Magnetics*, vol. 29, no. 6, pp. 2923–2925, 1993.

[10] Y. D. Yao et al., "The radial magnetic coupling studies of perpendicular magnetic gears," *IEEE Transactions on Magnetics*, vol. 32, no. 5, pp. 5061–5063, 1996.

[11] C. C. Huang, M. C. Tsai, D. G. Dorrell, and B. J. Lin, "Development of a magnetic planetary gearbox," *IEEE Transactions on Magnetics*, vol. 44, no. 3, pp. 403–412, 2008.

[12] L. Jian and K. T. Chau, "A coaxial magnetic gear with Halbach permanent-magnet arrays," *IEEE Transactions on Energy Conversion*, vol. 25, no. 2, pp. 319–328, 2010.

8

Prognostics and Health Management for Sustainable Electric Aircraft: Pathways to Reliability and Innovation

Md. Asif Hussain, Nishant Mishra, Ananda Shankar Hati, Altaf Q. H. Badar, and Prashant Kumar

8.1 Introduction of Health Management and Prognostics for Sustainable Electric Aviation

Influenced by an urge to lower its environmental impact and embrace sustainable development, the aviation sector sets off on a transforming trip. Fundamental in this transformation is the entrance of electric aviation, a paradigm change that promises to redefine the ideas of aircraft propulsion, design, and operation. By reducing greenhouse gas emissions, noise pollution, and reliance on fossil fuels, sustainable electric aviation – characterised by electric propulsion systems – aims to offset the environmental impact of air travel. This young discipline presents a set of new challenges, especially in ensuring the dependability and safety of these advanced systems, even if it also offers the possibility for improved operational efficiency and reduced costs [1]

Prognostics and Health Management (PHM), a multidisciplinary approach combining data analytics, predictive modelling, and engineering knowledge to monitor, diagnose, and manage important systems' health, is core to solving these issues. Within the framework of sustainable electric aviation, PHM becomes an indispensable tool since it offers the means to preserve the integrity of electric propulsion systems, which are naturally complex and lack great historical data accessible for conventional jet engines [2, 3]. This introduction looks at the symbiotic interaction between PHM and sustainable electric aviation, clarifying how PHM could help to enable the change to a safer, more ecologically friendly aviation environment [4].

Driven by battery technology and energy storage, where advances in lithium-ion batteries drive performance, central to electric aviation is health monitoring, degradation detection, prediction of remaining useful life, and failure prevention, all of which depend on PHM, guaranteeing dependability

DOI: 10.1201/9781003747994-8

and safety. Beyond the aircraft, PHM maximises infrastructure needs and maintenance ecosystems – including charging stations, battery swapping facilities, and specialised maintenance hangars – by means of PHM to lower downtime and costs. Even if PHM integration increases operational efficiency, dependability, availability, and maintainability also improve. Moreover, to lower aviation's environmental impact, PHM supports environmental sustainability by enhancing waste-minimising techniques and energy use. Economically, PHM lowers maintenance costs and guides market strategies to increase viability; conformance to regulatory criteria from agencies like the FAA and EASA guarantees certification and safety [5]. Leveraging sensor innovations, artificial intelligence, and machine learning, future directions in PHM promise even greater predictive accuracy, thus enabling a sustainable and efficient electric aviation future [6].

The evolution of electric propulsion systems forms the technological foundation upon which the next generation of aircraft is being developed, necessitating robust design frameworks that seamlessly integrate prognostics and health management (PHM) capabilities. As these frameworks mature, the role of advanced battery technologies and energy storage becomes pivotal for operational efficiency, real-time health monitoring, and predictive maintenance. This, in turn, demands a reimagining of supporting infrastructure and maintenance ecosystems capable of sustaining electric aviation. Integrating performance metrics with PHM systems ensures that operational reliability and safety are continuously optimised. Moreover, adopting PHM strategies enhances environmental sustainability by enabling efficient energy usage and reducing life cycle emissions. These advances contribute to the economic viability of electric aircraft by minimising downtime and maintenance costs, while offering valuable market insights. Finally, the ecosystem must align with evolving regulatory standards, ensuring compliance and fostering trust in PHM-integrated electric aviation.

8.2 Core Concepts of Electric Propulsion Systems

Electric propulsion systems turn electrical energy into mechanical thrust to move vehicles, usually spacecraft or electric cars. Instead of burning chemicals, electric propulsion thrusts by speeding up charged particles (ions) or using electromagnetic fields. It needs electrical power sources like batteries, solar panels, or nuclear reactors [2].

Different kinds of electric propulsion:

- *Ion Thrusters:* Use electrostatic or electromagnetic forces to speed up ions: low thrust, high efficiency. Hall-effect thrusters and gridded electrostatic ion thrusters are two examples.

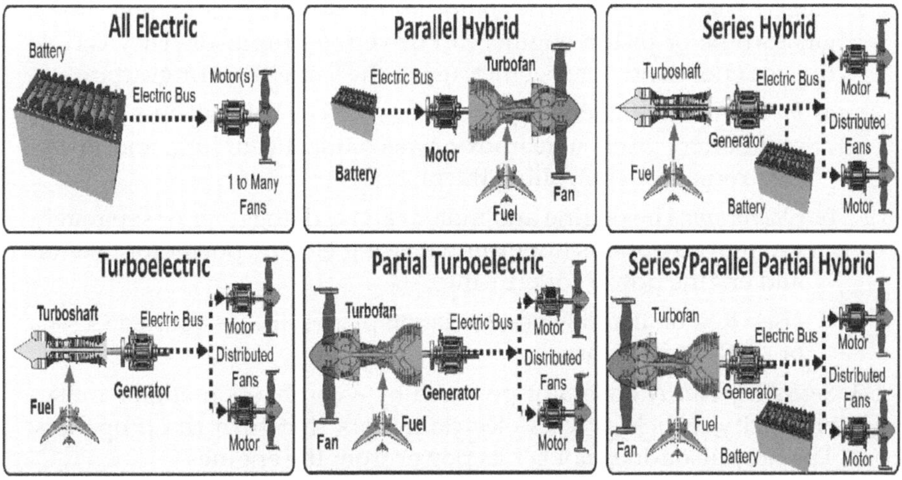

FIGURE 8.1
EAP architectures [2].

- *Electromagnetic Propulsion:* Uses magnetic fields to speed up plasma, like magneto plasma dynamic thrusters. More thrust, but more complicated.
- *Electrothermal Propulsion:* Heats the propellant with electricity and then expands it through a nozzle (like resistojets or arcjets). Not very efficient or powerful.

Electrified aircraft propulsion (EAP) systems use electric power to make aircraft propulsion more efficient, lower emissions, and allow for new aircraft designs. Electric motors, power electronics, and energy storage or generation systems are used in EAP architectures to power propulsors like fans and propellers [7]. Here are the main EAP architecture options, with a focus on their design, parts, and uses:

1. *All-Electric:* There is no combustion engine; electric motors powered by batteries on board do all the work.
    a. *Parts:* Batteries (like lithium-ion, ~200–300 Wh/kg), electric motors (like permanent magnet motors), power electronics (like inverters and controllers), and propulsors (like propellers and ducted fans).
    b. *Operation:* The battery's energy is turned into electricity, which powers motors that turn propellers.
    c. *Uses:* For urban air mobility, small, short-range planes (like Joby Aviation's eVTOLs or Pipistrel Alpha Electro's trainers).

2. *Parallel Hybrid:* An electric motor and a combustion engine (like a gas turbine or piston engine) can drive the propulsor. They can do this independently or together, using the same shaft or gearbox.
   a. *Parts:* A combustion engine (gas turbine or ICE), electric motor(s) and battery pack, a gearbox or mechanical coupling, and power electronics for controlling the motor.
   b. *Operation:* The engine and motor can work together or separately, giving you options (for example, using electric power for take-off and engine power for cruising).
   c. *Uses:* Regional and general aviation planes, like retrofitted Cessna or ATR-sized planes.
3. *Series Hybrid:* A generator powered by a combustion engine makes electricity, which powers electric motors that turn the propellers. The propulsor does not get its power from the engine.
   a. *Parts:* A combustion engine (like a gas turbine or piston engine), a generator to make electricity, batteries (for backup or peak power), electric motors, and power electronics.
   b. *Operation:* Motors are powered by electricity from the engine (or batteries). The engine runs at its best regardless of the speed of the propulsor.
   c. *Uses:* Regional and small commuter planes, like the NASA X-57 Maxwell.
4. *Turboelectric:* Gas turbines turn generators to make electricity, then power electric motors that turn propulsors. No batteries or other energy storage devices are on board; all power comes from burning fuel.
   a. *Parts:* Gas turbine(s), generator(s), electric motors, power distribution system, power electronics, and propulsors (fans, propellers).
   b. *Operation:* The electricity from the turbine is sent to the motors, which lets the propulsors be set up in different ways (for example, distributed propulsion).
   c. *Uses:* Large commercial planes and distributed propulsion designs like NASA STARC-ABL.
5. *Partial Turboelectric:* This type of propulsion uses both electric and mechanical thrust. It combines turboelectric propulsion (electric motors powered by turbine-driven generators) with regular turbine-driven propulsors (like direct-drive fans).
   a. *Parts:* Gas turbine(s) for both direct thrust and making electricity, a generator, electric motors, a power distribution system, and mixed propulsors (like turbine-driven fans and electric propellers).

b. *Operation:* Turbines give some direct thrust and make electricity for electric motors, which benefits hybrid propulsion.

c. *Uses:* Big commercial or regional planes with distributed propulsion.

6. *Series/Parallel Partial Hybrid:* This hybrid architecture uses both series and parallel configurations. The propulsion system can work in series (the engine drives the generator to power the motors), parallel (the engine and motors drive the propulsors directly), or in a mix of both (Table 8.1). Usually comes with batteries for extra flexibility.

a. *Parts:* A combustion engine (gas turbine or ICE), a generator and electric motors, batteries for storing energy or peak power, a gearbox or clutch for parallel operation, and an advanced power management system.

b. *Operation:* The system changes between series and parallel modes depending on the flight phase (for example, series for cruise efficiency and parallel for high-power take-off). Batteries can give you more power.

c. *Uses:* Regional or medium-range planes that need to be flexible, like retrofitted turboprops or new designs.

TABLE 8.1

Comparison of Propulsion Architectures for Electrified Aircraft [2]

Architecture	Power Source	Thrust Source	Applications	Emissions	Complexity	Range
All-Electric	Batteries	Electric motors	eVTOL, short-range	Zero	Low	Short
Parallel hybrid	Engine + batteries	Engine + motors	Regional, general aviation	Reduced	Medium	Medium
Series hybrid	Engine + generator (+batteries)	Electric motors	Regional, small commuters	Reduced	Medium	Medium
Turboelectric	Turbine + generator	Electric motors	Large aircraft, distributed	Same as turbine	High	Long
Partial turboelectric	Turbine (+generator)	Turbine + motors	Large/regional, distributed	Reduced	High	Long
Series/parallel partial	Engine + batteries + generator	Engine + motors	Regional, flexible designs	Reduced	Very high	Medium-long

Important trade-offs

- *Efficiency versus Range:* All-electric cars have no emissions but a short range. Hybrid and turboelectric cars have longer ranges but are more complicated and have some emissions.
- *Complexity:* All-electric is the easiest; series/parallel partial hybrid is the hardest because it can work in two modes.
- *Scalability:* Turboelectric and partial turboelectric work well on bigger planes, but all-electric only works on smaller, short-range platforms.
- *Distributed Propulsion:* Series, turboelectric, and series/parallel architectures allow for distributed propulsion, improving aerodynamics using more than one propulsor.

Electric propulsion systems are the foundational technology for developing next-generation aircraft, offering improved efficiency, reduced emissions, and simplified mechanical architectures. These core concepts directly influence the creation of design frameworks for electric aircraft that incorporate prognostics and health management (PHM) capabilities. By understanding electric propulsion components' unique operational characteristics and failure modes, designers can develop PHM-integrated architectures that enable real-time monitoring, fault prediction, and enhanced system reliability. This synergy ensures that the aircraft is optimised for performance and energy use and built with intelligent health-aware features from the ground up.

8.3 Design Frameworks for PHM-Enabled Electric Aircraft

Designing prognostics and health management (PHM)-enabled systems for electrified aircraft propulsion (EAP) requires a structured framework to monitor, diagnose, and predict subsystem degradation, faults, failure modes, and their effects. These frameworks enhance safety, reliability, and efficiency by leveraging real-time data, predictive modelling, and condition-based maintenance (CBM) [8]. Below is a comprehensive design framework for PHM-enabled electric aircraft, tailored to EAP subsystems, focusing on degradation, faults, failure modes, and effects, integrating insights from relevant research.

Objectives: Detect and mitigate faults to prevent catastrophic failures in EAP systems (e.g., electric motors, batteries, power electronics). Predict remaining useful life (RUL) to ensure operational integrity – optimise maintenance through CBM, reducing downtime and costs. Address unique EAP challenges, such as high-voltage systems, thermal management, and distributed propulsion.

EAP Subsystems and Failure Modes: EAP subsystems include batteries, electric motors, power electronics (inverters, converters), generators (for hybrid/turboelectric), and propulsors. Key failure modes and effects include the following:

- *Batteries:* Capacity fades and internal resistance increases due to cycling or thermal stress. Short circuits, thermal runaway, and cell imbalance. Reduced range, power loss, and fire risk.
- *Electric Motors:* Magnet demagnetisation, bearing wear, winding insulation breakdown. Turn-to-turn short circuits, eccentricity, and dry friction. Reduced torque, overheating, and mechanical failure.
- *Power Electronics:* Capacitor ageing, thermal fatigue in semiconductors. Switching failures, short/open circuits. Power delivery disruption, system instability.
- *Generators (Hybrid/Turboelectric):* Efficiency loss, rotor/stator wear. Diode failures, winding faults. Reduced power generation, voltage oscillations.
- *Propulsors:* Blade erosion, bearing wear. Imbalance, structural failure. Reduced thrust, vibration, and aerodynamic inefficiency.

8.3.1 PHM Framework Components

The PHM framework integrates sensing, diagnostics, prognostics, and decision-making, tailored to EAP architectures (all-electric, series/parallel hybrid, turboelectric, etc.).

1. **Data Acquisition and Sensing**
    (a) *Sensors:* Voltage, current, and temperature sensors for batteries and power electronics. Vibration, torque, and speed sensors for motors and propulsors. Frequency and power quality sensors for generators.
    (b) *Data Types:* Real-time operational data (e.g., motor current, battery state-of-charge). Environmental data (e.g., temperature, humidity). Historical data for baseline comparisons.
    (c) *Challenges:* Sensor noise and data imbalance (faults are rare). Limited sensor placement in lightweight designs.
2. **Diagnostics**
    (a) *Objective:* Detect, identify, and isolate faults in EAP subsystems.
    (b) *Methods:*
        i. *Physics-Based Models:* Simulate subsystem behaviour (e.g., battery electrochemical models, motor electromagnetic models) to detect deviations.

ii. *Data-Driven Approaches:* Use machine learning (e.g., neural networks, random forests) to classify faults based on sensor data patterns.

iii. *Hybrid Approaches:* Combine physics-based models with data-driven methods for improved accuracy (e.g., particle filtering for motor winding faults).

Key Techniques:

- Feature extraction (e.g., current harmonics, temperature gradients).
- Anomaly detection (e.g., comparing real-time data to healthy baselines).
- Fault classification (e.g., distinguishing short circuits from eccentricity).

Example: Motor current signature analysis for detecting turn-to-turn shorts in electric motors.

3. **Prognostics**
   a. *Objective:* Predict RUL and fault progression to enable proactive maintenance.
   b. *Methods:*
      i. *Model-Based Prognostics:* To estimate RUL, use physics-based degradation models (e.g., Paris' law for mechanical wear, Arrhenius models for battery ageing).
      ii. *Data-Driven Prognostics:* Employ algorithms like Gaussian process regression (GPR) or deep residual-recurrent neural networks (DR-RNN) to predict degradation trends.
      iii. *Metaheuristic Algorithms:* Use differential evolution (DE) or particle swarm optimisation (PSO) to optimise prognostic models under uncertainty.

Key Metrics:

- RUL estimation accuracy.
- Prediction horizon (time before failure).
- Confidence intervals for prognostic uncertainty.

Example: Particle filtering to predict battery capacity fade based on cycle data.

4. **Decision-Making and Maintenance**
   a. *Objective:* Optimise maintenance scheduling and system reconfiguration.
   b. *Strategies:*
      i. *Condition-Based Maintenance (CBM):* Repair or replace components based on PHM predictions, reducing unnecessary maintenance.

ii. *Adaptive Power Management:* Reconfigure power distribution in hybrid/turboelectric systems to mitigate faults (e.g., re-routing power to healthy motors).
iii. *Mission Reconfiguration:* Adjust flight profiles (e.g., reduce thrust demand) to extend component life during faults.

Tools:

- Digital twins for real-time simulation and decision support.
- Deep reinforcement learning (DRL) for optimising maintenance schedules.

Example: Tri-redundant power management system using FPGA-DSP controllers to ensure safe operation during generator faults.

8.3.2 Framework Implementation

1. **System Architecture**
   a. *On-Board PHM:* Embedded PHM computer (PHMC) in the avionics bay for real-time diagnostics/prognostics. Low-fidelity models for computational efficiency (e.g., 80 Hz down sampling).
   b. *Off-Board PHM:* Ground-based analysis using high-fidelity models and historical data. Cloud-based digital twins for fleet-wide monitoring.
   c. *Data Flow:*

   Sensor data → Preprocessing (e.g., exponential filtering) → Feature extraction → Diagnostics/prognostics → Decision-making, as shown in Figure 8.2.

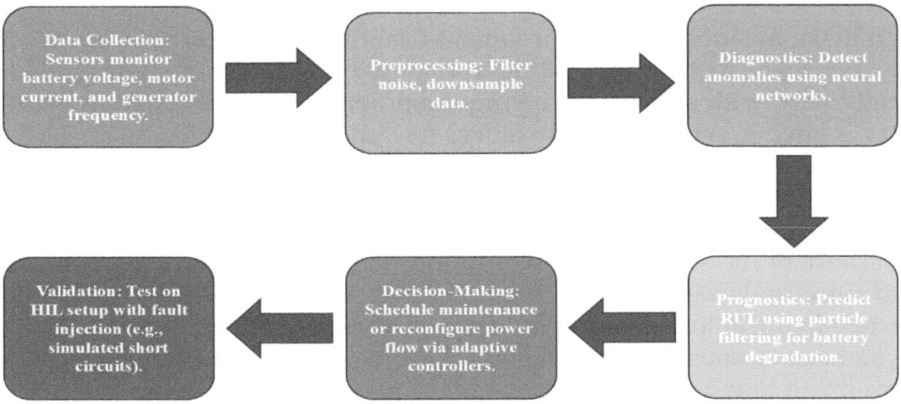

FIGURE 8.2
Flowchart for the PHM framework to EAP architectures.

2. **Validation and Testing**
   a. *Simulation:* Use physics-based models to simulate degradation (e.g., battery capacity fade, motor winding shorts). Inject faults (e.g., via fault injection mechanisms) to create failure databases.
   b. *Hardware-in-the-Loop (HIL):* Test PHM algorithms on iron bird setups or test stands with realistic conditions.
   c. *Flight Testing:* Validate PHM on prototype aircraft (e.g., NASA X-57 for series hybrid).

Metrics:

- Fault detection rate, false positive rate.
- RUL prediction accuracy (e.g., mean absolute error).
- Computational latency for real-time applications.

Design frameworks for PHM-enabled electric aircraft must strategically incorporate battery technology and energy storage systems as central elements, given their critical role in powering electric propulsion. These frameworks are shaped by performance and safety requirements and the need to embed health monitoring capabilities within the battery systems. Advanced energy storage solutions demand continuous tracking of parameters such as state-of-charge, state-of-health, and thermal behaviour, which are integral to PHM functionality. By aligning design principles with the operational characteristics and degradation patterns of battery systems, engineers can ensure that the aircraft maintains optimal performance, longevity, and safety throughout its life cycle.

8.4 Battery Technology, Energy Storage, and Health Monitoring

To make aviation more environment-friendly, battery technology needs to improve. Battery innovation and development are moving quickly right now, with a focus on solving the problems that the aviation industry is having. As the industry moves towards electrification, the need for high-performance batteries has grown significantly. Right now, lithium-ion batteries are the most promising and widely used type of battery for electric applications, although many problems still need to be solved. Because of the strict safety rules in the aviation industry, it is important to fully understand how lithium-ion batteries can overheat, especially when flying at low pressure [9]. To keep batteries safe in these situations, you need to know a lot about how cells break down, how to manage heat well, and how to use advanced battery management systems (ABMS) to lower the risks.

- *Edge Devices:* On-board computational units process sensor data in real time, enabling preliminary health assessments and reducing the need for constant data transmission to ground stations. This is particularly important for electric aircraft, where rapid response to battery or motor anomalies can prevent critical failures.
- *Communication Networks:* High-speed, low-latency networks, such as 5G or IoT protocols, facilitate the transfer of large volumes of data between the aircraft and ground-based PHM systems. These networks ensure seamless data flow for real-time monitoring and analysis, which is critical for fleet-wide operations.

8.5.1.2 Digital Twin and Simulation Platforms

- *Digital Twin Frameworks:* Virtual replicas of physical aircraft systems allow for predictive simulations and "what–if" scenario testing. Digital twins are particularly valuable for electric aircraft, where complex interactions between electrical and mechanical components require advanced modelling to predict performance and degradation [17, 18].
- *Simulation Infrastructure:* High-performance computing (HPC) environments are necessary to run complex PHM algorithms and stress scenarios. These systems support real-time monitoring and prediction, ensuring that PHM can handle the computational demands of electric aircraft operations.

8.5.1.3 Data Management and Cloud Infrastructure

- *Big Data Storage:* Scalable databases and data lakes are required to manage the massive volumes of operational data generated by electric aircraft, including sensor data, maintenance records, and simulation results. This infrastructure must support the storage and retrieval of large datasets for analysis [18].
- *Cloud Computing:* Centralised analytics, model training, and maintenance scheduling insights are performed in the cloud, providing a flexible and scalable solution for PHM. Cloud platforms enable data integration from multiple aircraft, facilitating fleet-level health management.

8.5.2 Maintenance Ecosystems Enabling PHM

PHM transforms traditional maintenance practices by enabling condition-based maintenance (CBM) and leveraging advanced technologies for decision-making and training. The following sections describe the key components of the maintenance ecosystem.

8.5.2.1 Condition-Based Maintenance (CBM) Facilities

- *CBM Approach:* CBM shifts maintenance from scheduled checks to need-based servicing, using real-time diagnostics to identify when components require attention. For electric aircraft, this is critical for battery systems, where subtle degradation can impact performance and safety. CBM reduces unnecessary maintenance, extending component lifespan and lowering costs [20].
- *Mobile Inspection Units:* Integration of mobile inspection units, such as drones or AR glasses, with augmented diagnostics allows for on-site, real-time assessments. These tools enhance efficiency by providing immediate feedback during inspections, reducing downtime for electric aircraft.

8.5.2.2 AI-Driven Decision Support Systems

- *Explainable AI (XAI):* XAI assists human operators in understanding predictive insights and maintenance recommendations, ensuring transparency and trust. This is particularly important for electric aircraft, where complex systems require clear decision-making processes [15, 17].
- *Integration with Planning Tools:* AI-driven systems integrate with maintenance planning tools and supply chain logistics to optimise scheduling and resource allocation. This minimises disruptions and ensures that spare parts and technicians are available when needed.

8.5.2.3 Augmented and Virtual Reality (AR/VR) for Training and Support

- *VR Training Modules:* Immersive virtual reality training prepares technicians for the unique challenges of maintaining electric aircraft, such as handling high-voltage systems or interpreting complex electrical schematics. These modules enhance skill development and safety [19].
- *AR Overlays:* Augmented reality provides in situ guidance during repair or inspection tasks, displaying real-time data and instructions via AR glasses. This improves accuracy and efficiency, particularly for complex electric aircraft systems.

A systematic flowchart depicting the levels of infrastructure requirements and maintenance ecosystem is shown in Figure 8.5. It shows different modules which need to be taken into account for the proper functioning of automated aircraft propulsion systems with the use of AI as a tool incorporated for operational and maintenance purposes.

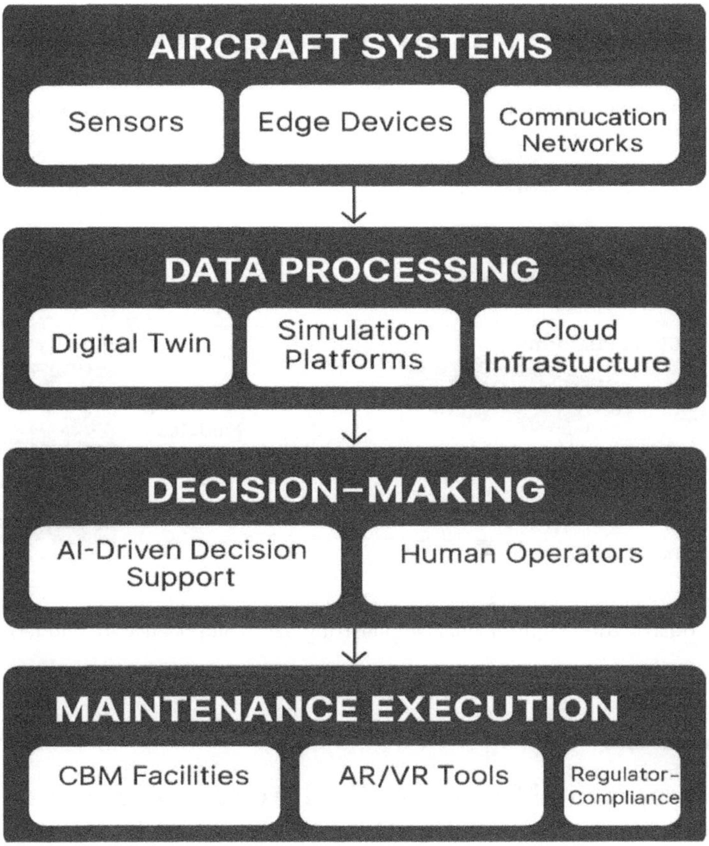

FIGURE 8.5
Flowchart depicting levels of infrastructure and maintenance ecosystem.

Figure 8.6 outlines the critical infrastructure requirements for modern aerospace systems, emphasising four key components: sensor systems, communication networks, cloud platforms, and AR/VR tools. Sensor systems are vital for monitoring battery health and motor performance in real time, ensuring the safety and efficiency of electric aircraft. Communication networks, particularly 5G and IoT, enable high-speed, low-latency data exchange between on-board systems and external platforms. Cloud platforms support big-data storage and cloud computing, facilitating predictive analytics, digital twin applications, and real-time diagnostics. Lastly, AR/VR tools enhance training and maintenance through immersive virtual simulations and real-time augmented reality overlays, streamlining operational workflows and boosting technician effectiveness. Together, these components form the backbone of a smart, connected, and efficient aviation ecosystem.

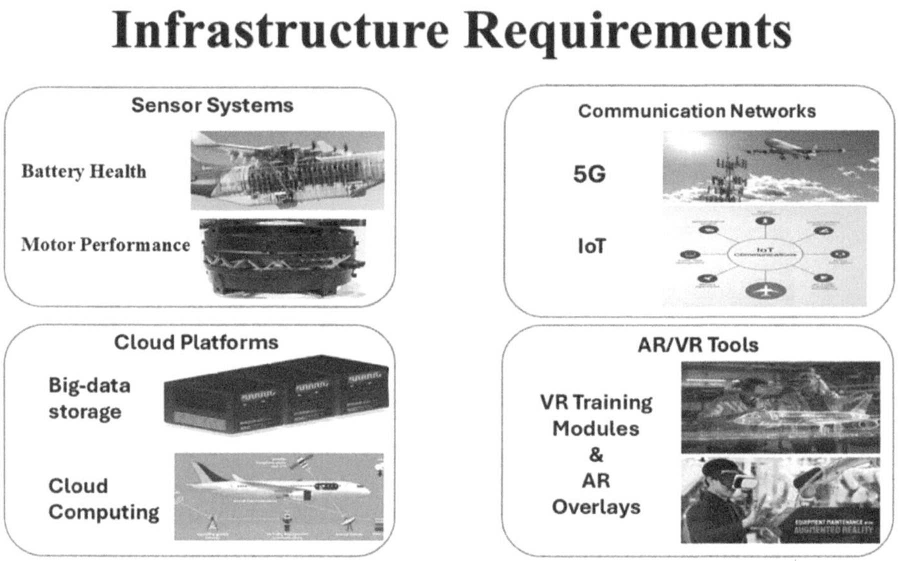

FIGURE 8.6
Critical infrastructure requirements emphasising key components for modern aerospace systems.

8.6 Environmental Sustainability through Advanced PHM Strategies for Sustainable Electric Aircraft

The aviation sector is at a critical juncture, compelled by global demands to curb greenhouse gas emissions, reduce ecological footprints, and transition to sustainable propulsion technologies [24]. Electric aircraft represent a transformative solution, delivering zero in-flight emissions, significantly reduced noise pollution, and lower operational costs than traditional fossil-fuel-powered aircraft [25]. Despite these advantages, challenges such as battery reliability, system safety, and energy efficiency must be addressed to ensure their scalability and commercial viability [27, 28].

Prognostics and health management (PHM) is a cornerstone in overcoming these hurdles. By leveraging real-time data, predictive analytics, and advanced diagnostics, PHM enhances electric aircraft systems' reliability, safety, and longevity. Furthermore, PHM contributes to environmental sustainability by optimising resource use, minimising waste, and supporting life cycle management, aligning with global efforts to achieve net-zero aviation [23, 26].

The study investigates how advanced PHM strategies can bolster the sustainability and operational efficiency of electric aircraft by achieving the following goals:

- *Minimising Operational Disruptions:* Reducing unscheduled maintenance and downtime through proactive fault detection [21, 22].
- *Extending Component Lifespan:* Enhancing the durability of critical systems, such as batteries and electric motors, to maximise service life.
- *Optimising Resource Utilisation:* Decreasing energy consumption and material waste through data-driven maintenance practices.
- *Reducing Environmental Impact:* Supporting lower emissions and sustainable practices throughout the aircraft's life cycle.
- *Enhancing Safety and Reliability:* Ensuring robust performance under diverse operating conditions to meet stringent aviation standards [24].

Figure 8.7 illustrates the integration of cutting-edge technologies in prognostics and health management (PHM) for modern aircraft systems. It highlights the role of digital twins and edge computing in real-time simulation and feedback, supporting predictive maintenance. The fusion of physics-based and data-driven methods is emphasised through hybrid PHM models, which enable more accurate and robust fault detection. Machine learning and explainable AI enhance decision-making transparency and reliability. Additionally, AR/VR-based PHM enhancements and battery

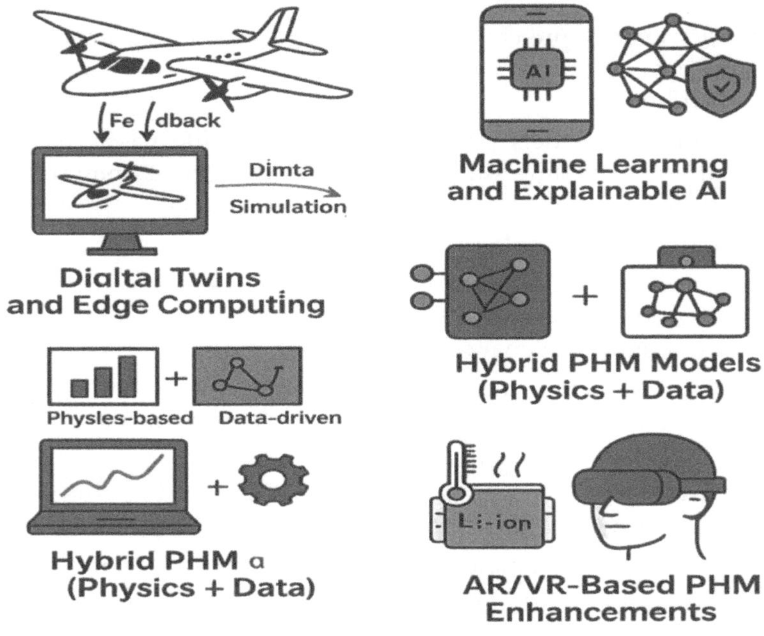

FIGURE 8.7
Advanced PHM technologies for next-generation aircraft systems.

health monitoring (e.g., Li-ion systems) are included, showcasing immersive and intelligent support tools for advanced diagnostics and maintenance planning.

8.6.1 Advanced PHM Techniques and Strategies

The following techniques form the backbone of PHM implementation in electric aircraft, driving both performance and sustainability:

(a) **Real-Time Condition Monitoring**

Implementation: Deploys advanced sensors and Internet of Things (IoT) technologies to monitor critical components, including lithium-ion batteries, electric propulsion systems, power electronics, and airframe structures.

Benefits: Detects early signs of degradation or anomalies, preventing system failures and improving safety.

Example: Battery health monitoring tracks voltage, temperature, and charge cycles to predict performance degradation.

(b) **Predictive Analytics and Machine Learning**

Implementation: Employs machine learning algorithms trained on historical and real-time data to forecast component failures and optimise operational parameters.

Benefits: Enables pre-emptive maintenance scheduling and enhances energy efficiency by optimising flight profiles and power distribution.

Example: AI models predict battery end-of-life based on usage patterns, enabling timely replacements without compromising safety.

(c) **Digital Twin Technology**

Implementation: Develops a real-time digital replica of the aircraft, integrating sensor data to simulate performance, environmental stress, and maintenance requirements.

Benefits: Facilitates scenario analysis, supports predictive maintenance, and informs sustainable design improvements.

Example: A digital twin of an electric propulsion system simulates thermal stress under varying flight conditions to optimise cooling strategies.

(d) **Data-Driven Maintenance and Life Cycle Optimisation**

Implementation: Shifts from traditional scheduled maintenance to condition-based and predictive maintenance, informed by real-time data and usage patterns.

Benefits: Reduces unnecessary part replacements, lowers spare part inventories, and minimises waste, contributing to cost-savings and sustainability.

Example: Usage-based maintenance for electric motors reduces overhaul frequency by aligning interventions with actual wear.

(e) **Integration with Sustainable Materials and Recycling**

Implementation: Tracks component health to support closed-loop recycling systems, identifying opportunities for repurposing or recycling materials like batteries and composites.

Benefits: Reduces environmental impact by extending material life cycles and minimising landfill waste.

Example: PHM data guides the repurposing of degraded batteries for secondary applications, such as ground-based energy storage.

8.6.2 Summary

Advanced PHM strategies are pivotal in realising the full potential of electric aircraft as a sustainable alternative to conventional aviation. PHM enhances system reliability, reduces operational costs, and minimises environmental impacts by integrating real-time monitoring, predictive analytics, digital twins, and life cycle optimisation. These strategies ensure electric aircraft meet rigorous safety and performance standards while aligning with global sustainability objectives. As the aviation industry evolves, PHM will play an indispensable role in fostering a resilient, efficient, and environmentally responsible aerospace ecosystem.

8.7 Regulatory Standards and Compliance for PHM Systems

As the aviation industry accelerates towards electrification to meet global climate commitments, the reliability and safety of electric aircraft systems have become paramount. These aircraft, powered by batteries, hybrid-electric systems, or fuel cells, operate under new thermal, electrical, and mechanical stress profiles that differ significantly from conventional propulsion [35]. Prognostics and health management (PHM) addresses these operational uncertainties by enabling real-time condition monitoring and predictive maintenance of key components such as electric motors, power electronics, energy storage systems, and thermal management subsystems.

Given the safety-critical nature of aviation, the deployment of PHM systems in electric aircraft must comply with stringent regulatory frameworks. Moreover, integrating novel technologies such as AI-based diagnostics and digital twins demands regulatory evolution. This chapter examines the current and emerging regulatory standards applicable to PHM in the context of sustainable electric aviation.

8.7.1 Regulatory Standards and Guidelines

8.7.1.1 Aerospace Safety Standards

Electric aircraft must meet the same airworthiness certification standards as conventional aircraft, with additional scrutiny on electrified components. Key standards include the following [29, 35]:

- RTCA DO-178C / DO-330: Software assurance for airborne systems.
- RTCA DO-254: Hardware assurance standards for complex electronics.
- SAE ARP 4754A / ARP 4761: System-level safety assessment and development processes.
- DO-160G: Environmental conditions for testing airborne equipment (especially relevant to battery thermal and vibration profiles).
- DO-311A: Safety standards for rechargeable lithium battery systems in aircraft.

8.7.1.2 PHM- and IVHM-Specific Standards

PHM systems are governed by additional guidance and recommended practices [30, 32]:

- *SAE ARP 6313:* Guidelines for integrated vehicle health management (IVHM) systems in aerospace platforms.
- *SAE JA6268:* Data interoperability and performance metrics for PHM.
- *FAA AC 33.76:* Advisory circular on engine health management systems for predictive maintenance.

8.7.1.3 Electric Propulsion and Sustainability-Focused Guidance

Emerging initiatives from EASA and FAA aim to provide updated guidance specific to electric and hybrid propulsion systems [31, 35]:

- *EASA Special Condition for Electric Propulsion (SC-E19):* Safety and compliance for electric propulsion in general aviation.
- *FAA MOSA (Modular Open System Approach):* Encouraging flexible PHM architectures for electric platforms.
- *SAE AIR6894:* Safety and risk-based electric/hybrid propulsion certification strategies.

8.7.1.4 Cybersecurity and Data Integrity

Cybersecurity becomes a compliance-critical area as electric aircraft incorporate connected health monitoring systems and cloud-based analytics. Regulatory frameworks relevant to PHM cybersecurity include the following [33, 34]:

- *DO-326A / ED-202A:* Airborne system security requirements.
- *NIST SP 800–82 / ISO 21434:* Cybersecurity for embedded and automotive systems, often leveraged for similar architectures in electric aviation.
- *EU Cyber Resilience Act (CRA):* Impacts digital platforms and onboard analytics used in European aircraft.

Securing sensor networks, edge processors, and data pipelines is essential to maintaining PHM system integrity and avoiding certification denial.

8.7.1.5 Trends towards Regulatory Evolution

The transition to sustainable aviation necessitates new approaches to compliance [29, 30, 31, 35]:

8.7.1.5.1 Digital Twin-Based Certification

Digital twins allow in situ PHM logic validation against physical asset behaviour, supporting virtual certification loops and reducing physical testing costs.

8.7.1.5.2 Certification of AI Models

Work is underway to formalise methods for verifying explainability, performance bounds, and robustness of AI models used in PHM, for example, via RTCA and EUROCAE working groups.

8.7.1.5.3 Life Cycle Compliance Tracking

Continuous airworthiness and sustainability metrics (e.g., battery state of health, carbon offset logs) are being considered for digital compliance records.

8.8 Future Directions: Innovations in PHM for Electric Aviation

As the aviation industry accelerates towards sustainable, electric, and autonomous air mobility, prognostics and health management (PHM) is emerging

TABLE 8.2

Challenges and Mitigation Techniques for the Aviation Industry

Challenge	Mitigation
Data scarcity	Semi-supervised learning, transfer learning
Power constraints	Low-energy edge AI systems
Certification complexity	Early collaboration with regulators
System integration	Interoperable, modular PHM frameworks

as a critical enabler of safety, reliability, and operational efficiency. Electric aircraft face unique challenges due to their reliance on high-power components such as battery systems, electric motors, and power electronics, all of which demand advanced, intelligent maintenance solutions (Table 8.2).

This report outlines key future directions and innovations in PHM tailored for electric aviation, highlighting the integration of cutting-edge technologies, adaptive methodologies, and regulatory alignment.

8.8.1 Digital Twins and Edge Computing

Digital twins provide virtual replicas of physical systems and are expected to become central in PHM through the following [36]:

- Real-time fleet monitoring of electric aircraft components (batteries, motors, etc.).
- Multi-physics simulations to predict failure modes (e.g., thermal/electrical stress).
- Edge computing integration enables on-board diagnostics with minimal latency for eVTOLs and autonomous aircraft.

Impact: Up to 20% reduction in unscheduled maintenance and enhanced aircraft availability.

8.8.2 Machine Learning (ML) and Explainable AI (XAI)

Future PHM will leverage advanced ML and DL techniques to manage complex data from electric systems [37]:

- Semi-supervised/unsupervised learning to handle limited failure data.
- Transfer learning from EV and UAV systems to electric aircraft.
- Explainable AI (XAI) for regulatory transparency in predictions.

Impact: ML-driven systems have shown up to 88% accuracy in predicting component failure.

8.8.3 Hybrid PHM Models (Physics + Data)

Combining physics-of-failure models with data-driven algorithms leads to the following [38]:

- More accurate RUL predictions for components like batteries and motors.
- Self-adaptive models that refine predictions using live data.
- System-wide prognostics for cascading failure detection.

Impact: Reduced prediction errors and a 10–20% decrease in maintenance costs.

8.8.4 PHM for Batteries and Fuel Cells

Battery systems and PEFCs are crucial in electric aviation. Key PHM advancements include the following [38]:

- Deep learning models for SoH prediction with minimal data.
- Hybrid models for PEFC degradation mechanisms (e.g., catalyst wear).
- Thermal management integration to detect heat-induced failures.

Impact: Up to 20% extended component life and enhanced safety.

8.8.5 Autonomous and Real-Time PHM

Autonomous aircraft require lightweight, real-time PHM for on-board decision-making [39]:

- Edge-enabled PHM systems for real-time diagnostics and RUL computation.
- Vehicle-to-cloud integration for continuous fault tracking.
- Autonomous flight re-routing or load balancing based on component health.

Impact: Significantly reduced in-flight failure risk and enhanced operational autonomy.

8.8.6 AR/VR-Based PHM Enhancements

AR and VR will revolutionise PHM in electric aviation maintenance [40]:

- AR-guided diagnostics overlay health data onto components in real time.

- VR training for technicians on PHM-enabled aircraft maintenance.
- Remote support with real-time visualisation for field servicing.

Impact: 30% reduction in maintenance time and improved repair accuracy.

8.8.7 Standardisation and Certification

Widespread adoption requires regulatory alignment:

- PHM certification frameworks in collaboration with the FAA and EASA.
- Standardised performance metrics (e.g., RUL accuracy, false-positive rate).
- Cross-industry cooperation for interoperability and validation.

Impact: Accelerates certification and builds stakeholder trust.

8.8.8 Economic and Operational Impact

Cost-benefit optimisation is vital for PHM adoption:

- Life cycle cost-benefit analysis shows improved net present value with PHM.
- Thin-haul applications (four to nine seaters) can achieve up to 30% DOC reduction.
- Scalable PHM solutions are adaptable from eVTOLs to regional aircraft.

Impact: 10–20% reduction in maintenance costs and increased utilisation.

8.9 Conclusion

The integration of prognostics and health management (PHM) is essential for the advancement and sustainability of electric aviation. The industry addresses critical challenges related to system reliability, safety, and environmental impact by embedding PHM into the core of electric aircraft design and operations. PHM ensures continuous health monitoring and predictive maintenance for complex electric propulsion systems, which lack extensive historical data compared to conventional aircraft engines. This proactive approach extends system lifespan, improves operational efficiency, and supports the reduction of life cycle emissions and maintenance costs, which are key contributors to sustainable aviation.

Furthermore, PHM enhances infrastructure readiness, regulatory compliance, and economic viability by facilitating real-time diagnostics and optimised maintenance strategies. Emerging technologies such as advanced sensors, artificial intelligence, and machine learning strengthen PHM's capabilities, offering greater predictive accuracy and a stronger foundation for the evolving electric aviation ecosystem. Addressing unique challenges, including data scarcity and certification complexity, PHM accelerates the path towards safer, more efficient, and environmentally responsible air travel.

Future PHM systems for electric aviation must integrate digital twins, AI/ML, hybrid models, and real-time edge computing to support the next generation of sustainable and autonomous aircraft. Standardisation, AR/VR support, and cost-optimised solutions will further accelerate the deployment of PHM, aligning with global goals for zero-emission aviation by 2050.

References

[1] L. Tang, K. Younsi, and A. Saxena, "Prognostics and health management for electrified aircraft propulsion: State of the art and challenges," *Journal of Engineering for Gas Turbines and Power*, pp. 1–16, Sep. 2024, https://doi.org/10.1115/1.4066598.

[2] D. L. Simon, and NASA Glenn Research Center, "Health management considerations for electrified aircraft propulsion systems," report, 2022. Available: https://ntrs.nasa.gov/api/citations/20220004260/downloads/EAP_Health MangementConsideration_DSimon_v2.pdf.

[3] B. Sarlioglu and C. T. Morris, "More electric aircraft: Review, challenges, and opportunities for commercial transport aircraft," *IEEE Transactions on Transportation Electrification*, vol. 1, no. 1, pp. 54–64, Jun. 2015, https://doi.org/10.1109/tte.2015.2426499.

[4] S. R. Hashemi *et al.*, "A fast diagnosis methodology for typical faults of a Lithium-Ion battery in electric and hybrid electric aircraft," *Journal of Electrochemical Energy Conversion and Storage*, vol. 17, no. 1, Sep. 2019, https://doi.org/10.1115/1.4044956.

[5] X. Hu, K. Zhang, K. Liu, X. Lin, S. Dey, and S. Onori, "Advanced fault diagnosis for Lithium-Ion battery systems: A review of fault mechanisms, fault features, and diagnosis procedures," *IEEE Industrial Electronics Magazine*, vol. 14, no. 3, pp. 65–91, Sep. 2020, https://doi.org/10.1109/mie.2020.2964814.

[6] V. Samavatian, M. Fotuhi-Firuzabad, P. Dehghanian, and F. Blaabjerg, "Reliability modeling of multistate degraded power electronic converters with simultaneous exposure to dependent competing failure processes," *IEEE Access*, vol. 9, pp. 67096–67108, Jan. 2021, https://doi.org/10.1109/access.2021.3075974.

[7] D. L. Simon, R. Thomas, and K. M. Dunlap, "Considerations for the extension of gas path analysis to electrified aircraft propulsion systems," *The American Society of Mechanical Engineering*, Jun. 2021, https://doi.org/10.1115/gt2021-58578.

[8] "Remaining useful life prediction and state of health diagnosis for Lithium-Ion batteries using particle filter and support vector regression," *IEEE Journals & Magazine | IEEE Xplore*, Jul. 1, 2018. Available: https://ieeexplore.ieee.org/document/8186223
[9] S. Saxena, C. L. Floch, J. MacDonald, and S. Moura, "Quantifying EV battery end-of-life through analysis of travel needs with vehicle powertrain models," *Journal of Power Sources*, vol. 282, pp. 265–276, Jan. 2015, https://doi.org/10.1016/j.jpowsour.2015.01.072.
[10] T. Pattanayak and D. Mavris, "Battery technology for sustainable aviation: A review of current trends and future prospects," *Applied Energy*, vol. 397, p. 126356, Jul. 2025, https://doi.org/10.1016/j.apenergy.2025.126356.
[11] C.-X. Zu and H. Li, "Thermodynamic analysis on energy densities of batteries," *Energy & Environmental Science*, vol. 4, no. 8, p. 2614, Jan. 2011, https://doi.org/10.1039/c0ee00777c.
[12] S. Saxena, C. L. Floch, J. MacDonald, and S. Moura, "Quantifying EV battery end-of-life through analysis of travel needs with vehicle powertrain models," *Journal of Power Sources*, vol. 282, pp. 265–276, Jan. 2015, https://doi.org/10.1016/j.jpowsour.2015.01.072.
[13] B. Saha and K. Goebel, "Modeling li-ion battery capacity depletion in a particle filtering framework," 2009. Available: https://papers.phmsociety.org/index.php/phmconf/article/view/1614.
[14] N. Kovvali, M. Banavar, and A. Spanias, "An introduction to Kalman filtering with MATLAB examples," *Synthesis Lectures on Signal Processing*, vol. 6, no. 2, pp. 1–81, Sep. 2013, https://doi.org/10.2200/s00534ed1v01y201309spr012.
[15] L. Tang, A. Saxena, and K. Younsi, "Prognostics and health management for electrified aircraft propulsion: State of the art and challenges," *Proceedings Of the ASME Turbo Expo*, vol. 147, no. 4, p. 041018, Aug. 2024. https://doi.org/10.2514/6.2019-4474.
[16] C. S. Kulkarni and M. Corbetta, "Health management and prognostics for electric aircraft powertrain," *AIAA Propulsion and Energy 2020 Forum*, Aug. 2019, https://doi.org/10.2514/6.2019-4474. I. Kabashkin, "Digital twin framework for aircraft lifecycle management based on data-driven models," *Mathematics*, vol. 12, no. 19, Art. 2979, Sep. 2024.
[17] B. Xiao *et al.*, "Digital twin-driven prognostics and health management for industrial assets," *Scientific Reports*, vol. 14, Art. 13443, 2024.
[18] "Prognostic and health management of critical aircraft systems and . . .," *Sensors*, vol. 23, no. 19, Art. 8124, 2023.
[19] "Prognostics and health management in automotive and aerospace," *International Journal of PHM*, ca. 2019.
[20] R. F. Orsagh, D. W. Brown, P. W. Kalgren, C. S. Byington, A. J. Hess, and T. Dabney, "Prognostics and health management for avionics systems," in: *2006 IEEE Aerospace Conference*. 2006, p. 7. https://doi.org/10.1109/AERO.2006.1656086.
[21] C. Kulkarni, and M. Corbetta, "Health management and prognostics for electric aircraft powertrain," in: *2019 AIAA/IEEE Electric Aircraft Technologies Symposium (EATS)*. Ames Research Center: AIAA, Aug. 2019, pp. 1–13. https://doi.org/10.2514/6.2019-4474.
[22] C. Cicek and B. R. Leao, "Digital twin driven sustainable product lifecycle," *Procedia CIRP*, 90, pp. 165–170, 2020.

[23] S. Baskar and E. Balaban, "Prognostics and health management in electrified aircraft," NASA Ames Research Center Technical Report, 2018.
[24] European Union Aviation Safety Agency (EASA), "Environmental report – the future of electric aviation," 2021.
[25] R. Zhao et al., "Health-aware and energy-efficient mission planning for electric aircraft," *Journal of Power Sources*, 476, p. 228654, 2020.
[26] B. Saha and K. Goebel, "Battery data set," NASA Ames Prognostics Data Repository, 2009.
[27] DNV-GL, "Battery safety and reliability for electric aircraft," Technical Brief, 2019.
[28] RTCA DO-178C, "Software considerations in airborne systems and equipment certification," 2011.
[29] SAE ARP6313, "IVHM for aerospace systems – performance and design guidelines," 2019.
[30] EASA, "Special condition for electric propulsion systems," 2020.
[31] FAA AC 33.76, "Engine health monitoring," Federal Aviation Administration, 2021.
[32] ISO 21434:2021, "Road vehicles – cybersecurity engineering".
[33] RTCA DO-326A, "Airworthiness security process specification," 2014.
[34] SAE AIR6894, "Safety considerations for electric propulsion systems in aircraft," 2021.
[35] L. C. Jaw, "Recent advancements in aircraft Engine Health Management (EHM) technologies and recommendations for the next step," in Volume 1: Turbo Expo 2005. ASMEDC, Jan. 2005, pp. 683–695, https://doi.org/10.1115/GT2005-68625.
[36] A. Ratner and Y. Edwards, "Digital twin-driven health monitoring for avionics systems," *Applied Sciences*, vol. 14, no. 14, p. 6120, Jul. 2024.
[37] J. Wang et al., "Explainable AI approaches for predictive maintenance: A review," *Sensors*, vol. 23, no. 19, p. 8124, Sep. 2023.
[38] Y. Zhang et al., "State-of-health estimation of lithium-Ion batteries using DLinear and hybrid modeling techniques," *IEEE Transactions on Industrial Informatics*, 2024.
[39] Cyient, "Prognostics and health management for safer electric vehicles," *Cyient Blog*, 2024.
[40] M. K. Jangra et al., "Augmented and virtual reality applications in aerospace maintenance: A review," *Applied Sciences*, vol. 14, no. 14, p. 6120, Jul. 2024.
[41] L. M. O'Sullivan and B. R. Harris, "Certification challenges for prognostics and health management systems in next-gen aircraft," *Journal of Aerospace Systems*, vol. 2, no. 3, pp. 101–110, 2023.

9

Electrical Insulation for Powertrain Components in Electric Aircraft

Shakthi Prasad D. and Shesha H. Jayaram

9.1 Introduction

Contemporary electric aircraft architectures demand insulation systems that transcend the conventional role of preventing short circuits. These systems must demonstrate reliable operation under extreme conditions: high-voltage busbars operating more than 1 kV, power densities exceeding 10 kW/kg, wide operating temperatures, while experiencing continuous vibration and mechanical shock. Within this operational context, insulation materials and designs must balance three fundamental and often conflicting requirements:

- *High Electrical Performance:* High dielectric strength capabilities to withstand transient overvoltages and rapid *dv/dt* switching without breakdown or accelerated ageing due to partial discharge phenomena.
- *Excellent Thermal Performance:* Low thermal resistance characteristics to facilitate heat extraction from WBG inverters and motors, preventing localised hotspots that degrade insulation.
- *Robust Mechanical Performance:* Resistance to vibration, thermal cycling, and mass constraints, while maintaining compliance with aerospace fire, smoke, and toxicity (FST) standards.

Electric aircraft encounter distinctive insulation challenges due to reduced atmospheric pressure at high altitudes, which significantly decreases the partial discharge inception voltage, necessitating careful insulation design and voltage derating. The electrification of aircraft, encompassing more-electric aircraft (MEA) and all-electric aircraft (AEA) configurations, demands high-power-density, lightweight electric power systems (EPS) with medium voltage levels to effectively transmit power.

A fundamental challenge within this operational context pertains to the mitigation of partial discharge (PD) phenomena, arcing events, and

 DOI: 10.1201/9781003747994-9

thermal stress accumulation resulting from elevated voltage requirements and reduced atmospheric pressure environments at cruise altitudes, which significantly complicate aircraft cable system design methodologies. Novel engineering solutions encompass the implementation of multilayer insulation architectures that simultaneously enhance electrical performance characteristics and thermal management capabilities while preserving optimal weight-to-volume ratios. These sophisticated insulation systems demonstrate the requisite capacity to withstand voltage and environmental stresses encountered during high-altitude operations.

High-temperature insulation materials, including fluorinated ethylene propylene (FEP), ethylene tetrafluoroethylene (ETFE), and polyether ether ketone (PEEK), have demonstrated efficacy in addressing challenges imposed by partial discharge activities under variable pressure conditions. These materials exhibit high resistance to surface discharge phenomena, rendering them suitable for medium-voltage direct current (MVDC) power distribution applications within aviation systems.

Sophisticated thermal management methodologies, such as the integration of liquid cooling circuits with ram air heat exchanger systems, are imperative for managing substantial thermal loads in AEA and MEA configurations. These approaches are critical for maintaining operational efficiency and system reliability of electric propulsion systems (EPS) within demanding operational environments.

Furthermore, the development of ultra-compact gate driver architectures incorporating integrated current sensing capabilities and optimised commutation pathways facilitates the management of high loop inductance challenges, partial discharge mitigation, and short-circuit protection in SiC-based propulsion inverters, thereby optimising power system design for high-altitude operational requirements. For aircraft propulsion applications, achieving superior power density while maintaining compliance with electromagnetic interference (EMI) regulatory standards remains paramount. The inverter configuration required to effectively mitigate the EMI emissions and bearing current issues, ensuring seamless integration of short-circuit protection mechanisms, gate driver circuits, commutation loops, and sensing systems. The implementation of multilayer, multifunctional insulation systems and advanced thermal management solutions represents critical enabling technologies for achieving requisite power-to-weight ratios and operational reliability throughout extended flight operations.

Electric aircraft insulation systems encounter distinctive challenges relative to terrestrial applications. The diminished atmospheric pressure at cruise altitudes (can go down till 5 mbar) substantially reduces partial discharge inception voltage thresholds, necessitating careful voltage derating protocols and insulation coordination strategies. Concurrently, the aviation industry's stringent mass constraints demand materials exhibiting exceptional power-to-weight characteristics while maintaining long-term operational reliability exceeding 20,000 flight hours.

This chapter examines the fundamental physics and failure mechanisms of insulation systems under these operational stresses, surveys material systems and design methodologies, reviews testing protocols and certification standards, and explores emerging technologies including nanocomposite insulators and self-healing coating systems.

9.2 Classification of Insulation Systems

Insulation in aircraft electric powertrains can be categorised by its primary application and material family.

9.2.1 Wire and Cable Insulation

High-performance polymers, including polyimide (Kapton®), ETFE, PEEK, and polytetrafluoroethylene (PTFE), are prominently utilised in aircraft wiring harnesses due to their excellent dielectric strengths, thermal ratings, and chemical resistance properties. These attributes are critical for withstanding the rigorous conditions of aircraft environments, including thermal cycling stresses.

Enamel and Extruded Jackets: Polyimide, ETFE, PEEK, and PTFE are employed for magnet wire and lead wire applications, offering dielectric strengths of 200–300 kV/mm and thermal ratings up to 200°C. These materials provide excellent chemical resistance and maintain flexibility at temperature extremes, critical for aircraft wiring harnesses subjected to thermal cycling.

Polyimide (PI): Polyimide exhibits high dielectric properties and exceptional thermal stability. The polymer maintains operational efficiency at temperatures up to 500°C, as demonstrated in various high-temperature industrial applications. Its dielectric constant can be optimised to enhance efficiency in aerospace applications. Polyimide with added fillers demonstrates notable resistance to partial discharges, a critical feature that enhances its longevity in electrical insulation systems.

Ethylene Tetrafluoroethylene (ETFE): This shares key characteristics with PTFE and other fluoropolymers, including excellent chemical resistance and the ability to maintain flexibility across a wide temperature range. This renders ETFE suitable for environments requiring robust mechanical performance without the degradation observed in other polymers under similar conditions.

Polyether Ether Ketone (PEEK): This possesses dielectric and mechanical properties that render it ideal for high-voltage applications within aircraft systems. PEEK's electrical insulation capability is exceptionally beneficial, given the high dielectric strength that supports high-voltage applications.

However, the dielectric strength can be influenced by the crystalline structure; increased crystallinity typically results in reduced breakdown strength. This aspect must be managed to optimise PEEK for specific applications in aircraft systems.

Polytetrafluoroethylene (PTFE): It is renowned for its low dielectric constant and excellent thermal resistance, traits that are significantly enhanced with fillers like boron nitride which improve both the dielectric strength and thermal conductivity. Studies have demonstrated that incorporating such fillers can substantially increase PTFE's resistance to moisture and improve electrical properties, a critical adaptation for maintaining performance in harsh environmental conditions encountered in aircraft wiring systems.

Braided Sleeving: Glass-fibre or aramid fibre braid impregnated with varnish provides abrasion and partial discharge protection. The braided construction distributes mechanical stress while the impregnated varnish provides corona resistance at high altitudes.

Glass-fibre braids are valued for their high tensile strength and corrosion resistance. These fibres can withstand significant mechanical loads without degradation, making them ideal for applications requiring long-term durability. The braided construction allows for the distribution of mechanical stress across the material, preventing localised failure and enhancing overall structural integrity. Their resistance to environmental factors contributes to their effectiveness in outdoor and high-altitude applications where exposure to the elements is a concern.

Aramid fibres, such as Kevlar, provide exceptional strength-to-weight ratios, superior to many traditional materials. These fibres are especially noted for their high tensile strength and resistance to impact and abrasion, making them suitable for applications requiring high protective qualities. The presence of varnish impregnated into the braid enhances these properties by offering additional resistance against abrasion and partial discharge. Aramid fibres also perform well under conditions of transverse mechanical stress, often providing more efficient performance compared to traditional materials like round fibres.

9.2.2 Busbars and Connectors

Solid Laminates: Epoxy–glass or mica–epoxy laminates providing 15–30 kV creepage per mm are frequently utilised in busbar insulation modules. Mica-based systems offer superior tracking resistance and maintain dielectric properties at elevated temperatures up to 180°C for continuous operation.

Epoxy–glass and mica–epoxy laminates are pivotal in electrical insulation materials due to their unique properties and applications. Epoxy–glass laminates typically consist of glass fibres embedded in an epoxy–resin matrix, which imparts high mechanical strength, excellent thermal stability, and notable dielectric properties. These laminates are primarily used in electrical applications where a high degree of insulation is required, such as in transformers, insulators, and as part of busbar insulation systems due to their reliable electric breakdown resistance and thermal stability over a range of temperatures.

Mica–epoxy laminates provide enhanced performance, particularly in conditions demanding high thermal resistance and excellent dielectric properties. Mica-based composites are recognised for their superior resistance to tracking and maintain their dielectric properties at elevated temperatures up to 180°C, crucial for applications involving high thermal operational conditions. The mica within these composites plays a critical role in providing increased thermal endurance and improved performance in high-voltage applications.

The epoxy and mica–epoxy composites provide electrical insulation while offering chemical, mechanical, and thermal stability that is vital for the operation of rotating machines under varying stress conditions such as electrical, thermal, and mechanical stresses.

Advanced manufacturing processes like selective laser sintering (SLS) have enabled the production of complex structures from these materials, essential for sophisticated electrical devices. Such innovations in fabrication and formulation of these composites ensure they meet the evolving demands of electrical insulation in modern applications.

Coatings: Silicone or epoxy-based conformal coatings applied by dip or spray for connectors and small components, achieving 20–40 μm thickness layers with >20 kV breakdown strength. These coatings provide environmental protection against moisture, salt spray, and contaminants while maintaining flexibility. Silicone and epoxy-based conformal coatings offer distinct advantages for protecting small components, but differ significantly in application methods, protection capabilities, flexibility, and breakdown strength.

Application Methods: Silicone coatings are commonly applied as single or duplex layers. They provide flexibility in release mechanisms, as demonstrated in fracture mechanics models. Epoxy coatings can be integrated with various fillers such as graphene and alumina to enhance properties, making them suitable for alternating multilayer structures designed for corrosion protection.

Protection Capabilities: Both coatings offer excellent environmental protection. Silicone–epoxy composite coatings exhibit outstanding superhydrophobic properties with remarkable chemical stability and self-cleaning ability. This ensures excellent protection against acid–base environments, saline immersion, and mechanical wear. Epoxy coatings can be tailored with additives to create effective anticorrosion barriers and to enhance adhesion and wear properties. The use of fillers like nano-silica enhances epoxy coatings' durability and anticorrosive properties.

Flexibility: Silicone-coated materials are known for their inherent flexibility and low surface energy, making them effective in applications requiring adhesion yet easy release properties. This flexibility is crucial in dynamic components where movement is expected. In contrast, epoxy coatings are more rigid but can achieve improved flexibility through modifications, such as the integration of polydimethylsiloxane (PDMS) brushes. These brushes enhance frictional properties and long-term stability in superlubricity coatings.

Breakdown Strength: Breakdown strength is a critical factor in electrical applications. Studies suggest that the interaction between silicone and epoxy affects their electric field breakdown strength. Silicone has been shown to influence breakdown characteristics positively, especially when functioning as an interface between different dielectric materials. Additionally, enhancing epoxy with nanoparticles can improve breakdown strength, although this can be complex due to the required balance between filler content and dispersal.

9.2.3 Stator and Rotor Insulation in Electric Machines

In electric vehicles, including electric aircraft, the motor insulation system plays a critical role in ensuring long-term performance, electrical reliability, and thermal stability under high-voltage and high-frequency operating conditions. Modern motor insulation systems are engineered to withstand partial discharge, thermal cycling, and mechanical vibration – challenges that are amplified in aerospace environments due to altitude, variable pressure, and compact form factors. Conventional insulation materials such as polyimide films and mica-based composites have been widely adopted due to their excellent dielectric strength (>20 kV/mm) and thermal endurance (Class H and above). However, with the increasing use of wide-bandgap (WBG) semiconductors like SiC and GaN, which introduce higher switching frequencies and steeper voltage gradients (dv/dt), traditional insulation materials face accelerated ageing and degradation. To address these demands, advanced insulation systems are being developed using nanocomposite technologies and multilayered architectures that combine high thermal conductivity, partial discharge resistance, and mechanical flexibility. For instance, epoxy-based nanocomposites with ceramic or boron nitride fillers have shown improved tracking resistance and thermal stability, while enamel-coated winding wires with optimised layering reduce corona discharge initiation at elevated voltages. Vacuum pressure impregnation (VPI) techniques are often employed to achieve uniform resin distribution and void-free encapsulation, critical for aerospace-grade reliability. Additionally, research into self-healing and adaptive insulation materials is gaining traction, aiming to extend the operational life and safety margins of electric propulsion motors in demanding environments.

9.2.4 Power Electronic Module Insulation

In power electronic modules, effective insulation is crucial to ensure device reliability and performance. Potting compounds and ceramic insulators are two vital components utilised for this purpose.

Potting Compounds: Silicone and epoxy potting compounds are extensively employed in power electronics, offering thermal conductivities ranging from 0.8 W/m·K to 1.5 W/m·K, along with dielectric strengths exceeding 15 kV/mm. These materials are utilised to fill gaps in silicon carbide (SiC) and gallium nitride (GaN) module assemblies, providing electrical insulation while accommodating thermal expansion mismatches between different materials. This attribute helps prevent cracking or other mechanical failures due to differing thermal expansion rates, particularly over wide temperature ranges. They are designed to maintain their properties under the harsh thermal conditions typical of high-power electronic applications.

Advanced nanocomposite formulations of epoxy materials exhibit improved dielectric breakdown strength and thermal conductivity compared to traditional formulations, making them suitable for high-voltage applications. Novel epoxy designs, including liquid crystalline networks, have exhibited significant enhancements in thermal and dielectric properties. The strategic use of fillers, such as graphene or ceramic-based additives, further enhances these properties, enabling efficient thermal management and reduced operating temperatures in electronic devices.

Ceramic Insulators and Substrates: Ceramic materials, such as aluminium nitride (AlN) and aluminium oxide (Al_2O_3), are utilised as substrates and insulators due to their excellent thermal conductivity (150–280 W/m·K) and electrical insulation properties. These substrates support the direct bonding of power dies, crucial for efficient heat spreading and maintaining electrical insulation within power electronic modules. Ceramic insulators have a pivotal role in enhancing the thermomechanical performance and reliability of power modules. They are often coupled with various bonding materials like silver–tin or copper to improve joint reliability despite high thermal stresses. The choice of ceramic substrate can significantly impact the fatigue life of a module by reducing stress and strain during temperature fluctuations.

9.3 Challenges in Designing Insulation for Electric Aircraft

9.3.1 Partial Discharge and Ageing

Partial discharge (PD) and material ageing are critical concerns in high *dv/dt* (typically 10–50 kV/μs) pulse applications, such as those involving wide-bandgap (WBG) inverters. These inverters produce significant electric fields concentrated at interfaces and voids within insulating materials, initiating partial discharge that can result in the degradation of polymer insulation through the formation of reactive species and ozone. Atmospheric conditions, particularly at high altitudes, further exacerbate these effects due to reduced air pressure, as described by Paschen's law, which causes PD inception voltages to drop below operational levels.

The degradation is facilitated by the production of ozone and free radicals, which chemically attack polymer insulation. This process gradually erodes dielectric materials, increasing the risk of electrical failure over time. High *dv/dt* stresses associated with such inverters lead not only to partial discharge but also to dielectric heating, which can contribute to thermal ageing if the generated heat exceeds the dissipation capacity. Insulating materials such as epoxy and silicone composites are often employed to mitigate these effects, but their performance is challenged under high frequency and *dv/dt* voltage conditions.

Materials like cross-linked polyethylene and silicone composites have shown promise in sustaining better performance under such stresses. For instance, incorporating fillers can enhance the dielectric and thermal properties of these materials, helping to resist the mechanical and chemical stressors that contribute to ageing. However, the intricacies of PD and thermal effects in these situations necessitate careful insulation design, often validated through diagnostics that consider non-ideal voltage conditions and harmonic content, which significantly affect PD behaviours.

Advancements in non-destructive testing, like terahertz spectroscopy, are being explored to better detect the early signs of material degradation and improve the reliability of insulation systems. This technology holds the potential for assessing both structural defects and the degree of ageing in insulating materials, potentially extending the operational life of high-voltage systems and reducing the risk of failure due to partial discharges.

9.3.2 Thermal Management and Heat Dissipation

The operational conditions of compact inverter and motor assemblies, which experience heat flux densities up to 100 kW/m^2, impose significant thermal and mechanical stresses on traditional polymer insulators. These insulators typically possess low thermal conductivity (0.2–0.4 W/m·K), resulting in the formation of hotspots that can substantially impact the longevity and effectiveness of the insulating materials.

The Arrhenius principle indicates that each 10°C rise in temperature approximately doubles the chemical reaction rate, leading to accelerated degradation of the insulation. Such rapid temperature increases can reduce the lifespan of insulators, as the higher rates of chemical reactions facilitate the thermal ageing process and exacerbate mechanical stresses.

The low thermal conductivity of polymeric materials presents a challenge as it leads to heat accumulation, increasing operational temperatures and accelerating insulation wear. Efforts to improve the thermal conductivity of polymer insulators focus on integrating conductive fillers to disperse heat efficiently, reduce insulation breakdown risk, and extend the operational life of the equipment. One approach is the development of polymer composites with high thermal conductivity, which can replace traditional polymer insulators, enabling better thermal management and durability in high-temperature environments.

In addition to enhancing thermal conductivity, advanced cooling techniques such as two-phase spray cooling are being explored to manage heat in power inverter modules. This method significantly reduces thermal resistance, achieving higher power densities and uniform device temperatures, thereby protecting the insulating materials from excessive heat exposure and potential failure.

9.3.3 Mechanical Vibration and Fatigue

Aircraft engines induce broad-spectrum vibration (20 Hz to 2 kHz) with acceleration levels up to 20*g* that can fatigue insulation layers, causing cracks and delamination. Multilayer tape wraps must employ high-elongation varnishes and fibre reinforcements to accommodate dynamic strain without compromising dielectric integrity. Fatigue testing according to DO-160G requires insulation systems to withstand 10^7 cycles at operational stress levels without degradation in dielectric properties. This necessitates careful material selection and interface design to prevent stress concentration and progressive failure.

Fibre-reinforced composites, particularly those using high-modulus fibres such as carbon and glass, are advantageous for this application due to their ability to provide increased strength and stiffness, as well as better toughness and dimensional stability. In the context of aircraft engine insulation, fibre reinforcements help to accommodate dynamic strain, thus reducing the likelihood of material fatigue and the associated deterioration.

Various fibre types can be used for these composites, each offering specific benefits. While carbon fibres are widely favoured in aerospace for their mechanical properties, alternatives like basalt and rockwool fibres may offer cost-effective solutions with adequate mechanical and thermal properties. Furthermore, hybrid fibre composites, which combine different fibre types, can enhance performance characteristics, including impact resistance and thermal stability. These hybrid composites exhibit a favourable balance of stiffness and flexibility, making them suitable for applications where both structural integrity and insulation are critical.

In terms of processing, incorporating fibres into polymer matrices, like epoxy or cyanate ester, allows for the manipulation of material properties such as thermal expansion and thermal conductivity. The mismatch in thermal expansion coefficients in traditional composites can lead to dimensional instability; however, the use of low thermal expansion matrices or hybrid composite configurations containing nanoparticles can mitigate these effects. Additionally, carbon aerogel composites reinforced with fibres like oxidised polyacrylonitrile (PAN) serve as effective thermal insulators with minimal cracking and good mechanical integrity, useful for high-temperature applications. Similarly, silica aerogel composites with anti-infrared radiation fibres offer superior thermal insulation and are adaptable for both low- and high-temperature conditions.

9.3.4 Fire, Smoke, and Toxicity Compliance

Aerospace Fire, Smoke, and Toxicity (FST) standards, such as FAR 25.855 and DO-160G, are critical for ensuring the safety of insulators used in aircraft by requiring them to self-extinguish within 15 seconds, produce smoke with an optical density of less than 100%, and avoid toxic gas emissions during combustion. Meeting these stringent criteria often necessitates the modification of high-performance polymers or the utilisation of non-halogenated copolymers.

Polymers commonly employed in aerospace applications may not inherently possess the required fire-retardant properties, necessitating the addition of flame-retardant additives. Flame retardants can be incorporated into polymers to enhance their resistance to ignition and slow down the spread of fire. These additives function by promoting char formation, which acts as a barrier to heat and mass transfer, or by releasing flame-inhibiting gases that help to extinguish the fire. In addition to using flame retardants, an alternative approach involves utilising inherently non-halogenated copolymers. Such polymers are designed to meet FST standards naturally without the need for additional toxic additives. This is particularly beneficial for reducing potential health and environmental impacts associated with the use of traditional halogen-based flame retardants.

9.4 Recent Advancements in Electrical Insulation

9.4.1 Nanocomposite Dielectrics

Incorporating nanoscale fillers such as SiO_2, TiO_2, and BN into epoxy–polymer matrices can significantly impact the properties of these composites, enhancing dielectric breakdown strength, minimising PD activity through trap sites creation, and boosting thermal conductivity without severely compromising flexibility. However, achieving good dispersion is a considerable difficulty in preparing nanocomposites. Recent studies demonstrate the enhancement of maximum dispersion of the filler material in the silicone rubber matrix using high shear and electrostatic dispenser techniques.

The addition of nanoscale fillers can enhance dielectric breakdown strength by up to 25%. TiO_2 contributes significantly to mechanical strength and stability due to enhanced interfacial interactions between the polymer matrix and the filler. The small size of fillers like TiO_2 increases the interfacial area, leading to a significant volume fraction of interfacial polymer that has distinct properties from the bulk polymer, even at low loadings.

Furthermore, nanoscale fillers like BN are particularly effective in improving the thermal conductivity of polymer composites. It was observed that the thermal conductivity increased by up to 189% with the inclusion of boron nitride (BN) in polymer matrices, showing that BN fillers can considerably enhance thermoconductive properties. This increase in thermal conductivity is critical for applications that require effective heat dissipation.

The introduction of hexagonal boron nitride (h-BN) fillers into insulating materials significantly improves their resistance to partial discharges. This improvement is critical because PDs are a primary cause of accelerated degradation and premature failure in organic electrical insulation. For instance, BN fillers incorporated into polyphenylsulfone (PPSU) matrices have been found to substantially enhance PD resistance, outperforming traditional insulating materials like corona-resistant Kapton®.

Several studies highlight the significant enhancements in mechanical properties such as tensile strength, Young's modulus, and impact strength when BN fillers are used in hybrid forms. For instance, epoxy nanocomposites reinforced with a hybrid of molybdenum disulfide (MoS_2) and hexagonal boron nitride (h-BN) showed a 95% increase in tensile strength and a 58% improvement in Young's modulus, alongside a massive enhancement in thermal conductivity by 203% compared to the pure polymer.

The optimal loading is typically 1–5 wt% to maximise benefits while avoiding agglomeration. Surface functionalisation of nanoparticles is critical to ensure uniform dispersion and strong polymer-filler interfaces.

9.4.2 Self-Healing Insulators

Microencapsulated healing agents embedded within epoxy coatings represent a prominent approach for autonomously repairing microcracks in dielectric materials, thereby prolonging the service life of coated surfaces, particularly in demanding, long-duration missions. The encapsulation technique involves incorporating microcapsules containing healing agents like cerium nitrate or epoxy–mercaptan mixtures into an epoxy matrix. These microcapsules rupture upon the formation of microcracks, releasing the healing agents, which then react to heal the damaged area and restore the material's structural and dielectric properties. The capsule-based self-healing approach, inspired by biological systems, addresses both mechanical and chemical damage without manual intervention, effectively maintaining the integrity of polymeric coatings.

One example of this technology is found in the development of anticorrosion epoxy coatings incorporated with a dual-capsule healing system. Microcapsules, created via the electrospray method, use polymers for the shell material and reactive chemicals like polyetheramine and isocyanate prepolymers for the core, demonstrating a healing efficiency of up to 85%. This shows the potential for such systems in industrial applications where corrosion resistance is crucial.

Microencapsulated healing agents embedded within epoxy coatings release monomers upon microcrack formation, autonomously repairing dielectric breaches and extending service life in long-duration missions. Current systems can heal cracks up to 50 μm width with 80% recovery of original dielectric strength.

9.4.3 Additive Manufacturing and Graded Dielectrics

The 3D printing of graded dielectric structures within insulation components offers several substantial advantages over traditional methods, primarily due to its ability to enable complex geometries with locally optimised properties. The use of functionally graded materials (FGMs) in insulators allows for spatial variations in the dielectric properties, which are particularly beneficial for improving insulation performance without complicating the structural design. This technique can enhance insulation performance significantly, as evidenced by improvements in surface flashover voltages by 23% in SF6 and 20% in vacuum environments.

Furthermore, the use of advanced 3D printing technologies, such as greyscale digital light processing (g-DLP), supports the creation of highly functionally graded materials with precise control over gradients, which in turn dictates the mechanical, thermal, and dielectric properties of the final product. This advanced control allows for the creation of materials that can better manage electric field distributions, thus reducing peak stress by 30–40% compared to uniform materials by smoothing out electric fields across the insulator.

3D printing of insulation components enables complex geometries with locally optimised properties. Graded dielectric structures with varying permittivity can smooth electric field distributions, reducing peak stress by 30–40% compared to uniform materials.

9.5 Testing, Standards, and Certification

9.5.1 Partial Discharge Testing

IEC 60270 governs PD measurement methods essential for assessing the insulation systems in electrical apparatus. It includes traditional methods like free-space, coupling capacitor, and high-frequency current transformer (HFCT) techniques, which are pivotal in detecting PD levels effectively down to 1 pC. These methods serve as a baseline for offline and online testing, particularly in high-voltage power systems like transformers and stator windings, ensuring both quality control and ongoing maintenance.

Advanced PD testing techniques provide deeper insights and overcome the limitations of traditional methods. Techniques like phase-resolved partial discharge (PRPD) analysis allow detailed defect identification by analysing discharge patterns and their relation to the phase of the power cycle, distinguishing among different types of discharges such as corona, void, and surface discharges. Additionally, ultra-high frequency (UHF) detection enhances sensitivity by capturing higher frequency emissions, thus improving the reliability of the PD measurements in the presence of noise. This technique is particularly beneficial in environments where noise can easily contaminate data. Furthermore, a combined acoustic-electrical PD localisation system with a non-iterative method focuses on the strategic placement of acoustic sensors, offering improvements in computational time over existing iterative methods.

9.5.2 Thermal Endurance and Burnthrough

FAR 25.853 annex tests are essential for evaluating the fire resistance of materials used in aircraft insulation. These tests involve exposing insulators to thermal radiation at 65 kW/m^2 and direct flames for 15 seconds to assess the material's burnthrough time and flaming dropout behaviour. This ensures that the materials can withstand severe fire conditions without compromising safety.

Moreover, these tests are complemented by mechanical load-cycle tests to validate the vibration resistance of the insulation. These tests adhere to the guidelines outlined in DO-160G Section 8, which sets the standards for environmental conditions and test procedures for airborne equipment, focusing on ensuring the material's durability under mechanical stress.

In summary, the combination of FAR 25.853 annex tests and advanced ageing analyses offers a robust method for evaluating insulator performance, balancing between immediate fire safety requirements and long-term durability predictions.

9.5.3 Flame Smoke and Toxicity (FST)

Ensuring the safety of cabin materials involves rigorous testing for smoke density, heat release, and toxicity, as outlined in standards such as FAR 25.853(d), ASTM E 1354, and ISO 5659. These standards are essential for maintaining safety in aerospace applications due to the critical environment in which aircraft operate.

Smoke density testing, as per FAR 25.853(d), is designed to assess the optical density of smoke produced by burning materials. This measurement is crucial because high smoke density can obstruct visibility during an emergency, hindering evacuation efforts and potentially leading to fatalities. Heat release rate testing, conducted in accordance with ASTM E 1354, evaluates how much energy is emitted by a material as it burns. A higher heat release rate indicates a more intense fire, which can be more challenging to control and extinguish. Thus, understanding and limiting the heat release rate of materials is imperative for fire safety in aviation.

ISO 5659 focuses on the toxicity of gases released when materials burn. During combustion, materials can emit a variety of harmful gases, including CO, HCN, HCl, HF, and NO_2. These gases can be deadly, even in small concentrations, and effective testing ensures that materials do not release these gases in quantities that exceed specified safety limits. For example, carbon monoxide (CO) exposure can lead to oxygen deprivation, while hydrogen cyanide (HCN) is a potent poison that can cause rapid incapacitation and death.

These tests help in setting parameters that materials must meet to be considered safe for use in the unique and sensitive environment of an aircraft cabin or bay. The assessment and regulation of these parameters ensure that in the event of a fire, the materials will not contribute to a catastrophic escalation and that passengers and crew will have the best possible chance of survival.

9.5.4 High-Voltage Testing Protocols

Specialised high-voltage testing in the aerospace industry is crucial to ensure the safety and reliability of electrical systems used in aircraft and spacecraft. Each testing method serves a distinct purpose in evaluating potential failure modes and environmental impacts. Here is an overview of some key testing methods.

Lightning Transient Simulation (DO-160G Section 23): This test simulates the effects of lightning on aerospace electrical systems, which is essential because high-altitude operations increase susceptibility to such electrical transients. By assessing the lightning transient simulation, engineers can ensure that the systems can withstand high voltage surges and remain operational during storm conditions.

Wet High-Voltage Testing for Moisture Ingress Effects: Due to varying environmental conditions, aerospace systems can potentially experience moisture ingress, which can lead to electrical insulation failures. Wet high-voltage tests are conducted to simulate these scenarios and evaluate the system's robustness against moisture, ensuring that electrical systems remain reliable under humid or wet conditions.

Corona Extinction Voltage at Reduced Pressure: At high altitudes, reduced atmospheric pressure can influence electrical discharge characteristics. Corona extinction voltage testing assesses the voltage level at which corona discharges cease, which is critical because corona discharge can lead to insulation degradation and system failures in aerospace applications. This test helps design systems that minimise the risk of electrical discharge under low-pressure environments.

Tracking Resistance as Per IEC 60112: This method evaluates the resistance of insulating materials against tracking, which occurs when conductive paths form across the surface of an insulator due to electrical discharge or humidity. Tracking resistance testing ensures that materials can withstand conditions that promote tracking and prevents potential electrical hazards.

These tests collectively address the challenges posed by high voltage levels and the unique environmental conditions experienced in aerospace operations. By implementing these testing methods, the aerospace industry can advance towards safer and more reliable electrified systems, accommodating higher voltages while managing risks associated with electrical discharges, moisture ingress, and environmental challenges.

9.6 Design Guidelines and Best Practices

9.6.1 Electric Field Management

Effective electric field management, particularly in high-performance systems, involves several strategies to mitigate risks associated with hotspots and field concentrations. Employing multi-physics simulations that integrate electric, thermal, and mechanical aspects in the early design phase is vital. This approach helps in identifying potential hotspots and field concentrations that could lead to system inefficiency or failure. These simulations are essential for designing robust systems that handle the physical complexities encountered during operation.

Maintaining peak electric stress under 50% of a material's breakdown strength is crucial. This involves accounting for altitude derating, which adjusts the permissible stress levels based on changes in environmental conditions such as altitude, where air pressure and density alter dielectric properties. This preventive measure significantly reduces the likelihood of dielectric breakdown and enhances system longevity.

Rounded edges can be effectively used in various high-voltage applications, including power electronic modules and HVDC cables, where electric field control is crucial. For example, the application of non-linear resistive field grading techniques, or the use of coatings, can mitigate the electric field within the system, reducing potential for discharge and improving operational integrity.

Additionally, experiments with functionally graded materials have demonstrated the potential for reducing electric field stress around sharp edges by varying the dielectric permittivity spatially, thereby lowering the risk of electrical breakdown and enhancing the long-term reliability of insulating material.

9.6.2 Partial Discharge Prevention

Selecting insulation materials with a partial discharge inception voltage (PDIV) exceeding 20% above the maximum operating voltage can enhance the reliability of electrical insulation systems. Materials such as nanocomposite-enamelled wires have been developed for their PD-resistant properties and are beneficial under repetitive surge voltages with steep wavefronts, as found in low-voltage inverter-fed motor applications.

Furthermore, it is important to understand the impact of environmental factors, such as pressure and ambient humidity, on PDIV. Research has shown that the partial discharge inception voltage can be modelled based on ambient humidity and temperature, which is crucial for achieving PD free designs. It is also noted that polyimide films are main insulating materials for inverter-fed motors, although their ageing mechanism is unique under pulse voltage conditions, highlighting the need for specific insulation strategies.

Moreover, implementing insulation systems that account for the effects of inter-harmonic AC voltages can further enhance insulation efficacy under fluctuating voltage conditions. This involves understanding the characteristics of insulation, which impacts PD inception voltage and related phenomena.

9.6.3 Thermal Design Optimisation

Thermal design optimisation is crucial in ensuring efficient heat management, particularly in electronic components and systems. The integration of thermally conductive fillers and heat spreaders adjacent to windings represents an effective strategy for enhancing heat dissipation. These materials facilitate the rapid transfer of heat away from critical areas, thereby reducing the risk of thermal hotspots and improving the overall performance and reliability of electronic systems. Minimising thermal interface resistances is another essential aspect of thermal design. Proper surface preparation can significantly reduce contact resistance, thereby ensuring better thermal conduction across interfaces. This can be achieved through techniques such as surface polishing, applying thermal interface materials (TIMs), and ensuring adequate pressure across interfaces to improve contact quality.

In conclusion, optimising thermal management involves a holistic approach that integrates advanced materials, proper surface preparation, efficient cooling system design, and sophisticated temperature monitoring. These strategies not only enhance the thermal performance of electronic systems but also improve their operational reliability and longevity.

9.7 Summary

Electric aircraft powertrains demand advanced insulation to withstand high voltages (>1 kV), extreme temperatures, low-pressure partial discharge, and intense vibration. High-performance polymers (PEEK, PTFE) and laminates (epoxy–glass, mica–epoxy) offer thermal stability, dielectric strength, and mechanical resilience. Potting compounds and ceramic substrates ensure heat spreading and stress relief in power modules, while flame-retardant, low-smoke materials meet stringent FST standards. Nanocomposite fillers and self-healing epoxies enhance breakdown strength, thermal conductivity, and autonomous crack repair. Rigorous multiphysics design, partial discharge testing, and graded insulation architectures guarantee lightweight, reliable electric aircraft systems.

Bibliography

1. *Aerospace Materials Handbook*, CRC, 2016. https://doi.org/10.1201/b13044.
2. R. Agarwal, H. Li, Z. Guo, and P. Cheetham, "The effects of PWM with high *dv/dt* on partial discharge and lifetime of medium-frequency transformer for Medium-Voltage (MV) solid state transformer applications," *IEEE Transactions on Industrial Electronics*, vol. 70, no. 4, pp. 3857–3866, 2023, https://doi.org/10.1109/tie.2022.3174243.
3. J. Ahmad and M. G. Niasar, "Aging behavior of PEEK, PTFE, and PI insulation materials under thermal oxidative and humid conditions for aerospace applications," *Journal of Applied Polymer Science*, vol. 142, no. 19, 2025, https://doi.org/10.1002/app.56858.
4. J. Andersson, S. M. Gubanski, and H. Hillborg, "Properties of interfaces between silicone rubber and epoxy," *IEEE Transactions on Dielectrics and Electrical Insulation*, vol. 15, no. 5, pp. 1360–1367, 2008, https://doi.org/10.1109/tdei.2008.4656245.
5. N. Andraschek, A. J. Wanner, C. Ebner, and G. Riess, "Mica/Epoxy-composites in the electrical industry: Applications, composites for insulation, and investigations on failure mechanisms for prospective optimizations," *Polymers*, vol. 8, no. 5, p. 201, 2016, https://doi.org/10.3390/polym8050201.
6. D. Antony and G. S. Punekar, "Noniterative method for combined acoustic-electrical partial discharge source localization," *IEEE Transactions on Power Delivery*, vol. 33, no. 4, pp. 1679–1688, 2018, https://doi.org/10.1109/tpwrd.2017.2769159.
7. M. U. Azam, I. Belyamani, A. Schiffer, S. Kumar, and K. Askar, "Progress in selective laser sintering of multi functional polymer composites for strain- and self-sensing applications," *Journal of Materials Research and Technology*, vol. 30, pp. 9625–9646, 2024, https://doi.org/10.1016/j.jmrt.2024.06.024.
8. A. Azizi and M. Ghassemi, "Design of high power density MVDC cables for wide-body all electric aircraft," *IEEE Transactions on Dielectrics and Electrical Insulation*, vol. 30, no. 5, pp. 2315–2324, 2023, https://doi.org/10.1109/tdei.2023.3285849.
9. A. Beigbeder, M. Devalckenaere, P. Dubois, M. Claes, P. Degée, R. Lazzaroni, M. Linares, and D. Beljonne, "CH-π interactions as the driving force for silicone-based nanocomposites with exceptional properties," *Advanced Materials*, vol. 20, no. 5, pp. 1003–1007, 2008, https://doi.org/10.1002/adma.200701497.

10. R. Benrashid, and G. C. Tesoro, "Effect of surface-limited reactions on the properties of Kevlar® fibers," *Textile Research Journal*, vol. 60, no. 6, pp. 334–344, 1990, https://doi.org/10.1177/004051759006000604.
11. H. Bostanci, B. A. Saarloos, L. C. Chow, E. Van Ee, and D. P. Rini, "Thermal management of power inverter modules at high fluxes via two-phase spray cooling," *IEEE Transactions on Components, Packaging and Manufacturing Technology*, vol. 2, no. 9, pp. 1480–1485, 2012, https://doi.org/10.1109/tcpmt.2012.2190933.
12. C. Cazan, A. Enesca, and L. Andronic, "Synergic effect of TiO2 filler on the mechanical properties of polymer nanocomposites," *Polymers*, vol. 13, no. 12, p. 2017, 2021, https://doi.org/10.3390/polym13122017.
13. E. G. Chatzi, and J. L. Koenig, "Morphology and structure of Kevlar fibers: A review," *Polymer-Plastics Technology and Engineering*, vol. 26, nos. 3–4, pp. 229–270, 1987, https://doi.org/10.1080/03602558708071938.
14. I. Cotton, D. Schweickart, D. Grosean, R. Gardner, and C. Severns, Design considerations for higher electrical power system voltages in aerospace vehicles, pp. 57–61, 2016, https://doi.org/10.1109/ipmhvc.2016.8012771.
15. M. Coutinho, D. Bento, A. Souza, R. Cruz, F. Afonso, F. Lau, A. Suleman, F. R. Barbosa, R. Gandolfi, W. Affonso, F. I. K. Odaguil, M. F. Westin, R. J. N. Dos Reis, and C. R. I. Da Silva, "A review on the recent developments in thermal management systems for hybrid-electric aircraft," *Applied Thermal Engineering*, vol. 227, p. 120427, 2023, https://doi.org/10.1016/j.applthermaleng.2023.120427.
16. L. Doyle, I. Weidlich, and E. Di Maio, "Developing insulating polymeric foams: Strategies and research needs from a circular economy perspective," *Materials*, vol. 15, no. 18, p. 6212, 2022, https://doi.org/10.3390/ma15186212.
17. B. Du, W. Yang, H. Sun, J. Jiang, and X. Kong, "Temperature-dependent electric field distribution in ±800 kV valve-side bushing insulation for a converter transformer," *High Voltage*, vol. 6, no. 1, pp. 106–115, 2020, https://doi.org/10.1049/hve.2019.0385.
18. B. Feichtenschlager, J. Svehla, G. Kickelbick, M. Sajjad, T. Koch, S. Pabisch, and H. Peterlik, "Epoxy resin nanocomposites: The influence of interface modification on the dispersion structure – a small-angle-X-ray-scattering study," *Surfaces*, vol. 3, no. 4, pp. 664–682, 2020, https://doi.org/10.3390/surfaces3040044.
19. Y. Feng, X. Zhao, X. Li, C. Liu, X. Zhou, Y. Ye, X. Xie, and H. Liu, "Synergetic improvement in thermal conductivity and flame retardancy of Epoxy/Silver nanowires composites by incorporating 'branch-like' flame-retardant functionalized graphene," *ACS Applied Materials & Interfaces*, vol. 10, no. 25, pp. 21628–21641, 2018, https://doi.org/10.1021/acsami.8b05221.
20. M. Florkowski, "Effects of partial discharge echo in motor-insulation systems of transportation power devices," *IEEE Transactions on Transportation Electrification*, vol. 11, no. 2, pp. 5387–5395, 2025, https://doi.org/10.1109/tte.2024.3480810.
21. H. Gao, and P. Liu, "High-temperature encapsulation materials for power modules: Technology and future development trends," *IEEE Transactions on Components, Packaging and Manufacturing Technology*, vol. 12, no. 11, pp. 1867–1881, 2022, https://doi.org/10.1109/tcpmt.2022.3225960.
22. M. A. Gharaibeh and J. Wilde, "Numerical characterizations of the thermo-mechanical response of power modules with ceramic substrates and lead-free bonds," *Multidiscipline Modeling in Materials and Structures*, vol. 20, no. 6, pp. 1384–1401, 2024, https://doi.org/10.1108/mmms-05-2024-0136.

23. M. Ghassemi, "Accelerated insulation aging due to fast, repetitive voltages: A review identifying challenges and future research needs," *IEEE Transactions on Dielectrics and Electrical Insulation*, vol. 26, no. 5, pp. 1558–1568, 2019, https://doi.org/10.1109/tdei.2019.008176.
24. C. Guo, M. Cai, X. Zhang, W. Xu, J. Fu, and S. Duan, "A review: Application of terahertz nondestructive testing technology in electrical insulation materials," *IEEE Access*, vol. 10, pp. 121547–121560, 2022, https://doi.org/10.1109/access.2022.3222860.
25. Z. Guo, Y. Pan, P. Zhao, J. He, R. Gao, J. Fu, and C. Zhang, "Molding of polyether ether ketone (PEEK) and its composites: A review," *Journal of Zhejiang University-SCIENCE A*, vol. 25, no. 10, pp. 788–823, 2024, https://doi.org/10.1631/jzus.a2400004.
26. Y. Hao, Y. Chen, Y. Chen, L. Yang, T. Huang, and J. Xiao, "A partial discharge detection system for XLPE cable terminals and acoustic wave sensing characteristics based on fiber optic Mach-Zehnder interferometer," *IEEE Sensors Journal*, vol. 21, no. 19, pp. 21782–21790, 2021, https://doi.org/10.1109/jsen.2021.3101835.
27. Z. He, Q. Li, Y. Zhao, N. An, X. Li, J. Wang, M. Kang, and S. Jiang, "Self-healing isocyanate microcapsules for efficient restoration of fracture damage of polyurethane and epoxy resins," *Journal of Materials Science*, vol. 54, no. 11, pp. 8262–8275, 2019, https://doi.org/10.1007/s10853-018-03236-3.
28. J.-H. Hong, S.-E. Shim, and D.-W. Park, "A review on thermal conductivity of polymer composites using carbon-based fillers: Carbon nanotubes and carbon fibers," *Carbon Letters*, vol. 11, no. 4, pp. 347–356, 2010, https://doi.org/10.5714/cl.2010.11.4.347.
29. Y. Huang, H. Peng, J. Cao, F. Ye, C. Li, Y. Li, P. Ai, Q. Chen, Q. Ji, and W. Wu, "Investigation on the preparation of a highly water-resistant emulsion coating for paper-based packaging materials," *Journal of Coatings Technology and Research*, vol. 22, no. 3, pp. 1159–1169, 2025, https://doi.org/10.1007/s11998-024-01040-0.
30. H. Illias, H. Mokhlis, T. Yuan, A. H. A. Bakar, P. L. Lewin, and G. Chen, Partial discharge patterns in high voltage insulation, pp. 750–755, 2012, https://doi.org/10.1109/pecon.2012.6450316.
31. K. A. Imran and K. N. Shivakumar, "Graphene-modified carbon/epoxy nanocomposites: Electrical, thermal and mechanical properties," *Journal of Composite Materials*, vol. 53, no. 1, pp. 93–106, 2018, https://doi.org/10.1177/0021998318780468.
32. A. A. Jaber, M. D. Judd, M. F. Q. Vieira, I. A. Glover, M. Moradzadeh, Z. D. Zaharis, P. I. Lazaridis, and R. C. Atkinson, "Calibration of free-space radiometric partial discharge measurements," *IEEE Transactions on Dielectrics and Electrical Insulation*, vol. 24, no. 5, pp. 3004–3014, 2017, https://doi.org/10.1109/tdei.2017.006730.
33. W. Jarrett, S. P. Jeffs, F. Korkees, and M. Rawson, "The opportunities and challenges of hybrid composite driveshafts and their couplings in the aerospace industry: A review," *Composite Structures*, vol. 320, p. 117203, 2023, https://doi.org/10.1016/j.compstruct.2023.117203.
34. Y. Jia, T. D. Ajayi, B. H. Wahls, K. R. Ramakrishnan, S. Ekkad, and C. Xu, "Multifunctional ceramic composite system for simultaneous thermal protection and electromagnetic interference shielding for carbon fiber-reinforced polymer composites," *ACS Applied Materials & Interfaces*, vol. 12, no. 52, pp. 58005–58017, 2020, https://doi.org/10.1021/acsami.0c17361.

35. J. Jiang, C. Zhang, X. Wei, W. Li, P. Ranjan, X. Zhang, and Z. Li, "A review on insulation challenges towards electrification of aircraft," *High Voltage*, vol. 8, no. 2, pp. 209–230, 2023, https://doi.org/10.1049/hve2.12304.
36. K. K. Khanum, A. M. Sharma, C. Angammana, S. H. Jayaram, and F. Aldawsari, "Influence of filler-polymer interface on performance of silicone nanocomposites," *IEEE Transactions on Industry Applications*, vol. 56, no. 1, pp. 686–692, 2019, https://doi.org/10.1109/tia.2019.294344.
37. M. S. Koochaki, A. Ashrafi, M. Magni, S. P. Trasatti, S. Nouri Khorasani, and R. Esmaeely Neisiany, "Facile strategy toward the development of a self-healing coating by electrospray method," *Materials Research Express*, vol. 6, no. 11, p. 116444, 2019, https://doi.org/10.1088/2053-1591/ab4d1b.
38. R. S. Krishna Moorthy, D. Rahman, M. Olimmah, L. Yang, B. Aberg, I. Husain, A. N. Lemmon, and W. Yu, "Estimation, minimization, and validation of commutation loop inductance for a 135-kW SiC EV traction inverter," *IEEE Journal of Emerging and Selected Topics in Power Electronics*, vol. 8, no. 1, pp. 286–297, 2019, https://doi.org/10.1109/jestpe.2019.2952884.
39. X. Kuang, J. Wu, K. Chen, Z. Zhao, Z. Ding, F. Hu, D. Fang, H. J. Qi, "Grayscale digital light processing 3D printing for highly functionally graded materials," *Science Advances*, vol. 5, no. 5, 2019, https://doi.org/10.1126/sciadv.aav5790.
40. A. Kumar, S. Sreenivasa, S. Dixit, P. S. Bains, R. Sharma, and S. Singh, "Recent developments in the mechanical properties and recycling of fiber-reinforced polymer composites," *Polymer Composites*, vol. 46, no. 5, pp. 3883–3908, 2024, https://doi.org/10.1002/pc.29261.
41. M. Kurimoto, M. Hanai, H. Okubo, Y. Hoshina, M. Takei, and K. Kato, "Application of functionally graded material for reducing electric field on electrode and spacer interface," *IEEE Transactions on Dielectrics and Electrical Insulation*, vol. 17, no. 1, pp. 256–263, 2010, https://doi.org/10.1109/tdei.2010.5412025.
42. H. Lee, H. Kim, J. Jeong, K. Lee, G. Stone, and S. B. Lee, "Inverter-embedded partial discharge testing for reliability enhancement of stator winding insulation in low voltage machines," *IEEE Transactions on Industry Applications*, vol. 58, no. 2, pp. 2088–2096, 2022, https://doi.org/10.1109/tia.2022.3142712.
43. K. H. Lee, A. Dahshan, A. Bibi, U. Zubair, E.-J. Lee, A. Javid, Z. Arshad, M. Alshareef, Q. A. Alsulami, and M. Nawaz, "Porous aerogel structures as promising materials for photocatalysis, thermal insulation textiles, and technical applications: A review," *Catalysts*, vol. 13, no. 9, p. 1286, 2023, https://doi.org/10.3390/catal13091286.
44. S. Li, G. Yin, S. Bai, and J. Li, "A new potential barrier model in epoxy resin nanodielectrics," *IEEE Transactions on Dielectrics and Electrical Insulation*, vol. 18, no. 5, pp. 1535–1543, 2011, https://doi.org/10.1109/tdei.2011.6032822.
45. X. Li, Z. Wang, B. Gao, Y. Yang, G. Wu, P. Xu, and Y. Li, "Partial discharge characteristics of oil-paper insulation for on-board traction transformers under superposed inter-harmonic AC voltages," *IEEE Transactions on Dielectrics and Electrical Insulation*, vol. 27, no. 1, pp. 240–248, 2020, https://doi.org/10.1109/tdei.2019.008404.
46. X.-R. Li, J.-Y. Xue, W.-D. Li, C. Wang, G.-Y. Sun, Z. Liu, and G.-J. Zhang, "3D printing fabrication of conductivity non-uniform insulator for surface flashover mitigation," *IEEE Transactions on Dielectrics and Electrical Insulation*, vol. 26, no. 4, pp. 1172–1180, 2019, https://doi.org/10.1109/tdei.2019.007938.

47. J. Liu, X. Zhang, W. Xi, M. Xu, and R. Zhang, "Progress of Porous/Lattice structures applied in thermal management technology of aerospace applications," *Aerospace*, vol. 9, no. 12, p. 827, 2022, https://doi.org/10.3390/aerospace9120827.
48. Y. Liu, X. Cao, W. Li, S. Zhang, and C. Zhang, "Simulation of electric field distribution in the XLPE insulation of a 320 kV DC cable under steady and time-varying states," *IEEE Transactions on Dielectrics and Electrical Insulation*, vol. 25, no. 3, pp. 954–964, 2018, https://doi.org/10.1109/tdei.2018.006973.
49. M. Lizcano, T. S. Williams, E.-S. E. Shin, D. Santiago, and B. Nguyen, "Aerospace environmental challenges for electrical insulation and recent developments for electrified aircraft," *Materials*, vol. 15, no. 22, p. 8121, 2022, https://doi.org/10.3390/ma15228121.
50. Y. Luo, P. Wang, G. Zhu, J. Peng, K. Cao, G. Wu, and J. Liu, "PD characteristics and microscopic analysis of polyimide film used as turn insulation in inverter-fed motor," *IEEE Transactions on Dielectrics and Electrical Insulation*, vol. 21, no. 5, pp. 2237–2244, 2014, https://doi.org/10.1109/tdei.2014.003868.
51. X. Meng, X. Wu, H. Yu, B. U. Amin, and L. Wang, "Recent progress on fabrication and performance of polymer composites with highly thermal conductivity," *Macromolecular Materials and Engineering*, vol. 306, no. 11, p. 2100434, 2021, https://doi.org/10.1002/mame.202100434.
52. Y. J. Min, D.-E. Kim, and K.-H. Kang, "Development of polyimide films reinforced with boron nitride and boron nitride nanosheets for transparent flexible device applications," *Nano Research*, vol. 11, no. 5, pp. 2366–2378, 2018, https://doi.org/10.1007/s12274-017-1856-0.
53. R. Mohanraman, C. Kocabas, M. A. Bissett, P. Steiner, I. A. Kinloch, "Synergistic improvement in the thermal conductivity of hybrid boron nitride nanotube/nanosheet epoxy composites," *ACS Applied Nano Materials*, vol. 7, no. 11, pp. 13142–13146, 2024. https://doi.org/10.1021/acsanm.4c01646.
54. G. C. Montanari, M. Lizcano, T. S. Williams, S. B. Myneni, and M. Shafiq, "Electrical characterization of boron nitride-filled insulation for aerospace and avionics applications," *Energies*, vol. 17, no. 12, p. 3016, 2024. https://doi.org/10.3390/en17123016.
55. H. Naderiallaf, P. Giangrande, M. Galea, and Y. Ji, "Modeling humidity impact on PDIV for turn-to-turn insulation of inverter-fed motors at different temperatures," *IEEE Transactions on Dielectrics and Electrical Insulation*, vol. 31, no. 3, pp. 1573–1582, 2024, https://doi.org/10.1109/tdei.2023.3338138.
56. H. Okubo, G. Montanari, and N. Hayakawa, "Technical development on partial discharge measurement and electrical insulation techniques for low voltage motors driven by voltage inverters," *IEEE Transactions on Dielectrics and Electrical Insulation*, vol. 14, no. 6, pp. 1516–1530, 2007, https://doi.org/10.1109/tdei.2007.4401236.
57. P. Patpatiya, A. Shastri, S. Sharma, and K. Chaudhary, "A review on polyjet 3D printing of polymers and multi-material structures," *Proceedings of the Institution of Mechanical Engineers, Part C: Journal of Mechanical Engineering Science*, vol. 236, no. 14, pp. 7899–7926, 2022, https://doi.org/10.1177/09544062221079506.
58. T. Petri, M. Keller, and N. Parspour, "The insulation resilience of inverter-fed low voltage traction machines: Review, challenges, and opportunities," *IEEE Access*, vol. 10, pp. 104023–104049, 2022, https://doi.org/10.1109/access.2022.3210348.
59. Y. Qin, J. Spencer, C. Dimarino, C. Buttay, Y. Zhang, B. Albano, M. Tadjer, J. S. Lundh, and B. Wang, "Thermal management and packaging of wide and ultra-wide

bandgap power devices: A review and perspective," *Journal of Physics D: Applied Physics*, vol. 56, no. 9, p. 093001, https://doi.org/10.1088/1361-6463/acb4ff.

60. M. A. Rahman, M. Ghassemi, and A. Saha, "Optimized fabrication process and PD characteristics of MVDC multilayer insulation cable systems for next generation wide-body all-electric aircraft," *Energies*, vol. 17, no. 12, p. 3040, 2024, https://doi.org/10.3390/en17123040.
61. A. Ramgobin, G. Fontaine, and S. Bourbigot, "Thermal degradation and fire behavior of high performance polymers," *Polymer Reviews*, vol. 59, no. 1, pp. 55–123, 2019, https://doi.org/10.1080/15583724.2018.1546736.
62. C. Reddy, and T. Ramu, "Polymer nanocomposites as insulation for HV DC cables – investigations on the thermal breakdown," *IEEE Transactions on Dielectrics and Electrical Insulation*, vol. 15, no. 1, pp. 221–227, 2008, https://doi.org/10.1109/t-dei.2008.4446754.
63. J.-R. Riba, A. Gómez-Pau, M. Moreno-Eguilaz, and S. Bogarra, "Arc tracking control in insulation systems for aeronautic applications: Challenges, opportunities, and research needs," *Sensors (Basel, Switzerland)*, vol. 20, no. 6, p. 1654, 2020, https://doi.org/10.3390/s20061654.
64. H. Ribeiro, M. C. Lopes, J. P. C. Trigueiro, W. M. Silva, J. M. Tour, R. Villegas Salvatierra, N. Chopra, J. J. Pedrotti, P. M. Ajayan, A. Cristian Chipara, G. G. Silva, I. N. Odeh, C. S. Tiwary, P. S. Owuor, and C. F. Woellner, "Hybrid MoS2/h-BN nanofillers as synergic heat dissipation and reinforcement additives in epoxy nanocomposites," *ACS Applied Materials & Interfaces*, vol. 11, no. 27, pp. 24485–24492, 2017, https://doi.org/10.1021/acsami.7b09945.
65. J. Rubio-Serrano, J. A. Garcia-Souto, J. Posada, M. V. Rojas-Moreno, G. Robles, and J. M. Martínez-Tarifa, "Electro-acoustic detection, identification and location of partial discharge sources in oil-paper insulation systems," *IEEE Transactions on Dielectrics and Electrical Insulation*, vol. 19, no. 5, pp. 1569–1578, 2012, https://doi.org/10.1109/tdei.2012.6311502.
66. V. S. Saji, "2D hexagonal boron nitride (h-BN) nanosheets in protective coatings: A literature review," *Heliyon*, vol. 9, no. 9, p. e19362, 2023, https://doi.org/10.1016/j.heliyon.2023.e19362.
67. T. Sathishkumar, J. Naveen, and S. Satheeshkumar, "Hybrid fiber reinforced polymer composites – a review," *Journal of Reinforced Plastics and Composites*, vol. 33, no. 5, pp. 454–471, 2014, https://doi.org/10.1177/0731684413516393.
68. A. Sezer Hicyilmaz, and A. Celik Bedeloglu, "Applications of polyimide coatings: A review," *SN Applied Sciences*, vol. 3, no. 3, 2021, https://doi.org/10.1007/s42452-021-04362-5.
69. T. Shahsavarian, C. Lents, D. Zhang, C. Li, Y. Cao, and X. Wu, "Temperature-dependent partial discharge characteristics of high temperature materials at DC voltage for hybrid propulsion systems," *High Voltage*, vol. 6, no. 4, pp. 590–598, 2021, https://doi.org/10.1049/hve2.12110.
70. L. Simoni, "A general approach to the endurance of electrical insulation under temperature and voltage," *IEEE Transactions on Electrical Insulation*, vol. EI-16, no. 4, pp. 277–289, 1981, https://doi.org/10.1109/tei.1981.298361.
71. I. L. Singer, J. G. Kohl, and M. Patterson, "Mechanical aspects of silicone coatings for hard foulant control," *Biofouling*, vol. 16, nos. 2–4, pp. 301–309, 2000, https://doi.org/10.1080/08927010009378453.
72. J. Singh, M. Kumar, S. Kumar, and S. K. Mohapatra, "Properties of glass-fiber hybrid composites: A review," *Polymer-Plastics Technology and Engineering*,

vol. 56, no. 5, pp. 455–469, 2016, https://doi.org/10.1080/03602559.2016.1233271.

73. R. R. Singh, I. Vairavasundaram, G. Bhatti, D. Kalel, and F. Alsaif, "Building a digital twin powered intelligent predictive maintenance system for industrial AC machines," *Machines*, vol. 11, no. 8, p. 796, 2023, https://doi.org/10.3390/machines11080796.
74. G. C. Stone, D. G. Dunn, and M. K. W. Stranges, "Common questions on partial discharge testing: A review of recent developments in IEEE and IEC standards for offline and online testing of motor and generator stator windings," *IEEE Industry Applications Magazine*, vol. 22, no. 1, pp. 14–19, 2016, https://doi.org/10.1109/mias.2015.2458337.
75. Q. Sun, R. Tao, S. Zhou, X. Dong, J. Qi, Q. Shen, and G. Zhi, "Advanced design and manufacturing approaches for structures with enhanced thermal management performance: A review," *Advanced Materials Technologies*, vol. 9, no. 15, https://doi.org/10.1002/admt.202400263.
76. T. Takahashi, T. Okamoto, K. Shibata, and Y. Ohki, "Breakdown strength at the interface between epoxy resin and silicone rubber – a basic study for the development of all solid insulation," *IEEE Transactions on Dielectrics and Electrical Insulation*, vol. 12, no. 4, pp. 719–724, 2005, https://doi.org/10.1109/tdei.2005.1511097.
77. T. Tanaka, "Aging of polymeric and composite insulating materials: Aspects of interfacial performance in aging," *IEEE Transactions on Dielectrics and Electrical Insulation*, vol. 9, no. 5, pp. 704–716, 2002, https://doi.org/10.1109/tdei.2002.1038658.
78. M. M. Tousi and M. Ghassemi, "Characterization of nonlinear field-dependent conductivity layer coupled with protruding substrate to address high electric field issue within high-voltage high-density wide bandgap power modules," *IEEE Journal of Emerging and Selected Topics in Power Electronics*, vol. 8, no. 1, pp. 343–350, 2019, https://doi.org/10.1109/jestpe.2019.2953145.
79. O. Ugwueze, N. Horri, M. Bromfield, M. Innocente, and T. Statheros, Investigation of a mission-based sizing method for electric VTOL aircraft preliminary design, 2022, Jan. 3, https://doi.org/10.2514/6.2022-1931.
80. H. Ullah, K. A. M. Azizli, Z. B. Man, M. B. C. Ismail, and M. I. Khan, "The potential of microencapsulated self-healing materials for microcracks recovery in self-healing composite systems: A review," *Polymer Reviews*, vol. 56, no. 3, pp. 429–485, 2016, https://doi.org/10.1080/15583724.2015.1107098.
81. C. Wang, A. Roy, Z. Chen, and V. V. Silberschmidt, "Modelling of damage evolution in braided composites: Recent developments," *Mechanics of Advanced Materials and Modern Processes*, vol. 3, no. 1, 2017, https://doi.org/10.1186/s40759-017-0030-4.
82. P. Wang, Z. Li, S. Akram, J. Castellon, G. C. Montanari, S. Ma, and P. Meng, "Design of an effective antenna for partial discharge detection in insulation systems of inverter-fed motors," *IEEE Transactions on Industrial Electronics*, vol. 69, no. 12, pp. 13727–13735, 2022, https://doi.org/10.1109/tie.2021.3130335.
83. T. Wang, Y. Sun, S. Chen, J. Yang, X. Jia, and H. Song, "Long-term stable superlubricity coatings enabled by the interaction between the polydimethylsiloxane brush and silicone oil," *ACS Applied Materials & Interfaces*, vol. 16, no. 8, pp. 11134–11144, 2024, https://doi.org/10.1021/acsami.4c00734.

84. Z. Wang, Y. Cheng, T. Zhang, Y. Zhou, M. Li, G. Yang, X. Cao, and C. Liu, "Dielectric and thermal properties characterisation and evaluation of novel epoxy materials for high-voltage power module packaging," *High Voltage*, vol. 9, no. 5, pp. 1021–1032, 2024, https://doi.org/10.1049/hve2.12443.
85. G. Wu, J. Wu, B. Gao, X. Guo, K. Cao, K. Zhou, and L. Zhou, "Microscopic view of aging mechanism of polyimide film under pulse voltage in presence of partial discharge," *IEEE Transactions on Dielectrics and Electrical Insulation*, vol. 17, no. 1, pp. 125–132, 2010, https://doi.org/10.1109/tdei.2010.5412010.
86. X. Wu, H. Yu, Z. Huang, L. Wang, X. Meng, X. Gong, J. Liu, and X. Liu, "Enhancing thermal conductivity of epoxy composites via f-BN@f-MgO hybrid fillers assisted by hot pressing," *Polymer Composites*, vol. 44, no. 5, pp. 2966–2976, 2023, https://doi.org/10.1002/pc.27294.
87. K. Wynne, G. Swain, R. Fox, S. Bullock, and J. Uilk, "Two silicone nontoxic fouling release coatings: Hydrosilation cured PDMS and CaCO3 filled, ethoxysiloxane cured RTV11," *Biofouling*, vol. 16, nos. 2–4, pp. 277–288, 2000, https://doi.org/10.1080/08927010009378451.
88. R. Yang, Y. Pei, X. Zhang, X. Lin, H. Gong, M. Sheng, and Y. Zhang, "Thermal and dielectric properties of epoxy-based composites filled with flake and whisker type hexagonal boron nitride materials," *High Performance Polymers*, vol. 33, no. 4, pp. 417–428, 2020, https://doi.org/10.1177/0954008320959413.
89. Z. Yuan, D. Huitink, Y. Wang, A. I. Emon, F. Luo, Z. Wang, and M. Ul-Hassan, "Insulation and switching performance optimization for partial-discharge-free laminated busbar in more-electric aircraft applications," *IEEE Transactions on Power Electronics*, vol. 37, no. 6, pp. 6831–6843, 2022, https://doi.org/10.1109/tpel.2021.3137839.
90. J. Zhang, T. Ding, Y. Li, H. Liu, J. Liu, J. Fan, W. Zhang, L. Pu, L. Wei, and Z. Guo, "Alternating multilayer structural epoxy composite coating for corrosion protection of steel," *Macromolecular Materials and Engineering*, vol. 304, no. 12, p. 1900374, 2019, https://doi.org/10.1002/mame.201900374.
91. P. Zhang, P. Neti, and K. Younsi, "A novel online stator ground-wall insulation monitoring scheme for inverter-fed AC motors," *IEEE Transactions on Industry Applications*, vol. 51, no. 3, pp. 2201–2207, 2015, https://doi.org/10.1109/tia.2014.2385937.
92. X. Zhao, C.-W. Chang, P. Asfaux, D. Dong, S. Uicich, R. Phukan, and R. Burgos, "Design of ultracompact gate driver integrated with current sensor and commutation path for a 211-kW three-level SiC aircraft propulsion inverter," *IEEE Journal of Emerging and Selected Topics in Power Electronics*, vol. 11, no. 4, pp. 4077–4094, 2023, https://doi.org/10.1109/jestpe.2023.3247378.
93. X. Zhao, A. Plat, D. Dong, P. Asfaux, S. Uicich, R. Phukan, M. Debbou, R. Burgos, and C.-W. Chang, "Design and implementation of SiC-based 200-kW high-density high-speed high-altitude electric propulsion AC drive system," *IEEE Journal of Emerging and Selected Topics in Power Electronics*, vol. 12, no. 5, pp. 5176–5199, 2024, https://doi.org/10.1109/jestpe.2024.3419149.
94. C. Zhi, Y. Bando, C. Tang, T. Terao, H. Kuwahara, and D. Golberg, "Towards thermoconductive, electrically insulating polymeric composites with boron nitride nanotubes as fillers," *Advanced Functional Materials*, vol. 19, no. 12, pp. 1857–1862, 2009, https://doi.org/10.1002/adfm.200801435.

10

Quality Assurance Tests for Motors Used in Electric Aircraft

Shesha H. Jayaram and Shakthi Prasad D.

10.1 Electrical Tests for Aircraft Motor Insulation Systems

10.1.1 Overview

Electrical testing of insulation systems in rotating electrical machines provides quantitative and diagnostic evidence of their ability to withstand both operational and transient electric stresses. In aircraft propulsion and auxiliary drives, such evaluation is particularly critical because the service environment combines reduced pressure, wide temperature variation, and exposure to steep-front converter voltages.

The sequence of tests used for insulation qualification includes the power-frequency dielectric withstand and breakdown verification, impulse-voltage withstand testing, partial-discharge measurement under both sinusoidal and repetitive-impulse voltages, insulation-resistance and polarisation-index evaluation, surge comparison for turn-to-turn integrity, partial-discharge endurance testing under repetitive impulses, and dissipation-factor (tan δ) measurement.

Each test contributes a complementary diagnostic parameter within a comprehensive quality-assurance framework based on IEC 60034, IEC 60060–1, IEEE Std. 4, IEEE 43, IEEE 522, and ASTM D149/D150.

10.1.2 Dielectric Withstand and AC Breakdown Verification

The dielectric-withstand test, commonly referred to as the Hi-Pot test, confirms that the stator insulation can tolerate its designated service and overvoltage stresses without electrical failure. It employs power-frequency alternating voltage (50 or 60 Hz) applied between the stator winding and the grounded frame. All phases are tied together and energised to the test potential specified by IEC 60034–1, generally 1.5 times the rated voltage plus 1,000 V rms, for machines up to 1 kV, and higher values as prescribed for

 DOI: 10.1201/9781003747994-10

machines in the 2.3–15 kV range. The voltage is increased gradually and held for 1 minute while the leakage current is monitored. Absence of flashover, puncture, or rising current indicates satisfactory integrity.

Closely related in concept, the AC breakdown or dielectric-strength test establishes the ultimate capacity of the insulation to withstand electric stress. In this destructive test, performed mainly for material qualification, the voltage is increased continuously until electrical failure occurs, and the breakdown field in kV/mm is recorded. The result defines the strength of the insulation system and provides a design reference for determining permissible operating stresses, typically limited to one-third of the measured breakdown value. Together, these tests confirm that the insulation possesses both adequate steady-state endurance and ultimate dielectric strength, forming the foundation for all subsequent evaluations of insulation quality.

10.1.3 Impulse Voltage Withstand Test

The impulse-voltage test determines the capability of the insulation system to resist transient overvoltages such as lightning surges, switching impulses, or commutation transients. It defines the basic impulse insulation level (BIL) of the machine.

A calibrated impulse generator produces standardised waveforms, commonly the 1.2/50 µs lightning impulse or the 250/2500 µs switching impulse, in accordance with IEC 60060-1. The test object is connected through a measuring divider to record the waveform applied between the winding and ground. Voltage is applied in successive shots up to the specified BIL level, and the insulation is deemed to have passed if no internal breakdown or flashover occurs and the recorded waveform conforms to standard parameters. Depending on the test object, either a simple up-and-down method or an elaborative statistical analysis is used to find the 50% breakdown voltage.

The impulse test thus assesses dynamic dielectric strength, complementing the Hi-Pot test by verifying that the insulation can also survive rapid, high-field transients without partial breakdown or reversal stress.

10.1.4 Partial Discharge Measurement

Partial-discharge (PD) measurement provides insight into localised dielectric imperfections that may not manifest during a conventional dielectric-withstand test. Microscopic gas voids, delamination, or interfacial defects within the insulation can undergo partial breakdown when the local electric field exceeds the ionisation strength of the trapped gas. Each discharge releases a small charge and produces chemical and thermal effects that cumulatively erode the insulation, eventually leading to electrical treeing or complete failure.

Under power-frequency excitation, PD is quantified as apparent charge, expressed in picocoulombs (pC), representing the equivalent charge that would produce the same voltage step across the measuring impedance as the actual discharge. The test set-up typically includes a coupling capacitor in series with the winding, a high-frequency detection impedance, and a calibrated PD measuring instrument conforming to IEC 60034-27-1.

As the applied sinusoidal voltage is gradually increased, a voltage is reached at which repetitive discharges first appear within the insulation system. This level is defined as the partial discharge inception voltage (PDIV), the lowest applied voltage, referred to the test object terminals, at which partial discharges of specified magnitude (commonly 10 pC) are observed under stated environmental and frequency conditions. When the voltage is subsequently reduced, the discharges cease at a lower level known as the partial discharge extinction voltage (PDEV), the highest voltage, during a downward voltage sweep, at which discharges no longer occur above the measurement threshold.

For aircraft-motor insulation systems, PDIV and PDEV serve as quantitative markers of insulation quality and stability. Type I systems (hairpin or random-wound, designed to be PD-free) are required to exhibit no measurable PD, typically below 10 pC, at rated phase-to-ground voltage plus 10%. Type II systems (form-wound, PD-tolerant) may display finite PD activity, but the PDIV and PDEV values must remain stable within established limits, showing no systematic reduction over repeated tests or during life cycle qualification.

In converter-fed applications, additional PD evaluation under repetitive-impulse excitation is performed in accordance with IEC TS 61934. In this case, the measuring system employs wide-band current transformers and the IEC-specified apparent-charge calibration adapted to the short-duration pulses typical of PWM inverters. The combination of power-frequency and impulse-mode PD tests provides a comprehensive assessment of both intrinsic insulation soundness and susceptibility to converter-induced degradation.

10.1.5 Insulation Resistance and Polarisation Index

Insulation-resistance testing characterises the bulk resistivity of the winding system and detects contamination or moisture ingress. A DC voltage, typically 500 V for low-voltage and 5 kV for high-voltage windings, is applied between the winding and ground, and the resistance after 1 minute is recorded. Continuing the same voltage for 10 minutes yields the Polarisation Index (PI), the ratio of the 10-minute to 1-minute resistance. A high PI (≥2) denotes clean, dry insulation; a low value indicates contamination or ageing. Results are corrected to 40°C and trended over time to monitor insulation health (IEEE 43; IEC 60034-27-3; IEC 60034-1).

10.1.6 Surge (Turn-to-Turn) Test

The surge test examines the strength of turn insulation within each phase. A steep-front oscillatory pulse with rise time typically ≤0.1 μs is applied, and the resulting voltage waveform is compared with that of a reference coil or phase. A deviation in oscillation frequency or waveform shape indicates interturn weakness or shorted turns.

For low-voltage windings, the test voltage corresponds to roughly twice the rated line-to-line voltage; for high-voltage form-wound coils, the levels follow IEC 60034–15. Consistent waveforms across all phases confirm uniform manufacturing quality and adequate turn insulation.

10.1.7 Partial-Discharge Endurance under Repetitive Impulses

The partial-discharge (PD) endurance test, sometimes referred to as electrical endurance under repetitive impulses, evaluates how an insulation system withstands continuous PD activity and high-frequency voltage stress over time. The purpose is not merely to determine whether PD occurs, but to measure the insulation's resistance to the erosive effects of sustained discharges and the cumulative heating and chemical degradation they cause.

Testing is carried out in a temperature-controlled environment using a repetitive-impulse generator that reproduces inverter-like voltage pulses with rise times shorter than 500 ns and frequencies of several kilohertz. The applied voltage is typically equal to the peak of the intended service voltage or a defined multiple thereof. PD magnitude and pattern are monitored throughout the test in accordance with IEC 60034-18-41 for PD-free Type I systems and IEC 60034-18-42 for PD-tolerant Type II systems.

For Type I insulation, the acceptance criterion is that no measurable PD should appear during the endurance period and that the system should exhibit no visible erosion or tracking upon post-test inspection. For Type II systems, controlled PD activity is permitted provided that the endurance time to failure equals or exceeds that of the qualified reference system established during type testing.

The PD-endurance test therefore serves as a lifetime-simulation assessment, correlating repetitive-stress conditions in the laboratory with expected service life under actual converter operation. It provides one of the most realistic indications of long-term insulation robustness.

10.1.8 Dissipation-Factor (tan δ) Measurement

The dissipation-factor (tan δ) test quantifies the dielectric losses within an insulation system when it is subjected to an alternating electric field at power frequency. It provides a sensitive measure of polarisation phenomena, dielectric heating, and contamination effects that may not yet be observable as partial discharge or breakdown. The parameter tan δ represents the tangent of the loss angle between the total current and the purely capacitive current through the insulation; it expresses the ratio of resistive (loss) current to reactive (capacitive) current.

In practice, tan δ is measured using a precision capacitance bridge or an automated dielectric-loss analyser, following the procedures specified in IEC 60034-27-3 for machine windings or ASTM D150 for insulating materials. The test is conducted at power frequency (50 or 60 Hz) with the voltage applied between the stator winding and the frame. Measurements are taken at several voltage levels, commonly from 0.2 to 1.0 times the rated phase-to-ground voltage, to establish a voltage-loss characteristic sometimes called the "tip-up" curve. The specimen should be thermally stable and at a recorded temperature, since dielectric losses increase with temperature due to enhanced dipolar relaxation and conduction.

A new or well-preserved insulation system exhibits a low and nearly constant tan δ value, typically below 0.5 % across the measured voltage range. A progressive rise in tan δ with increasing voltage, referred to as tan δ tip-up, signifies the onset of localised discharge, moisture absorption, or contamination that introduces resistive leakage paths. The difference in tan δ between low- and high-test voltages thus serves as a diagnostic indicator of partial discharge initiation and insulation ageing. For high-voltage form-wound stators, the tan δ tip-up between 0.5 and 1.0 times the rated voltage should not exceed approximately 0.5 percentage points; larger increases warrant further PD investigation.

As tan δ directly relates to dielectric loss power ($P = \omega C V^2 \tan \delta$), trending its value over repeated tests provides valuable information about the gradual deterioration of the insulation system. A steady upward drift of tan δ over time, even at constant voltage, suggests chemical or thermal ageing of the resin system or accumulation of conductive by-products. When correlated with other diagnostic results, particularly PD magnitude and PDIV/PDEV stability, tan δ offers a quantitative measure of dielectric health and complements the overall assessment of insulation quality assurance.

Summary

This sequence of standardised tests provides a comprehensive evaluation of insulation performance across all electrical stress regimes. The power-frequency dielectric withstand and breakdown tests confirm macroscopic integrity; the impulse test verifies transient robustness; partial-discharge measurement and endurance tests probe microscopic imperfections and long-term resistance to erosion; insulation-resistance and polarisation-index measurements assess cleanliness and moisture state; surge comparison verifies turn insulation; and dissipation-factor measurement monitors dielectric losses.

Together, these evaluations form a coherent quality-assurance methodology applicable in the qualification of aircraft electrical machines. Consistent adherence to these standardised tests ensures that insulation systems meet the stringent reliability and safety requirements of aerospace applications.

All tests described require strict adherence to laboratory safety protocols and accurate recording of environmental parameters. Temperature, humidity, and air pressure affect both breakdown and discharge inception levels; therefore, results must always be interpreted with reference to the prevailing ambient conditions.

10.2 Electro-Thermal Tests for Insulation Systems

10.2.1 Overview

Electro-thermal tests assess the performance and endurance of insulation systems when electrical and thermal stresses act simultaneously. They complement purely electrical tests by reproducing the combined service conditions that govern long-term reliability. These methods evaluate thermal ageing, electrical-thermal synergy, and surface degradation, ensuring that insulation materials maintain integrity under realistic operating environments.

10.2.2 Thermal–Electrical Coupled Tests

Thermal–electrical coupled testing evaluates the combined influence of temperature and electric stress on insulation performance. In real operation, electrical machines rarely experience purely electrical or purely thermal loading; instead, voltage stress, current-generated heating, and ambient temperature variations act simultaneously. The coupled test reproduces this condition by applying both electrical excitation and controlled thermal loading to the specimen while monitoring its dielectric behaviour.

A typical arrangement uses a representative coil or motorette equipped with embedded thermocouples, placed in a thermostatically controlled oven or environmental chamber. The winding is energised either at its rated power-frequency voltage or at an accelerated level, and the temperature is maintained at the desired value, often corresponding to the upper limit of the material's temperature class plus an additional safety margin (for example, 180°C for a Class F system). Leakage current, insulation resistance, and partial-discharge activity are recorded continuously. The combined electrical and thermal stress produces degradation mechanisms such as resin softening, gas-void expansion, and microcrack propagation that cannot be reproduced by either stress alone.

The rate of property decline, observed as a reduction in insulation resistance or an increase in tan δ and PD magnitude, provides an estimate of the system's thermal–electrical endurance. Results are compared with reference specimens aged electrically at ambient temperature or thermally without voltage to separate the synergistic component of degradation.

Thermal–electrical coupled testing thus offers a realistic simulation of service conditions in inverter-fed aircraft motors, where localised heating and steep voltage transients coexist and accelerate insulation ageing.

10.2.3 Thermal Index Testing

(IEEE 117, IEC 60085 – Determination of Insulation System Temperature Class)

The thermal-index (TI) test establishes the temperature rating, or *thermal class*, of an insulation system, the maximum temperature at which the system can operate for 20,000 hours without unacceptable deterioration. The procedure, standardised in IEEE 117 and IEC 60085, is an accelerated ageing test based on the Arrhenius relationship between temperature and rate of degradation.

Representative specimens, such as twisted-pair wires, small coils, or insulation test object, are subjected to continuous ageing at three or more elevated temperatures, each sufficiently high to produce measurable deterioration within a reasonable time (commonly 155°C, 180°C, and 200°C for Class F materials). At scheduled intervals, the specimens are removed and evaluated by an *electrical proof test*, such as a dielectric-withstand or surge test, to determine whether the insulation still meets functional requirements. The time-to-failure at each temperature is plotted on a logarithmic scale against the reciprocal of absolute temperature ($1/T$ in K^{-1}). A straight-line regression through these points extrapolates the time corresponding to 20,000 hours. The temperature at this intersection defines the system's thermal index and thereby its classification, for example, 155, 180, or 200°C class.

The method distinguishes between material ageing and system ageing, since the complete insulation system, including conductors, varnish, tapes, and other impregnating materials, is tested as an integrated assembly. The TI test, therefore, provides the fundamental data used to assign insulation classes and to verify that aircraft-motor insulation systems possess adequate long-term thermal endurance for continuous service at their rated temperature rise.

10.2.4 Tracking and Erosion Resistance

(ASTM D2303, IEC 60587 – Evaluation under Contaminated Conditions)

Tracking and erosion tests assess an insulating material's resistance to surface degradation when exposed to simultaneous electrical stress and contamination. Under service conditions, especially in humid or polluted atmospheres, conductive films or droplets can form on the surface of insulators, allowing leakage current to flow. Localised heating at discontinuities in these films can produce carbonised tracks or eroded craters, eventually leading to surface flashover.

The standardised test procedure of ASTM D2303 or IEC 60587 quantifies this behaviour. A planar or inclined specimen is placed between two electrodes, and a controlled electrolyte, usually an aqueous ammonium chloride or ammonium fluoride solution of defined conductivity, is applied intermittently to the surface through a dripping or spray mechanism. An AC voltage at power frequency is applied between the electrodes. As the liquid evaporates and re-forms, dry bands appear, across which arcing occurs. The test continues for a specified duration (for example, 4 hours) or until a continuous track bridges the electrodes.

Performance is expressed either as the Comparative Tracking Index (CTI), representing the highest voltage at which no tracking occurs under the prescribed conditions, or as a tracking and erosion resistance class, as defined in IEC 60587. Visual inspection and weight-loss measurements are used to assess erosion. Materials with high CTI values and minimal surface damage are preferred for components exposed to condensation, salt fog, or oil mist in aircraft environments.

Tracking and erosion resistance testing therefore complements bulk dielectric tests by evaluating the insulation's behaviour at its most vulnerable interface, the solid–air boundary, where surface contamination and electric stress interact. The results guide the selection of resins, coatings, and encapsulants for reliable performance under environmental exposure.

Summary

The electro-thermal evaluations discussed in this chapter extend the scope of insulation quality assurance beyond purely electrical phenomena. By addressing the couplings between temperature, electric field, and surface environment, they provide a realistic assessment of how insulation systems age and degrade under service-relevant stresses. Together, the thermal–electrical coupled tests, thermal-index determination, and tracking-and-erosion evaluations define the material's endurance envelope and confirm that laboratory data can be extrapolated confidently to field operation. Whether applied to low-voltage hairpin windings or high-voltage propulsion machines, these standardised tests provide the objective evidence required to ensure safety and reliability in aircraft electrical systems.

10.3 Analytical and Material Characterisation Tests

10.3.1 Overview

This section provides a comprehensive overview of advanced analytical techniques used for characterising materials and its degradation. It explains thermal analysis methods such as thermogravimetric analysis (TGA), which quantifies mass changes with temperature to reveal decomposition and stability, differential scanning calorimetry (DSC) for measuring heat flow associated with material transitions, and dynamic mechanical analysis (DMA) for assessing viscoelastic properties and thermal transitions. The section also introduces key spectroscopic and microscopic tools, including Fourier transform infrared spectroscopy (FTIR) for molecular structure identification, X-ray photoelectron spectroscopy (XPS) for analysing surface composition and chemical states, scanning electron microscopy with energy-dispersive X-ray analysis (SEM-EDAX) for high-resolution imaging and elemental analysis, and atomic force microscopy (AFM) for nanometre-scale surface and mechanical measurements. Additionally, it highlights electrical property measurements, such as volume and surface resistivity (ASTM D257) and dielectric spectroscopy, to understand the molecular mobility, and phase transitions. Together, these methods enable in-depth insights into chemical composition, structure, morphology, thermal behaviour, and electrical properties of a wide range of insulating materials.

10.3.2 Thermal Analysis

10.3.2.1 Thermogravimetric Analysis (TGA)

Thermogravimetric analysis continuously measures mass changes as a function of temperature under controlled atmospheric conditions, providing quantitative information about physical phenomena such as phase transitions, absorption, and desorption, and chemical processes including decomposition and oxidation reactions. The procedure involves suspending a sample in a small crucible from a sensitive microbalance positioned inside a programmable furnace. Under controlled temperature programmes, typically 10–20°C/min linear heating ramps, the balance records mass changes with microgram-level precision. These variations correspond to thermal events, revealing insights into thermal stability and composition. TGA results display as thermograms plotting mass percentage against temperature. Initial plateaus represent thermally stable regions; stepwise mass losses indicate distinct decomposition or volatilisation events; final residues represent non-volatile components such as fillers or ash. Advantages include small sample requirements, minimal preparation, applicability to solids and liquids, and high precision. However, the TGA detects only mass changes and cannot identify specific chemical species without coupling to spectroscopic techniques, it cannot detect phase transitions without a mass change, and materials with similar decomposition temperatures cannot be differentiated.

10.3.2.2 Differential Scanning Calorimetry (DSC)

Differential scanning calorimetry measures the difference in heat flow between a sample and an inert reference under controlled temperature programmes, providing quantitative and qualitative information about endothermic or exothermic processes and changes in heat capacity. Both sample and reference undergo identical temperature programmes. When thermal events such as melting, crystallisation, or chemical reactions occur, energy differences arise and are measured as heat flow. Unlike differential thermal analysis (DTA), DSC directly quantifies heat flow, enabling accurate enthalpy measurements. Common thermal events detected include melting (endothermic), crystallisation (exothermic), glass transitions, curing reactions (exothermic), and decomposition. DSC thermograms plot heat flow against temperature or time. By convention, endothermic peaks typically appear as downward deflections, while exothermic peaks appear as upward deflections. Critical information extracted includes peak temperatures characterising transitions, enthalpy changes (peak areas) quantifying energy in J/g, baseline shifts indicating glass transitions, and peak shape revealing material purity. Advantages include small sample requirements, rapid analysis (under 30 minutes), minimal preparation, broad applicability, and excellent quantitative capability. Limitations include the inability to identify specific chemical species, complications arising from overlapping events, dependence on critical experimental conditions, and potential baseline drift.

10.3.2.3 Dynamic Mechanical Analysis (DMA)

Dynamic mechanical analysis measures viscoelastic properties by applying oscillating force or deformation and monitoring material response as a function of temperature, time, or frequency. This technique provides critical information about material stiffness, damping behaviour, and thermal transitions essential for understanding polymer structure–property relationships. When sinusoidal stress is applied to viscoelastic materials, the resulting strain exhibits a phase lag (δ) between stress and strain. Purely elastic materials show $\delta = 0°$, purely viscous liquids show $\delta = 90°$. Viscoelastic polymers exhibit intermediate behaviour ($0° < \delta < 90°$), displaying both solid-like and liquid-like characteristics. DMA measurements yield three fundamental parameters: storage modulus (E' or G'), representing energy stored and recovered; loss modulus (E'' or G''), characterising energy dissipated as heat; and loss tangent ($\tan \delta = E''/E'$), providing insight into viscous–elastic balance. DMA exhibits exceptional sensitivity for measuring glass transition temperature (T_g), where storage modulus decreases dramatically (10–1,000 times) and loss modulus and $\tan \delta$ exhibit distinct maxima. Advantages include exceptional transition sensitivity, small sample requirements (10–50 mg), direct measurement of application-relevant mechanical properties, and broad temperature–frequency ranges. Limitations include parameter optimisation complexity and potential tip wear.

10.3.3 Spectroscopic and Chemical Characterisation

10.3.3.1 Fourier Transform Infrared Spectroscopy (FTIR)

Fourier transform infrared spectroscopy measures infrared radiation absorption to provide detailed molecular structure and chemical composition information. This versatile technique has become widely practised due to its sensitivity, flexibility, specificity, and minimal sample preparation requirements. Chemical bonds vibrate at characteristic frequencies determined by atom types and bonding arrangements. When samples are exposed to infrared radiation, molecular bonds absorb specific frequencies matching their natural vibrational frequencies. Different bond types (C—H, C=O, O—H, N—H, C=C) exhibit distinct vibrational frequencies, creating unique spectral fingerprints. Molecular vibrations occur as stretching (bond length changes) or bending (bond angle changes). For infrared activity, vibrations must change the molecular dipole moment. FTIR accommodates diverse sample types through transmission (infrared passes directly through sample) and attenuated total reflectance (ATR) sampling (total internal reflection at crystal–sample interface with 0.5–2 μm penetration depth). ATR requires minimal preparation and solid samples are simply pressed against the crystal surface. Spectra display infrared absorption versus wavenumber (cm^{-1}), typically 4,000–400 cm^{-1}. The spectrum divides into the Functional Group Region (4,000–1,500 cm^{-1}) containing characteristic bands and fingerprint region (1,500–400 cm^{-1}) enabling molecular identification. Advantages include rapid analysis, minimal preparation, non-destructive measurement, and extensive spectral databases. Limitations include water intolerance and limited penetration depth in ATR mode.

10.3.3.2 X-Ray Photoelectron Spectroscopy (XPS)

X-ray photoelectron spectroscopy, also known as electron spectroscopy for chemical analysis (ESCA), measures elemental composition, oxidation state, and chemical environment of atoms within the top 5–10 nm of material surfaces by analysing photoelectron kinetic energies through the photoelectric effect. XPS operates on the photoelectric effect: X-rays strike sample surfaces causing core-level electron ejection. Photoelectron kinetic energy follows: $KE = h\nu - BE - \varphi$, where KE is the photoelectron kinetic energy, $h\nu$ is the incident X-ray photon energy (typically Al Kα at 1,486.6 eV or Mg Kα at 1,253.6 eV), BE is the binding energy, and φ is spectrometer work function. Since X-ray source energy is precisely known and kinetic energy is directly measured, binding energy, a chemical fingerprint specific to each element and chemical environment can be uniquely determined, enabling elemental identification and chemical state analysis. XPS spectra display binding energy (eV) versus signal intensity (counts). Advantages include non-destructive surface analysis, exceptional surface sensitivity (5–10 nm), and quantitative elemental composition determination. Limitations include inability to detect hydrogen and helium and vulnerability to sample charging artifacts.

10.3.3.3 Scanning Electron Microscopy and Energy-Dispersive X-Ray Analysis (SEM-EDAX)

Scanning electron microscopy combined with energy dispersive X-ray analysis represents an integrated platform for simultaneous high-resolution morphological characterisation and quantitative elemental analysis at micrometre to nanometre scales. This technique provides capabilities for examining surface topography, identifying chemical composition, and understanding material structure across diverse applications. SEM generates a finely focused electron beam, typically accelerated to 1–40 keV, which scans the sample surface in raster patterns. When high-energy electrons strike atoms, they penetrate to 0.1–1 micrometre depths, creating interaction volumes generating multiple signals. Secondary electrons (SE) provide exceptional surface sensitivity and topographic contrast with resolution below 1 nm. Backscattered electrons (BSE) provide compositional contrast – heavier elements appear brighter. Characteristic X-rays are unique to each element, enabling chemical identification.

Energy-dispersive X-ray spectroscopy (EDAX/EDS) transforms SEM into a quantitative chemical analyser. Primary electrons eject inner-shell electrons; outer-shell electrons fall into vacancies, releasing characteristic X-rays. Silicon drift detectors detect individual X-ray photons; specialised software analyses spectra by plotting X-ray energy versus count frequency. Qualitative analysis matches detected peak positions to characteristic X-ray energies in spectral libraries and generates element mapping. Quantitative analysis determines elemental concentrations by measuring peak areas and applying matrix effect corrections, achieving ±2–5% relative error for major components (>1% concentration).

SEM-EDAX requires minimal sample preparation. Conductive samples (metals, alloys) can often be analysed directly. Non-conductive samples (insulators, ceramics, polymers) require conductive coatings to prevent charging and reduce beam damage. Sputter coating applies 2–20 nm metal layers (gold, platinum, chromium), while carbon evaporation deposits 5–50 nm carbon coatings ideal for EDAX analysis. Advantages include exceptional resolution, minimal preparation, non-destructive analysis, and rapid measurement. Limitations include inability to detect hydrogen and helium and light element detection limitations.

10.3.3.4 Optical Microscopy and Cross-Sectional Analysis

Optical microscopy remains a fundamental technique for high-resolution morphological characterisation. Cross-sectional analysis, examination of carefully sectioned and polished material surfaces, enables direct visualisation of internal structures, layer architecture, material composition, and structural defects invisible to surface examination. Optical microscopy illuminates specimens with visible light and collects transmitted or reflected light using precisely aligned lenses. The objective lens forms magnified real images further enlarged by eyepieces or cameras. Image quality depends on numerical aperture (NA), defined as NA = n sin(α), where n is the medium's refractive index and α is half the objective's angular aperture. Theoretical lateral resolution follows Abbe's formula: $d = \lambda/(2NA)$, where λ is the wavelength. With green light (550 nm) and oil-immersion objectives (NA ≈ 1.4), resolution approaches approximately 200 nm laterally. In practice, optical microscopy achieves resolution from approximately 200 nm to several micrometres. Cross-sectional optical microscopy serves critical functions: detecting and quantifying defects (voids, cracks, delamination, inclusions) for identifying manufacturing failures; measuring coating thickness and verifying layer adhesion; examining grain size and morphology; assessing microstructural features; and measuring critical dimensions with calibrated images.

10.3.3.5 Atomic Force Microscopy (AFM)

Atomic force microscopy, also known as scanning force microscopy, is a scanning probe technique achieving subnanometre lateral resolution and measuring nanoscale mechanical, electrical, and magnetic surface properties without requiring conductive samples or special preparation. AFM has evolved from simple topographic imaging into a sophisticated platform offering over 30 measurement modes. AFM uses flexible microfabricated silicon or silicon nitride cantilevers with sharp probe tips. As tips scan surfaces, intermolecular forces (van der Waals, electrostatic, magnetic, adhesive) cause cantilever bending. Cantilever deflection is measured using optical lever systems: lasers focus on reflective cantilever backs; reflected light directs onto position-sensitive photodetectors. Minute cantilever movements cause large laser spot displacements, enabling extraordinarily sensitive force detection. Three primary imaging modes dominate AFM use: contact mode maintains continuous cantilever–sample contact while scanning with feedback maintaining constant force. Tapping mode oscillates cantilevers at resonance frequencies (50–400 kHz) with 20–100 nm amplitude, touching surfaces only at oscillation bottoms. Non-contact mode oscillates with <10 nm amplitude, remaining above surfaces while sensing long-range van der Waals forces. AFM characterises nanoparticles, thin films, polymer properties, composite materials, coatings, and surface roughness. Advantages include non-contact capability, operation in diverse environments, minimal preparation, and quantitative nanoscale mechanical property measurement. Limitations include complexity in parameter optimisation and potential tip wear.

10.3.3.6 Volume and Surface Resistivity (ASTM D257)

Volume and surface resistivity represent complementary electrical property measurements standardised by ASTM D257 that quantify insulating materials' resistance to electrical current flow through bulk volume versus along surfaces. These measurements are fundamental for designing electrical insulation systems, qualifying materials for high-voltage applications, ensuring electrostatic dissipation compliance, and predicting long-term performance.

Volume resistivity (ρv) is defined mathematically as electrical resistance through 1 cm cubes of insulating material with uniform electric fields applied perpendicular to opposite faces. Volume resistivity is expressed in ohm-centimetres (Ω·cm) or ohm-inches (Ω·in) and represents bulk resistive properties. Insulating materials typically exhibit volume resistivity from 10^9 to 10^{20} Ω·cm, spanning approximately 11 orders of magnitude. Polymeric insulation systems typically exhibit 10^{16} to 10^{18} Ω·cm depending on resin chemistry, filler content, moisture absorption, and processing history. Surface resistivity (ρs) is defined as electrical resistance between electrodes placed on material surfaces, expressed uniquely in ohms per square (Ω/sq). Surface resistivity commonly ranges from 10^6 Ω/sq to 10^{18} Ω/sq for insulating materials. Moisture contamination significantly reduces surface resistivity, making it a sensitive quality control indicator. Both measurements apply known DC voltages across electrodes and measure resulting leakage currents using Ohm's law ($R = V/I$) with electrometer or picoammeter instrumentation detecting nanoampere to picoampere currents.

10.3.3.7 Dielectric Spectroscopy

Dielectric spectroscopy measures frequency-dependent electrical material responses across broad frequency ranges from microhertz (10^{-6} Hz) to terahertz (10^{12} Hz), characterising molecular dynamics, material structure, composition, and phase transitions across approximately 18 frequency decades. Dielectric spectroscopy rests on measuring how dielectric materials respond to alternating electric fields – specifically, their ability to store and dissipate electromagnetic energy. When oscillating electric fields are applied to dielectric materials, molecular dipoles attempt alignment. At low frequencies, dipoles have sufficient time to reorient, producing nearly perfect alignment; at high frequencies, dipole motion cannot keep pace. Frequency-dependent response is characterised mathematically by complex relative permittivity: $\varepsilon^*(\omega) = \varepsilon'(\omega) - j\varepsilon''(\omega)$, where ε' (real component) quantifies energy storage capacity and ε'' (imaginary component) measures energy dissipation as heat. Permittivity spectra plot ε' and ε'' versus frequency, revealing relaxation dynamics. Cole–Cole plot ε'' versus ε' provides an alternative visualisation. Modulus representation using electric modulus $M^* = 1/\varepsilon^*$ provides complementary information. Dielectric spectroscopy enables characterisation across diverse material types: polymers and glass transitions detect α-relaxations and β-relaxations through dramatic ε'' peaks, pinpointing glass transition temperatures with superior sensitivity compared to DSC or DMA. Ionic conductors and electrolytes reveal electrode polarisation and bulk conductivity from mobile ions, critical for battery electrolyte development. Composite materials exhibit multiple relaxation processes corresponding to individual phases, revealing phase interactions and interfacial effects. Quality control and material certification monitor composition, contamination, moisture content, and processing-induced changes in real-time industrial settings.

Summary

Thermal, spectroscopic, and electrical analysis techniques form the foundation of modern materials characterisation, with thermal methods like TGA, DSC, and DMA revealing decomposition behaviour, heat transitions, and mechanical properties, while spectroscopic techniques including FTIR, XPS, SEM-EDAX, and AFM provide molecular, surface, and nanoscale structural information, and electrical measurements such as resistivity and dielectric spectroscopy assess insulating performance and molecular dynamics across broad frequency ranges.

Bibliography

1. American Society for Testing and Materials, "Standard test method for dielectric breakdown voltage and dielectric strength of solid electrical insulating materials at commercial power frequencies (ASTM D149-13)," ASTM International, 2013.
2. American Society for Testing and Materials, "Standard test method for water absorption of plastics (ASTM D570-98(2018))," ASTM International, 2018a.
3. American Society for Testing and Materials, "Standard test methods for AC loss characteristics and permittivity (dielectric constant) of solid electrical insulation (ASTM D150-18)," ASTM International, 2018b.
4. American Society for Testing and Materials, "Standard test methods for DC resistance or conductance of insulating materials (ASTM D257-21)," ASTM International, 2021.
5. American Society for Testing and Materials, "Standard test method for thermal diffusivity by the flash method (ASTM E1461-22)," ASTM International, 2022a.
6. American Society for Testing and Materials, "Standard test methods for AC loss characteristics and permittivity (dielectric constant) of solid electrical insulation (ASTM D150-22)," ASTM International, 2022b.
7. J. F. J. Boggs and R. B. Prime, "Thermosets: Structure, properties, and applications," in: *Thermal Analysis of Polymers* (2nd ed.). Hanser, 2014.
8. D. Briggs and J. T. Grant (Eds.), *Surface Analysis By Auger and X-Ray Photoelectron Spectroscopy*, IM Publications, 2003.

9. M. E. Brown, *Introduction to Thermal Analysis: Techniques and Applications* (2nd ed.), Springer, 2001.
10. R. M. Chatfield, "Optical microscopy of fiber-reinforced composites," in: *Handbook of Composites* (2nd ed.). Springer, 1998.
11. B. D. Cullity, and S. R. Stock, *Elements of X-Ray Diffraction* (3rd ed.), Prentice Hall, 2001.
12. V. Gavrilenko, A. Leonov, A. Bukharkin, S. Hlioui, and S. Lefebvre, "A method for endurance testing of enameled round and rectangular wires for motors controlled by SiC-based inverters," *IEEE Transactions on Dielectrics and Electrical Insulation*, vol. 28, no. 6, pp. 2091–2098, 2021, https://doi.org/10.1109/TDEI.2021.009587.
13. M. Ghassemi, "Accelerated insulation aging due to fast, repetitive voltages: A review identifying challenges and future research needs," *IEEE Transactions on Dielectrics and Electrical Insulation*, vol. 26, no. 5, pp. 1558–1568, 2019, https://doi.org/10.1109/TDEI.2019.008176.
14. J. I. Goldstein, D. E. Newbury, P. Echlin, D. C. Joy, A. D. Romig, Jr., C. E. Lyman, C. Fiori, and E. Lifshin, *Scanning Electron Microscopy and X-Ray Microanalysis* (4th ed.), Springer, 2018.
15. P. R. Griffiths and J. A. de Haseth, *Fourier Transform Infrared Spectrometry* (2nd ed.), Wiley, 2007.
16. F. Guastavino, L. Briano, F. Gallesi, and E. Torello, "A review on the comparison of conventional and corona resistant nanofilled enamels [Paper presentation]," 2021 IEEE Workshop on Electrical Machines Design, Control and Diagnosis (WEMDCD), Modena, Italy, 2021 Apr. https://doi.org/10.1109/WEMDCD51469.2021.9425646.
17. T. J. A. Hammarstrom, "Partial discharge characteristics at ultra-short voltage risetimes," *IEEE Transactions on Dielectrics and Electrical Insulation*, vol. 25, no. 6, pp. 2241–2249, 2018, https://doi.org/10.1109/TDEI.2018.007445.
18. T. J. A. Hammarstrom, "Multilevel PWM: A tool to explore insulation PD characteristics," *IEEE Transactions on Instrumentation and Measurement*, vol. 68, no. 10, pp. 4075–4084, 2019, https://doi.org/10.1109/TIM.2018.2884636.
19. T. J. Hammarstroem, "Evaluation of different waveforms to study the insulation performance on electrical vehicle motors [Paper presentation]," 2022 IEEE Conference on Electrical Insulation and Dielectric Phenomena (CEIDP), Denver, CO, USA, 2022 Apr. https://doi.org/10.1109/CEIDP55452.2022.9985269.
20. Institute of Electrical and Electronics Engineers, "IEEE recommended practice for testing insulation resistance of rotating machinery (IEEE Std 43-2013)," IEEE, 2013a.
21. Institute of Electrical and Electronics Engineers, "IEEE standard techniques for high-voltage testing (IEEE Std 4-2013)," IEEE, 2013b.
22. Institute of Electrical and Electronics Engineers, "IEEE guide for testing and monitoring partial discharge in rotating machinery (IEEE Std 522-2016)," IEEE, 2016.
23. International Electrotechnical Commission, "High-voltage test techniques – Part 1: General definitions and test requirements (IEC 60060-1:2010)," IEC, 2010.
24. International Electrotechnical Commission, "Electric strength of insulating materials – test methods – Part 1: Tests at power frequencies (IEC 60243-1:2013)," IEC, 2013.

25. International Electrotechnical Commission, "Rotating electrical machines – Part 18-41: Partial discharge free electrical insulation systems (Type I) used in rotating electrical machines fed from voltage converters (IEC 60034-18-41:2014)," IEC, 2014.
26. International Electrotechnical Commission, "Rotating electrical machines – Part 27-3: Dielectric dissipation factor measurement on stator winding insulation of rotating electrical machines (IEC 60034-27-3:2015)," IEC, 2015.
27. International Electrotechnical Commission, "Rotating electrical machines (IEC 60034 series)," IEC, 2022.
28. International Electrotechnical Commission, "Electrical insulating materials and systems – electrical measurement of partial discharges (PD) under short rise time and repetitive voltage impulses (IEC TS 61934:2024)," IEC, 2024.
29. Y. Ji et al., "Partial discharge investigation under humidity conditions via dissipation factor and insulation capacitance tip-up test," *IEEE Transactions on Dielectrics and Electrical Insulation*, vol. 29, no. 4, pp. 1483–1490, 2022, https://doi.org/10.1109/TDEI.2022.3171197.
30. S. A. Jenkins, "X-ray microtomography in materials science," *Materials Characterization*, vol. 61, no. 3, pp. 215–233, 2010.
31. F. Kremer and A. Schönhals (Eds.), *Broadband Dielectric Spectroscopy*, Springer, 2003.
32. S. B. Lee, A. Naeini, S. Jayaram, G. C. Stone, and M. Šašić, "Surge test-based identification of stator insulation component with partial discharge activity for low voltage AC motors," *IEEE Transactions on Industry Applications*, vol. 56, no. 3, pp. 2541–2549, 2020, https://doi.org/10.1109/TIA.2020.2976719.
33. P. Mancinelli, S. Stagnitta, and A. Cavallini, "Qualification of hairpin motors insulation for automotive applications," *IEEE Transactions on Industry Applications*, vol. 53, no. 3, pp. 3110–3118, 2017, https://doi.org/10.1109/TIA.2016.2619670.
34. B. Mirafzal, G. L. Skibinski, and R. M. Tallam, "A failure mode for PWM inverter-fed AC motors due to the antiresonance phenomenon," *IEEE Transactions on Industry Applications*, vol. 45, no. 5, 1697–1705, 2009, https://doi.org/10.1109/TIA.2009.2027524.
35. D. E. Moghadam, J. Speck, S. Grossmann, and J. Stahl, "Parameters affecting the turn insulation lifetime and durability," *IEEE Transactions on Dielectrics and Electrical Insulation*, vol. 25, no. 2, pp. 516–523, https://doi.org/10.1109/TDEI.2017.006905.
36. G. C. Montanari, P. Seri, and F. Negri, "Partial discharge and aging phenomena in insulation systems of rotating machines fed by power electronics [Paper presentation]," 2017 International Symposium on Electrical Insulating Materials (ISEIM), 2017 Sept. https://doi.org/10.23919/ISEIM.2017.8088714.
37. T. Tanaka, G. C. Montanari, and R. Mülhaupt, "Polymer nanocomposites as dielectrics and electrical insulation – perspectives for processing technologies, material characterization and future applications," *IEEE Transactions on Dielectrics and Electrical Insulation*, vol. 11, no. 5, pp. 763–784, 2004.
38. B. Wunderlich, *Thermal Analysis of Polymeric Materials*, Springer, 2005.
39. S. Zeynalova, E. Cepparrone, E. Roffino, and D. Barbero, "Effect of thermal aging on electrical performance of perfluoroalkoxy- and polyamide-imide-coated magnet wire," *IEEE Electrical Insulation Magazine*, vol. 40, no. 3, pp. 6–14, https://doi.org/10.1109/MEI.2024.10508410

11

Electrical Machine Insulation Systems for Aerospace Electrification

Shesha H. Jayaram

11.1 Introduction

The rapid expansion of electrification in aerospace propulsion and actuation has intensified the demand for compact, efficient, and highly reliable electric machines. The electrical insulation system defines the ultimate reliability boundary of each machine: it governs dielectric withstand, thermal endurance, and mechanical cohesion under combined electrical, thermal, mechanical, and environmental stress.

In aerospace applications, stringent weight and efficiency constraints have accelerated the adoption of low-voltage random-wound and hairpin-type AC motors for auxiliary drives and actuation, and high-voltage form-wound propulsion machines for megawatt-class systems. These windings must tolerate high rates of voltage rise (*dv/dt*), repetitive pulse width–modulated (PWM) transients, significant thermal stresses arising from compact, high power-density configurations, and operation at reduced atmospheric pressure. Understanding insulation behaviour under these coupled electrical, thermal, mechanical, and environmental stresses is essential for designing machines that remain safe and durable throughout their service life.

This chapter outlines insulation system configurations, materials, and degradation mechanisms for low-, medium-, and high-voltage stator windings, emphasising aerospace-specific constraints and the diagnostic concepts used in quality-assurance testing and qualification.

11.1.1 Power Scaling in Electric Aircraft

Electric flight demonstrators validate electric propulsion with small aircraft that use hundreds of kilowatts, but scaling to large transport aircraft highlights current energy-storage limitations. Representative fully electric scaling illustrates the gap:

DOI: 10.1201/9781003747994-11

- *Airbus A380-Class:* average cruise power ≈ 80 MW for 14 hours → ~1,120 MWh total energy. At 250 Wh/kg, battery mass would be ~4,480 tonnes, far exceeding maximum take-off weight.
- *Boeing 787-Class:* average cruise power ≈ 45 MW for 11 h → ~495 MWh total energy. At 250 Wh/kg, battery mass would be ~1,980 tonnes, far exceeding maximum take-off weight.

These estimates indicate that full battery-electric long-haul flight is not feasible with current battery specific energy. Nevertheless, future architectures (hybrid-electric, turbo-electric, distributed propulsion) still require multi-megawatt electrical power transfer, which drives voltage escalation and elevates insulation requirements.

11.1.2 Voltage Scaling and the Insulation Implication

For a representative 20 MW propulsion motor operating at 20 kV line-to-line, the current is of order 0.6–0.7 kA for realistic power factor and efficiency, whereas low-voltage operation would demand impractically high currents and conductor mass. This motivates higher-voltage distribution and stator windings (up to ~20 kV class). However, increasing voltage directly increases

- groundwall and turn to turn electric-field stresses,
- local field enhancement at slot exits and end windings,
- high-frequency voltage stress and associated transient overvoltages, and
- partial-discharge susceptibility under reduced pressure.

Consequently, insulation design becomes a primary enabling technology across increasing power and voltage levels in electric machines.

11.2 Insulation for Low-Voltage Random-Wound Motors

11.2.1 Overview and Applications

Low-voltage random-wound AC motors typically operating below 1 kV are widely used in more-electric aircraft for actuators, pumps, compressors, and environmental-control subsystems. Their simplicity and manufacturability support compact, low-mass implementations. However, continuous winding of magnet wire into the stator slots, without controlled geometric alignment, produces inherently non-uniform electrical and mechanical stress distributions, thereby increasing the importance of material selection and impregnation quality under vibration, thermal cycling, and inverter-fed transients.

The electrification of flight control and actuation has expanded the operating envelope for low-voltage machines. They now experience elevated temperatures approaching 180°C insulation classes, high-frequency PWM voltages, and repetitive surge stress that accelerates insulation ageing. Therefore, an integrated electro–thermal–mechanical design philosophy is required.

11.2.2 Winding Configuration and Slot Geometry

Random-wound stators employ enamel-coated round copper conductors distributed irregularly within stator slots lined by thin dielectric barriers made from aramid paper, polyester-nonwoven laminates, or polyimide films. Windings are consolidated using trickle impregnation or dip-and-bake varnishing (rather than full vacuum-pressure impregnation) with solvent-free epoxy or polyester-imide resins having high glass transition temperature ($T_g > 180°C$) and low volatile emissions characteristics suitable for aerospace use.

Magnet-wire enamel systems typically use dual layers: a polyester-imide base for adhesion and a polyimide or polyamide-imide topcoat for thermal endurance and corona resistance. Impregnation fills residual voids and secures the winding mechanically, minimising movement under vibration and thermal cycling.

Slot insulation systems may incorporate multilayer composites, polyimide–glass laminates, or mica–paper reinforcements, depending on mechanical loading and expected service life. For small, weight-constrained motors, thin films (≤0.1 mm) are used to maximise copper fill while maintaining minimum clearances consistent with safety standards. Dimensional stability and mechanical integrity of liners during impregnation are essential to prevent delamination or resin starvation.

11.2.3 Electrical Stress under Inverter-Fed Operation

Although rated below 1 kV, the system experiences high electric-field stress due to steep voltage fronts often exceeding 5 kV/ms ($dv/dt > 5$ kV/µs), which is caused by wide-bandgap silicon-carbide (SiC) or gallium-nitride (GaN) switching devices. Such fast edges create several hundred volts to kilovolt-level potential differences between adjacent turns, producing localised electric field intensification at wire crossings and slot exits. The unequal voltage distribution arises from randomness in winding distribution, parasitic capacitances, and steep-front transients, particularly in short-lead configurations common in aerospace drives. Also, turn-to-turn spacing depends on wire diameter, enamel thickness, and fill factor (typically 45–55%). Consequently, turn insulation in random-wound coils becomes the first line of defence against premature dielectric breakdown.

When ambient pressure decreases at high altitude, the partial discharge inception voltage (PDIV) decreases. Voids, resin-starved regions, and airgaps at interfaces become discharge sites. The partial discharge extinction voltage (PDEV) is the voltage below which discharge ceases when the applied voltage is reduced. The PDIV–PDEV gap provides a measure of discharge stability and recovery margin under repetitive stresses. Finite-element analyses and experience indicate that local electric fields concentrate at wire crossings and slot exits and may reach ~15–20 kV/mm, sufficient to trigger partial discharges if defects exist. Therefore, materials with high PDIV and stable PDEV, along with uniform impregnation, are critical.

11.2.4 Thermal and Mechanical Considerations

Thermal cycling, vibration, and humidity contribute additional degradation pathways. Differences in coefficients of thermal expansion among copper, resin, and liner materials induce microcracks that may propagate into partial-discharge channels. A well-designed impregnation resin provides mechanical damping and improves heat transfer from the winding to the core. Thermally conductive fillers such as boron nitride or alumina can improve heat spreading while maintaining dielectric integrity. Flexible high-T_g resin systems mitigate brittleness under low-temperature starts.

11.2.5 Effects of Fast-Switching Inverter Pulses

Modern SiC and GaN power electronics enable increased switching frequencies and rapid turn-on and turn-off transitions. While beneficial for efficiency and control, these characteristics impose greater dielectric stress on motor insulation systems. Fast voltage edges can produce standing-wave effects along feeder cables and generate transient overvoltages at motor terminals. The resulting repetitive electrical stress promotes partial-discharge activity even at modest supply voltages, leading to enamel erosion and eventual turn-to-turn short circuits.

Mitigation strategies include enhancing corona resistance through the incorporation of nano-additives or inorganic fillers in enamels and impregnation resins. The use of output filters, shielded cables, and optimised PWM modulation schemes can further reduce electrical stress; however, these measures add mass and volume. Consequently, robust insulation design becomes essential in aircraft applications where weight and space constraints limit external mitigation approaches.

11.2.6 Failure Mechanisms and Degradation Modes

Failure in random-wound insulation typically originates from one or a combination of the following:

1. Thermal ageing governed by Arrhenius kinetics, reducing dielectric strength.
2. Electrical ageing due to voltage overshoot and the associated PD, eroding enamel surfaces and carbonising resin interfaces.
3. Mechanical fatigue induced by vibration leading to abrasion between wires.
4. Environmental factors, including humidity, contamination, and altitude-related pressure variation.

Failure analyses frequently reveal tree-like erosion channels along enamel boundaries and resin micro-voids near crossover points. Corona-resistant enamels and improved impregnation extend life, but aerospace qualification still requires accelerated endurance testing that reproduces combined stress conditions.

11.2.7 Testing and Qualification Methods

Validation of insulation integrity involves a hierarchy of tests:

- Surge comparison testing for turn-to-turn insulation.
- PD measurements (PDIV/PDEV) under ambient and reduced pressure.
- Thermal endurance testing (e.g., IEC 60034-18-21 or equivalents).
- Impulse and voltage endurance tests using repetitive PWM waveforms.
- Vibration and humidity cycling per RTCA DO-160 categories.

Comprehensive data feed reliability models predicting mean time to failure as functions of temperature, voltage, *dv/dt*, and switching frequency.

11.2.8 Diagnostic Concepts

Bulk insulation quality is commonly characterised by insulation resistance (IR) and polarisation index (PI):

- IR measures leakage current under DC voltage and indicates dryness/cleanliness.
- PI = R10/R1 (10-minutes to 1-minute IR ratio). PI $\geq$ 2 is generally associated with clean, dry insulation.

High IR and stable PI, together with high PDIV and low dissipation factor ($\tan \delta$), indicate robust insulation.

11.2.9 Summary

Low-voltage random-wound motor insulation must balance weight, compactness, and reliability. The unstructured winding geometry necessitates reliable enamels and impregnation systems capable of suppressing partial discharges under inverter-induced stress. As aerospace systems adopt faster SiC- and GaN-based drives, insulation must evolve towards enhanced corona-resistant coatings, void-free impregnation, and predictive health monitoring. High-thermal-conductivity resins and advanced nanodielectrics are promising directions for sustaining performance in more-electric aircraft.

11.3 Insulation for Low-to-Medium-Voltage Hairpin Motors

11.3.1 Overview and Motivation

Hairpin-type windings represent an evolutionary step bridging random-wound simplicity and form-wound robustness. Hairpin windings employ rectangular conductors bent into U-shaped pins inserted into stator slots and joined by laser or tungsten inert gas welding at the end turns. Geometric regularity enables high slot-fill factors (typically 70–80%) and improved heat removal, supporting high power density and efficiency. These benefits motivate adoption in electrified aerospace systems operating approximately from 400 V to ~3 kV.

The transition from round-wire to rectangular conductors introduces insulation challenges. Sharp edges, minimal inter-turn spacing, and welding processes create local field intensification that can trigger partial discharges under inverter-driven voltage stress. In aerospace platforms requiring compactness, low mass, and high reliability, the insulation system must have high dielectric strength, strong thermal endurance, and resistance to vibration and moisture within stringent dimensional constraints. Variations in atmospheric pressure also affect electric-field distribution and discharge behaviour, necessitating stable dielectric performance across the flight envelope.

11.3.2 Winding Configuration and Insulation System

Hairpin windings employ rectangular copper conductors bent into U-shaped pins that are inserted into stator slots and welded at the ends to form continuous phase windings. Each hairpin element consists of an enamel-coated rectangular copper bar (typically 1–3 mm × 3–6 mm) placed within a slot. After insertion, the protruding ends are bent and welded; however, the local curvature at bends and welded joints introduces electric field intensification, necessitating insulation reinforcement to prevent PD inception and ensure long-term dielectric reliability.

The slot insulation system combines a slot liner (polyimide–glass laminate or aramid paper) with optional corona-suppression layers. Inter-turn insulation is primarily provided by enamel that must maintain uniform thickness despite edge curvature. After assembly, controlled resin infiltration or RTM secures conductors and fills microscopic voids. These techniques are optimised for narrow clearances but do not always involve full VPI. High-T_g epoxy or cyanate-ester resins are selected for low moisture uptake and stable dielectric behaviour over temperature.

Hairpin windings enable excellent heat transfer, due to high copper cross section and close slot contact, but this also increases thermomechanical coupling. During take-off and descent, temperature gradients and vibration generate cyclic stresses at welds and end turns. Resins with intermediate elastic modulus (approximately 3–5 GPa) provide compliance, while thermally conductive fillers such as boron nitride or aluminium nitride improve heat spreading. Humidity, salt-fog, and condensation cycles are addressed through low-moisture-uptake resins and surface treatments that maintain dielectric strength after environmental exposure.

11.3.3 Electrical Stress and Partial-Discharge Behaviour

The rectangular geometry of hairpin conductors produces localised electric-field intensification at corners, weld regions, and slot exits. Under inverter-fed excitation with high dv/dt, transient voltage distribution between adjacent turns can result in several hundred volts of turn-to-turn stress within compact slot geometries. Local electric fields may exceed 15–20 kV/mm at sharp edges, increasing susceptibility to partial-discharge (PD) initiation in the presence of voids or insulation defects.

As Type I insulation systems, hairpin windings depend primarily on enamel-based turn insulation and slot liner ground insulation. Multilayer enamel coatings, typically combining polyester–imide base layers with polyamide–imide or polyimide topcoats, are employed to enhance thermal endurance and PD resistance. Advanced formulations incorporating inorganic nanofillers further improve corona resistance and suppress discharge-driven erosion.

Reduced atmospheric pressure lowers PD inception voltage (PDIV), increasing vulnerability to discharge activity under high-frequency operation. PD in hairpin systems typically initiates at turn-to-turn interfaces or conductor-to-slot insulation boundaries, leading to progressive enamel erosion and resin carbonisation. Mitigation therefore relies on optimised conductor edge geometry, uniform void-free impregnation, and corona-resistant enamel systems to ensure long-term dielectric reliability.

11.3.4 Manufacturing and Process Challenges

Narrow clearances make resin impregnation and void elimination difficult. Voids become PD focal points under repetitive voltage stress. Vacuum pressure impregnation (VPI), which uses vacuum and pressure cycles to drive resin into porous structures, and resin transfer moulding (RTM), a closed-mould injection process, are employed to achieve complete resin impregnation of the insulation system, for selected voltage levels (>1 kV) motors. Welding can degrade enamel if not shielded; local re-insulation or over-moulding using high-temperature silicone or polyimide film is common. Process control (cleanliness, viscosity, cure profile, bake) directly affects reliability.

Digital quality assurance, such as X-ray, CT, and ultrasonic inspection, verifies void-free impregnation and uniform insulation thickness for safety-critical aerospace components. Online PD monitoring and impedance spectroscopy enable early detection. Reliability models correlate PD intensity and thermal cycling to lifetime for condition-based maintenance.

11.3.5 Summary

Medium-voltage hairpin-type motors enable compact, high-efficiency actuation in aerospace platforms. Compared with random-wound motors, they provide 30–50% higher slot utilisation and 10–15°C lower steady-state temperatures, supporting increased torque density and improved thermal management. These performance gains are accompanied by more demanding insulation requirements arising primarily from their higher operating voltages relative to random-wound machines, which are typically rated below 1 kV. Elevated electric-field gradients within compact slot geometries, limited insulation thickness, and rectangular conductor profiles further intensify local dielectric stress. Since insulation thickness cannot be substantially increased within constrained slot dimensions, dielectric strength must be achieved through enhanced multilayer enamel systems, optimised slot liner insulation, and void-free resin impregnation rather than added bulk. As voltage ratings extend into the multi-kilovolt range, improved corona-resistant enamels, filled resin systems, and high-performance composite liners become increasingly important to maintain long-term reliability.

11.4 Insulation for High-Voltage Form-Wound Propulsion Motors

11.4.1 Overview and Motivation

High-voltage (HV) form-wound machines represent the backbone of emerging megawatt-class electric propulsion systems envisioned for hybrid-electric and fully electric aircraft. These machines employ Type II insulation systems, characterised by multilayer turn insulation and groundwall constructions designed to withstand repetitive partial-discharge activity and elevated electric field stress. In addition to turn-to-turn and ground insulation, effective slot corona suppression and stress grading at slot exits and end-winding regions are essential to control electric-field distribution and prevent localised discharge initiation. Unlike low- and medium-voltage random-wound or hairpin-type machines, form-wound coils are individually fabricated, shaped, and fully insulated before insertion into stator slots. Following coil insertion, the stator assembly is typically subjected to vacuum pressure impregnation (VPI) to eliminate residual voids, improve mechanical integrity, and enhance long-term dielectric endurance.

Operating voltages range from several hundred volts to tens of kilovolts (line-to-line), requiring insulation systems capable of tolerating high electric-field stress, steep inverter-induced dv/dt transients, reduced-pressure discharge susceptibility associated with altitude variation, mechanical vibration, cyclic thermal loading, and repeated pressure fluctuations across the flight envelope. Insulation reliability defines system safety. A single breakdown can cause catastrophic failure of both the motor and the associated drive converter. Therefore, insulation design must integrate multilayer electrical, mechanical, and thermal protection using advanced polymer composites, inorganic fillers, semiconductive slot coatings, stress-grading systems, and pressure-stable materials while remaining lightweight and manufacturable.

11.4.2 Coil Construction and Groundwall Architecture

Each form-wound coil is built from bundles of rectangular copper conductors, individually insulated with turn tape and wrapped with multilayer groundwall insulation. The groundwall, typically 1–3 mm thick, forms the principal dielectric barrier between the high-voltage winding and grounded stator core. Mica-paper–epoxy composites remain the industrial benchmark due to high dielectric strength (>20 kV/mm) and resistance to PD erosion.

A typical HV coil structure includes the following:

1. *Turn Insulation:* Polyimide or mica–polyester tapes.
2. *Main Groundwall:* Alternating mica-paper layers with epoxy/polyester resin and glass reinforcement.
3. *Slot Corona Protection:* Semiconductive coating along the slot portion for potential grading to the grounded core.
4. *End-Winding Stress Grading:* High-resistive or non-linear grading layers at slot exits to smooth the field transition and suppress corona inception.

Nanocomposite insulation systems containing silica, alumina, titania, or boron nitride fillers can increase PD endurance and thermal conductivity while limiting mass.

11.4.3 Environment- and Altitude-Related Effects

HV propulsion motors operate under simultaneous electrical, thermal, and mechanical stresses. Electrical stress originates from high fundamental voltage and steep inverter pulses ($dv/dt > 10$ kV/µs), generating strong local fields at slot exits and between adjacent turns. Thermal stress arises from current heating (>180°C local temperatures) combined with constrained cooling. Differential thermal expansion induces interfacial delamination, promoting partial-discharge (PD) inception and surface tracking. Mechanical stress results from vibration and high-speed operation, leading to microcracking in resin-rich regions.

A distinguishing challenge for aerospace HV machines is operation under variable pressure and temperature. At altitudes exceeding 10 km, reduced air density decreases PD inception voltage and alters thermal convection. Exposure to ozone, ultraviolet radiation, and humidity cycling further modifies surface conductivity and dielectric behaviour.

Material volatile release under reduced pressure may create micro-voids or bubbles within resin systems, encouraging PD initiation. Therefore, resins and insulation tapes must comply with low-volatile emission requirements (ASTM E595) and maintain dielectric stability across approximately 30–100 kPa. Thermal-vacuum cycling tests evaluate structural and dielectric integrity after repeated decompression and heating cycles typical of flight operation.

Over time, PD erosion, thermal oxidation, and mechanical fatigue interact to produce electrical treeing. Once such trees bridge groundwall insulation, dielectric breakdown follows. Achieving balanced mechanical compliance and dielectric robustness requires carefully tailored resin chemistry and controlled curing processes.

11.4.4 Quantitative Scaling Example: 20 MW Motor at 20 kV Class

For 20 kV line-to-line operation, the phase-to-ground voltage is

$$V_\phi = \frac{V_{LL}}{\sqrt{3}} = \frac{20\,\text{kV}}{\sqrt{3}} \approx 11.55\,\text{kV}$$

Average groundwall field:

$$E_{\text{avg}} = \frac{V_\phi}{t}$$

Representative values:

- $t = 3\text{ mm} \rightarrow E_{avg} \approx 3.85\text{ kV/mm}$
- $t = 2\text{ mm} \rightarrow E_{avg} \approx 5.78\text{ kV/mm}$

Local enhancement factors of ~2–4 × (slot geometry, tape steps, termination edges) yield local fields of approximately:

- 3 mm → ~7–15 kV/mm
- 2 mm → ~12–23 kV/mm

Under reduced pressure, PD inception thresholds decrease and the margin to PD-free operation becomes dominated by local field peaks and defect populations rather than average field alone. Weight-driven reductions in groundwall thickness therefore require improved stress grading, void control, and PD endurance materials.

11.4.5 Converter Switching Frequency, *dv/dt*, and Insulation Stress (40 kHz Context)

Wide-bandgap power electronics enable higher switching frequency and faster turn-on and turn-off transitions than conventional silicon. Hence, in insulation ageing, *dv/dt*, repetitive impulse exposure, overshoot, and local field enhancement often dominate over switching frequency alone. Increasing switching frequency (e.g., towards 40 kHz) increases the number of stress cycles per unit time, but insulation severity also depends strongly on edge rate, pulse shape, and overvoltage.

At 20 kV and multi-megawatt ratings, practical switching frequency is constrained by switching loss, EMI, device packaging, and thermal limits. Therefore, achieving high effective waveform quality at high voltage often requires multilevel converter topologies and/or soft-switching approaches to limit *dv/dt* per step and reduce insulation stress. Since the achievable switching frequency for MW-class, 20 kV–class propulsion converters is topology and device dependent, qualification should be based on representative *dv/dt* and repetitive impulse waveforms rather than assuming a single universal switching frequency.

11.4.6 High-Frequency Dielectric Behaviour, Qualification, and Diagnostics

In high-voltage inverter-fed operation, high-frequency dielectric loss becomes a significant consideration in addition to partial-discharge (PD) endurance. Under PWM excitation, dielectric heating scales approximately as follows:

$$P_{loss} \propto \omega C V^2 \tan\delta$$

At elevated voltage levels, the quadratic dependence on voltage dominates, making insulation susceptible to localised heating even when dissipation factor (tan δ) remains modest. Increased switching frequency and temperature can further amplify dielectric loss, particularly in compact stator geometries where cooling is constrained. Consequently, characterisation of tan δ as a function of frequency and temperature becomes an important evaluation criterion for inverter-fed HV stator insulation.

Qualification procedures for Type II insulation systems draw on IEC 60034-18-42 for inverter-fed voltage endurance, IEC 60060-1 for impulse withstand capability, and RTCA DO-160 for vibration, thermal, and environmental exposure. Key assessments include PD inception and extinction mapping (PDIV/PDEV), repetitive impulse endurance testing, thermal cycling, humidity exposure, and altitude simulation. Diagnostic tools such as dielectric spectroscopy and dissipation-factor measurement provide non-destructive monitoring of insulation condition and ageing progression.

11.4.7 Advanced Materials for High-Voltage Aerospace Motors

To achieve high dielectric strength with low mass, modern aerospace insulation systems employ multifunctional materials:

- Mica–epoxy composites (benchmark PD resistance but higher density).
- Polyimide–nanofiller systems (nano-SiO_2, Al_2O_3, BN) improving breakdown strength and thermal conductivity.
- Cyanate-ester resins (high thermal stability >250 °C, low moisture uptake).
- Flexible high-T_g epoxies (toughness and strain accommodation).
- Grading coatings and semiconductive layers (SiC- or ZnO-based) for uniform field distribution.

Emerging approaches include epoxy–boron nitride nanocomposites with improved PD endurance and self-healing dielectric polymers capable of recovering from localised discharge damage.

11.4.8 Testing, Qualification, and Standards

Qualification of HV form-wound insulation follows aerospace protocols derived from IEC 60034-18-42 and RTCA DO-160. Representative tests include the following:

- PD and endurance testing under repetitive inverter pulses.
- Impulse withstand per IEC 60060.

- Thermal cycling between −55 °C and +200 °C to evaluate delamination resistance.
- Vibration and shock testing for in-flight mechanical loads.
- Humidity and salt-fog exposure for environmental resilience.
- High-altitude simulation (PD and breakdown under controlled pressure).

11.4.9 Design Considerations for Electric Propulsion Integration

In integrating form-wound propulsion motors into aircraft power systems, insulation design interacts with thermal management, EMC, and packaging. Insulation thickness impacts slot fill and cooling efficiency; compromise between dielectric margin and thermal conductivity is critical. Direct-liquid cooling channels within stator slots impose additional dielectric demands on barriers separating coolant and copper. Specialised barrier coatings and resin-sealed cooling interfaces are required to prevent leakage-induced breakdown. Insulation must remain stable under high-frequency common-mode currents and EMI generated by inverter operation. Coupled electro-thermal modelling guides geometry optimisation to ensure uniform field distribution and adequate creepage while minimising mass.

11.4.10 Summary

High-voltage form-wound motors employing Type II insulation systems support increasing power and voltage levels in advanced electric and hybrid-electric aircraft architectures. Compared with Type I random-wound and hairpin machines, form-wound designs accommodate higher operating voltages through multilayer turn insulation, robust groundwall systems, and integrated slot corona suppression and stress-grading structures. These motors must withstand combined electrical, thermal, mechanical, and environmental stress across repeated flight cycles, including high *dv/dt* inverter excitation and reduced-pressure operation.

At elevated voltage levels, dielectric integrity directly governs propulsion system safety. Insulation design therefore relies on mica-based and advanced polymer composite groundwall constructions, carefully controlled vacuum pressure impregnation, and pressure-stable material systems. As voltage and power densities continue to increase, improvements in corona-resistant materials, thermally conductive yet electrically stable composites, and integrated diagnostic capabilities become increasingly important to ensure long-term reliability and condition-based monitoring.

11.5 Cross-Cutting Design Considerations

11.5.1 Altitude and Pressure Effects on Partial Discharge

Reduced atmospheric pressure lowers partial-discharge inception voltage (PDIV) in voids and interfacial regions, increasing susceptibility to discharge activity under inverter-driven excitation. Insulation systems across both Type I and Type II constructions must therefore exhibit high intrinsic PD resistance, low gas permeability, and stable dielectric behaviour under pressure variation. Testing under simulated flight pressure conditions identifies voltage ranges for corona-free operation and verifies performance across the aircraft flight envelope.

11.5.2 Thermal Management and Dielectric Heating

In compact aerospace machines, insulation plays a dual electrical and thermal role. Thermally conductive fillers such as boron nitride (BN), alumina, or aluminium nitride enhance heat transfer while preserving dielectric integrity. Under pulse width–modulated (PWM) operation, high-frequency dielectric loss, scaling approximately with $\omega C V^2 \tan\delta$, can contribute to localised heating, particularly at elevated voltage levels. Consequently, dissipation-factor characterisation as a function of frequency and temperature becomes important in inverter-fed systems. When cooling channels or liquid jackets are integrated near stator slots, hermetic dielectric barriers are required to prevent leakage-induced breakdown, validated through combined thermal and vacuum cycling.

11.5.3 Weight, Space, and Reliability Trade-Offs

Increasing insulation thickness improves dielectric margin but reduces copper fill and overall machine efficiency. High-performance dielectrics with strong breakdown strength per unit thickness, including polyimide-based systems, mica composites, and nanofilled polymers, enable compact, lightweight designs. Reliability across voltage classes is achieved through appropriate material selection, controlled processing, effective stress management, and environmental qualification rather than excessive bulk insulation.

11.5.4 Standards and Qualification Framework

Qualification of aerospace motor insulation integrates electrical machine standards and aerospace environmental requirements. IEC 60034-18 defines voltage endurance and inverter-fed evaluation for Type I and Type II systems, while IEC 60034-27-1 and IEC 60060-1 address PD measurement and impulse withstand capability. ASTM D149 and D150 support dielectric characterisation, and ASTM E595 defines low volatile emission requirements under reduced pressure. RTCA DO-160 governs vibration, temperature, humidity, and altitude testing. Current practice emphasises PD endurance evaluation under repetitive PWM excitation and reduced-pressure conditions to replicate realistic flight operation.

11.5.5 Emerging Materials and Integrated Diagnostics

Advanced polymer composite systems and nanofilled insulation materials enhance partial-discharge endurance, thermal transport, and mechanical compliance. High-temperature resin systems improve structural stability under combined electrical and thermal stress. Additive manufacturing techniques enable graded insulation geometries and integrated cooling features with reduced void content. Embedded fibre-optic and capacitive sensors support real-time monitoring of temperature, electric field, and discharge activity, facilitating predictive maintenance strategies in electrically driven aircraft systems.

11.6 Summary

Electrical insulation in aerospace electric machines has evolved from a passive dielectric barrier into a multifunctional engineered system integrating electrical strength, thermal conduction, mechanical integrity, and environmental resilience. Different motor topologies serve distinct functional roles within aircraft systems, and their insulation requirements reflect these operational differences.

Low-voltage random-wound motors remain well suited for auxiliary and actuation functions, where compactness, simplicity, and manufacturability are primary considerations. As voltage and power requirements increase beyond conventional low-voltage ranges, hairpin-type motors provide improved slot utilisation and thermal performance while maintaining Type I insulation systems based on enhanced enamel and slot liner constructions. For higher-voltage propulsion-class applications, form-wound machines employing Type II multilayer turn and groundwall insulation systems accommodate elevated electric field stress through integrated slot corona suppression and stress grading.

Across all voltage classes and motor functions, insulation must withstand coupled electrical, thermal, mechanical, and environmental stresses, including converter-generated fast repetitive voltage transients with high *dv/dt*, compact thermal loading, vibration, and reduced-pressure operation. Reliability, therefore, depends on appropriate material selection, precise process control, and rigorous environmental qualification. As aircraft architectures incorporate increasing levels of electrical power, insulation performance is increasingly influenced by local field enhancement, partial-discharge behaviour under altitude conditions, high-frequency dielectric heating, and the integrity of cooling–insulation interfaces. Continued advances in dielectric materials, manufacturing processes, and integrated sensing are enabling lightweight, high-endurance insulation systems that support the safe electrification of propulsion and auxiliary power systems alike.

Bibliography

1. IEC 60034–18, *Rotating Electrical Machines – Functional Evaluation of Insulation Systems.*
2. IEC 60034-27-1, *Off-line Partial Discharge Measurements on Stator Winding Insulation of Rotating Electrical Machines.*
3. IEC 60060–1, *High-Voltage Test Techniques – Part 1: General Definitions and Test Requirements.*
4. ASTM D149, *Standard Test Method for Dielectric Breakdown Voltage and Dielectric Strength of Solid Electrical Insulating Materials at Commercial Power Frequencies.*
5. ASTM D150, *Standard Test Methods for AC Loss Characteristics and Permittivity (Dielectric Constant) of Solid Electrical Insulation.*
6. ASTM E595, *Standard Test Method for Total Mass Loss and Collected Volatile Condensable Materials from Outgassing in a Vacuum Environment.*
7. RTCA DO-160, *Environmental Conditions and Test Procedures for Airborne Equipment.*
8. G. C. Stone, E. A. Boulter, I. Culbert, and H. Dhirani, *Electrical Insulation for Rotating Machines* (2nd ed.), IEEE Press, 2004.
9. L. A. Dissado and J. C. Fothergill, *Electrical Degradation and Breakdown in Polymers*, Peter Peregrinus Ltd., 1992.
10. T. Tanaka, G. C. Montanari, and R. Mülhaupt, "Polymer nanocomposites as dielectrics and electrical insulation – perspectives for processing technologies, material characterization and future applications," *IEEE Transactions on Dielectrics and Electrical Insulation*, vol. 11, no. 5, pp. 763–784, 2004.
11. S. B. Lee, A. Naeini, S. H. Jayaram, G. C. Stone, and M. Šašić, "Surge test-based identification of stator insulation component with partial discharge activity for low-voltage AC motors," *IEEE Transactions on Industry Applications*, vol. 56, no. 3, pp. 2541–2549, 2020.
12. S. Zeynalova, E. Cepparrone, E. Roffino, and D. Barbero, "Effect of thermal aging on electrical performance of perfluoroalkoxy- and polyamide-imide-coated magnet wire," *IEEE Electrical Insulation Magazine*, vol. 40, no. 3, pp. 6–14, 2024.
13. A. K. Khiz and S. H. Jayaram, "Effects of rise time and overshoot of WBG device-based power converters on hairpin winding with corona-resistant and non-corona-resistant wires in EVs," *IEEE Transactions on Dielectrics and Electrical Insulation*, vol. 32, no. 4, pp. 2334–2343, 2025.

14. A. Naeini, E. A. Cherney, and S. H. Jayaram, "Effect of conductivity on the thermal and electrical properties of the stress grading system of an inverter-fed rotating machine," *IEEE Transactions on Dielectrics and Electrical Insulation*, vol. 23, no. 1, pp. 29–35, 2019.
15. F. T. Emery, "The application of conductive and stress grading tapes to vacuum pressure impregnated high-voltage stator coils," *IEEE Electrical Insulation Magazine*, vol. 12, no. 4, pp. 15–22, 1996.
16. E. Sharifi, S. H. Jayaram, and E. A. Cherney, "Analysis of thermal stresses in medium-voltage motor coils under repetitive fast pulse and high-frequency voltages," *IEEE Transactions on Dielectrics and Electrical Insulation*, vol. 17, no. 5, pp. 1378–1384, 2010.
17. ASTM D1868, *Standard Test Method for Detection and Measurement of Partial Discharge (Corona) Pulses in Evaluation of Insulation Systems*, 2007.
18. F. P. Espino-Cortes, E. A. Cherney, and S. H. Jayaram, "Impact of inverter drives employing fast-switching devices on form-wound AC machine stator coil stress grading," *IEEE Electrical Insulation Magazine*, vol. 23, no. 1, pp. 16–28, 2007.
19. M. S. Moonesan, S. H. Jayaram, and E. A. Cherney, "Time to failure of medium-voltage form-wound machine turn insulation stressed by unipolar square waves," *IEEE Transactions on Dielectrics and Electrical Insulation*, vol. 22, no. 6, pp. 3118–3125, 2015.
20. I. M. Jarrar, E. A. Cherney, and S. H. Jayaram, "Switching wave characteristics affecting the life of Type II machine turn insulation," *IEEE Transactions on Dielectrics and Electrical Insulation*, vol. 29, no. 5, pp. 2000–2007, 2022.
21. A. Cavallini, D. Fabiani, and G. C. Montanari, "Power electronics and electrical insulation systems – Part 2: Life modeling for insulation design," *IEEE Electrical Insulation Magazine*, vol. 26, no. 4, pp. 33–39, 2010.
22. A. K. Khiz and S. H. Jayaram, "Effects of rise time and overshoot of WBG device-based power converters on hairpin winding with corona-resistant and non-corona-resistant wires in EVs," *IEEE Transactions on Dielectrics and Electrical Insulation*, vol. 32, no. 4, pp. 2334–2343, 2025.
23. T. Petri, M. Keller, and N. Parspour, "The insulation resilience of inverter-fed low-voltage traction machines: Review, challenges, and opportunities," *IEEE Access*, vol. 10, pp. 104023–104049, 2022.
24. E. Kuffel, W. S. Zaengl, and J. Kuffel, *High Voltage Engineering: Fundamentals* (2nd ed.), Butterworth-Heinemann, 2000.
25. A. Cavallini, G. C. Montanari, and D. Fabiani, "Advanced insulation systems for inverter-fed motors," *IEEE Electrical Insulation Magazine*.
26. Pipistrel Aircraft, https://www.pipistrel-aircraft.com.
27. NASA, https://www.nasa.gov.
28. Rolls-Royce Media, https://www.rolls-royce.com/media.

Index

Note: Page numbers in *italics* and **bold** indicate figures and tables on the corresponding page.

B

C

D

L

M

For Product Safety Concerns and Information please contact our EU representative GPSR@taylorandfrancis.com Taylor & Francis Verlag GmbH, Kaufingerstraße 24, 80331 München, Germany